Technology
and East-West Trade

U. S. Congress.

Office of Technology Assessment
CONGRESS OF THE UNITED STATES

ALLANHELD, OSMUN / GOWER

Published in the United States of America in 1981
by Allanheld, Osmun & Co. Publishers, Inc.
6 South Fullerton Avenue, Montclair, New Jersey 07042

Published in Great Britain in 1981
by Gower Publishing Co. Limited
Westmead, Farnborough, Hampshire, England
ISBN 0 566 00436 4

Library of Congress Cataloging in Publication Data

United States. Congress. Office of Technology
    Assessment.
    Technology and East-West trade.

    Reprint of the 1979 ed. published by The Office,
Washington, D.C.
    Includes bibliographical references.
    1. East-West trade (1945-   ) 2. United
States—Commerce—Communist countries. 3. Communist
countries—Commerce—United States. 4. Technology
transfer. I. Title.
[HF1411.U615 1981]   382′.091713′01717   80-26121
ISBN 0-86598-041-1

British Library Cataloguing Data
United States. Congress. Office of Technology Assessment.
Technology and East-West trade.

1. East-West trade (1945–)   2. Technology transfer.
3. United States—Commerce—Communist countries.
4. Communist countries—Commerce—United States.
Title.
382′. 091713′ 01717   HF 1411 80-26121
ISBN 0-566-00436-4.

Printed in the United States of America

Foreword

This assessment was made in response to requests from the House Committee on Foreign Affairs and the Senate Committee on Commerce, Science, and Transportation. Its purpose is to examine the economic and national security implications of the transfer of technology between the United States and the Communist world.

The study identifies and, where possible, evaluates the economic, political, and military costs and benefits that accrue to the United States in its trade with the Soviet Union, Eastern Europe, and the People's Republic of China, taking account of the complex ways in which these factors interrelate. It sets forth a spectrum of policy options which could potentially affect these relationships, and explains the difficulties in projecting their consequences.

The report also provides background information on the functioning and the implications of U.S. trade policy vis-a-vis the Communist world, including the areas of tariff and credit policy and export control, both in the United States and in selected allied nations. Finally, it surveys the past and potential contributions of Western technology to the economies of the Soviet Union and China.

The Director of this project is Dr. Ronnie Goldberg of OTA's International Security and Commerce program. That program is managed by Dr. Peter Sharfman (who succeeded Dr. Henry Kelly in September 1979) and is part of OTA's Energy, Materials, and International Security Division headed by Assistant Director Lionel S. Johns.

OTA is grateful for the assistance of its Technology and East-West Trade Advisory Panel, chaired by Mr. McGeorge Bundy, as well as for the assistance provided by the Central Intelligence Agency and the Departments of Defense, State, and Commerce. It should be understood, however, that OTA assumes full responsibility for its report, which does not necessarily represent the views of individual members of the Advisory Panel or of any Government agency.

JOHN H. GIBBONS
Director

Technology and East-West Trade Advisory Panel

Technology and East-West Trade Project Staff

Lionel S. Johns, *Assistant Director*
Energy, Materials, and International Security Division

Henry Kelly, *Program Manager* (to September 1979)
Peter Sharfman, *Program Manager* (from September 1979)
International Security and Commerce Program

Ronnie Goldberg, *Project Director*

Pamela Baldwin   Robert Rarog
Marvin Ott   Irene Szopo

Administrative Staff

Dorothy Richroath   Helena Hassell

OTA Publishing Staff

John C. Holmes, *Publishing Officer*
Kathie S. Boss   Joanne Heming

Acknowledgements

This report was prepared by the International Security and Commerce Program staff of the Office of Technology Assessment. The staff wishes to acknowledge the assistance of the following contractors in the collection, analysis, and preparation of material for the report:

Michael Checinski
Robert Fraser
Seymour Goodman
Mark Kuchment
Phillips Kuhl

Stephen Sternheimer
Richard P. Suttmeier
Rudi Volti
Angela Stent Yergin

Business Experience International
International Economic Studies Institute

The following Government agencies also provided assistance:

The Central Intelligence Agency
The Department of Commerce

The Department of Defense
The Department of State

Contents

LIST OF ACRONYMS

ACEP — Advisory Committee on Export Policy
AWG — Aussenwirtschaftsgesetz (West Germany)
BGW — Bundesamt fuer gewerbliche Wirtschaft (Ministry of Industrial Economy, West Germany)
CCC — Commodity Credit Corporation
CCL — Commodity Control List
CDC — Control Data Corporation
CDU — Christian Democratic Party (West Germany)
CG — Consultative Group
ChinCom — China Committee
CIA — Central Intelligence Agency
CMEA or COMECON — Council for Mutual Economic Assistance
CoCom — Coordinating Committee for Multilateral Export Controls
COFACE — Compagnie Francaise d'Assurance pour le Commerce Exterieur (France)
CTEGs — Critical Technology Expert Groups
DOD — Department of Defense
EARB — Export Administration Review Board
EDAC — Economic Defense Advisory Committee
EE — Eastern Europe
EEC — Euopean Economic Community
Eximbank — Export-Import Bank
FRG — Federal Republic of Germany (West Germany)
GAO — General Accounting Office
GATT — General Agreement on Trade and Tariffs
GDR — German Democratic Republic (East Germany)
GNP — gross national product
HVEC — High Voltage Engineering Corporation
IAEA — International Atomic Energy Agency
ICs — integrated circuits
IREX — International Research and Exchanges Board
ITC — International Trade Commission
I.W. — industrialized world
JDA — Japan Defense Agency

JERC — Japanese Economic Research Center
LDCs — less developed countries
LIBOR — London Interbank Borrowing Rate
LSI — large-scale integration
MFN — most-favored-nation
MFT — Ministry of Foreign Trade (PRC)
MIRVs — multiple independently targetable reentry vehicles
MITI — Ministry of International Trade and Industry (Japan)
NAS — National Academy of Sciences
NRC — Nuclear Regulatory Commission
NSC — National Security Council
OC — Operating Committee
OEA — Office of Export Administration
OECD — Organization for Economic Cooperation and Development
PRC — People's Republic of China
PRM 31 — Presidential Review Memorandum on East-West Technology Transfer
R&D — research and development
SCCC — State Capital Construction Commission (PRC)
SCST — Soviet Committee on Science and Technology
SEC — State Economic Commission (PRC)
SITC — Standard International Trade Classification
SPC — State Planning Commission (PRC)
SPD — Social Democratic Party (West Germany)
S&T — science and technology
STC — State Science and Technology Commission (PRC)
sub-ACEP — sub-Advisory Committee on Export Policy
TACs — Technical Advisory Committees
TECHIMPORT — China National Technical Import Corporation
TTG — Technical Task Group
U.S.S.R. — Union of Soviet Socialist Republics
VLSI — very large-scale integration

CHAPTER I

Summary: Issues and Options

CONTENTS

Summary: Issues and Options

INTRODUCTION

Trade and commerce between nations is a necessary, if not sufficient, requisite for achieving peaceful relationships. On the other hand, trade with a potential adversary will inevitably, to some extent, strengthen the economy and the military capability of the trading partner. It is in the context of this dilemma that present debates over the value and wisdom of selling U.S. goods and technology to the Communist world take place. This study has been undertaken at the request of the House Committee on Foreign Affairs and the Senate Committee on Commerce, Science, and Transportation to help provide Congress with the capability to address the complex issues raised by this trade, including the extent to which international trade in high technology endangers the national security of the United States. It addresses the controversies that surround the issue of East-West trade and technology transfer, i.e., the costs and benefits of the United States' selling technology to and expanding its commercial relations with the Soviet Union, Eastern Europe, and the People's Republic of China.

This subject is complicated by both conceptual problems and disagreements about the nature and future of U.S.-Soviet relations. The conceptual problems concern the difficult task of defining and measuring technology. These problems are dealt with in chapter VI. The disagreements are manifested in the divisiveness and ambivalence which surround the question of the appropriate nature and extent of U.S. trade with the East. At the center of these disagreements seems to lie an even more fundamental difference of views about the basic strategies that the United States should employ in its dealings with the Communist world.

From one perspective, technology transfer is a necessary part of a policy of expanded contacts with the Communist world. Out of this policy of detente arises a series of international and interpersonal relationships which, over time, could contribute to a lasting structure of peace. Those that argue from this perspective assert that present policies that restrict U.S. exports are both politically and economically ill-advised. The risk of some erosion of U.S. technical leadtime incurred by trade in technology is justified by the economic benefits of trade. Moreover, strict export controls are unworkable given the availability of much comparable technology abroad and the inability of the United States to obtain adequate cooperation from its allies for a restrictive trade policy. From this perspective, the denial of all but a small and specialized category of military technology is practically impossible. The safest policy therefore becomes one of vigorous promotion of 1) all U.S. exports to reduce balance-of-trade deficits, and 2) U.S. research and development to maintain a technological lead over friends and adversaries alike, thus minimizing the national security risks entailed in technology transfer. Consistent

4 • Technology and East-West Trade

with this view is the argument that corporate interest should be more than adequate to protect the United States from suffering substantial economic losses through trade in technology; it is, after all, in the interest of every corporation to protect its position of technical leadership.

Others view the basic nature of East-West confrontation in more Manichean terms, arguing that the fundamentally adversary relationship between East and West is unlikely to be changed in the near future through any gradual relaxation of tension brought about by trade. Trade is not seen as an opportunity for strengthening peace; rather it is contended that the West is being slowly bled of its most important assets by nations it has every reason to distrust. From this perspective, present policy is not restrictive enough. The only safe course is to deny assistance to our adversaries wherever possible, using trade only as necessary to extract political concessions. The difficulty of obtaining cooperation from our allies is acknowledged but countered by the argument that opinion in Western Europe and Japan is not monolithic and present official sentiments are not fixed for all time. With sufficient determination, funds, and energy it could be possible both to strengthen Western military alliances and to convince our allies to restrict trade in a common front. From this perspective even relatively passive aspects of trade in technology assume a strategic significance; programs allowing a constant interchange between Eastern and Western technologies, for instance, can gradually deplete advantages in technology, management skills, and other areas in which the West now enjoys substantial superiority.

The middle ground is occupied by those who feel that no judgments need necessarily be made about the prospects for detente. They argue that while existing policies may require adjustments to increase the efficiency and reliability of their administration, no basic reformulation is required. From this point of view, the objective of the export administration system is to maintain the mili-

tarily relevant technological leads that the West presently holds relative to the Communist world. The system, it is argued, is functioning properly so long as it delays the acquisition in the East of technologies that could close these gaps or ensures that their acquisition is relatively difficult and costly. Realistically, a Communist nation can ultimately acquire any item it prizes highly enough; either alternative suppliers will be found outside the United States, or it will be developed indigenously, at greater cost perhaps than if it were purchased from the West. It is acknowledged that the licensing system as it is presently administered may occasionally err either in subjecting harmless technologies to excessive and needless delay, or, less often, in allowing items of military significance to slip through the net. These defects can be remedied without altering the fundamental premises of the policy, however. Attempts to weight the policy on the side of economic advantage may have serious national security implications; efforts in the opposite direction must contend with economic and political realities. The United States is not the sole supplier of most of the technologies desired in the Communist world and U.S. allies in Western Europe and Japan are not likely to concur in more restrictive policies.

To discuss and evaluate these positions in a meaningful way requires the review of a host of complex economic, political, and military benefits and liabilities that may not be quantifiable, but which nevertheless must enter into any calculation of the risks inherent in all trading relationships with the Union of Soviet Socialist Republics (U.S.S.R.), People's Republic of China (PRC), and Eastern Europe. It is the goal of this assessment to present these and related points of view as clearly as possible, acknowledging the uncertainties which exist and which will continue to exist, but providing material that will allow a better analysis of the kinds of military, political, and economic costs and benefits that any program affecting East-West trade and technology transfer is likely to incur.

ISSUES AND FINDINGS

The following is a capsule summary of the major issues arising from U.S. trade—particularly in technology—with the Communist world. The discussion addresses the economic, military, and political concerns related to this trade and its role in relations between the United States and its major allies.

ECONOMIC

How Important Is Trade With the Communist World to the U.S. Balance of Trade?

East-West trade is a relatively small component of U.S. foreign trade, and a minor component of the overall American economy. Although trade with Communist States has grown rapidly since the beginnings of detente, and although trade with the People's Republic of China increased dramatically during 1979, trade with Communist nations is not expected to become a critical factor in the U.S. balance of trade in the foreseeable future.

Total turnover of U.S. trade with the East in 1978 was $6.3 billion—4.1 percent of U.S. world trade. The United States had a balance of trade surplus with the Communist world of $2.6 billion in 1978, as compared to a U.S. worldwide trade deficit of approximately $28.4 billion. Moreover, East-West trade remains a relatively small part of the Organization for Economic Cooperation and Development (OECD) trade as a whole, and the United States has captured only a minor share of this limited market. In no instance is the United States the major Western trading partner of a nonmarket economy.

What Are the Major Barriers to Continued Growth in U.S. Trade With the East?

Three factors are commonly cited as inhibiting the expansion of American trade with the Communist world: 1) the lack of official credits and guarantees to finance U.S. ex-

Photo credit: World-Wide Photo
Coca Cola is now available in the PRC

Photo credit: TASS from SOVFOTO
Pepsi Cola production line at Novorossiisk, U.S.S.R.

ports; 2) the lack of normal trading relations, including extension of most-favored-nation (MFN) status to Communist countries, notably the U.S.S.R. and PRC; and 3) U.S. export controls.

In fact, the primary obstacle to rapid growth of trade with the Communist world is the Communists' inability and/or unwillingness to export on a competitive basis to Western markets. Consequently, a shortage of hard currency inhibits Communist imports from the West. Credits that supply hard currency would attack this shortage directly; extension of MFN would facilitate some Communist exports; direct export controls are significant only in certain industries to which Communist nations accord priority in their allocation of hard currency (e.g., computers or oil extraction technology in the case of the U.S.S.R.).

Therefore, credit is and will continue to be a major factor influencing the growth of East-West trade. Subsidized credits and/or loan guarantees, especially for the U.S.S.R., are far more readily available from America's Western allies—West Germany, France, Great Britain, and Japan—than from the United States. In 1977, for instance, West Germany and France supplied on the order of $7 billion in official export credits to the U.S.S.R. and Eastern Europe; Japan provided nearly $5 billion; and the United States $945 million. There are strong indications that the availability of official credits substantially affects the choice of Western suppliers. For instance, all Soviet orders for American turnkey plants came during the brief period in which the U.S.S.R. was eligible for U.S. Export-Import Bank (Eximbank) credits. Contracts are now often concluded by American multinational firms with subsidiaries in countries that provide the U.S.S.R. with more competitive financing. This means that although American firms supply the technology, the United States does not receive the economic benefits of major equipment orders. It is unlikely to do so until Eximbank financing is once again available to the U.S.S.R.

Lack of MFN status appears hitherto to have had greater symbolic than practical impact on the volume of U.S. imports from the East.

Owing to the commodity composition of trade and the demand elasticity characteristics of Eastern products, absence of MFN status has had a relatively minor effect on the largest nonmarket exporters (U.S.S.R. and PRC, although the situation with respect to the latter may now be changing).

The existing tariff schedule has, however, probably had a relatively greater effect on the volume of U.S. imports from—and ability to export to—East Germany and Czechoslovakia.

Many U.S. firms contend that U.S. export controls are a serious barrier to expansion of trade with the East. Careful analysis does not support this proposition. While export controls may affect the U.S. market share of present trade, significant growth is retarded more by chronic hard-currency shortages and deliberate policy decisions in the Communist world. Even if U.S. trade with the PRC continued to grow at its present rate, for instance, changes in the Communist world's share of U.S. foreign trade would be incremental. Major increases in the volume of East-West trade can only occur if the Communist world alters certain of its fundamental policies regarding the degree of worldwide economic interdependence acceptable to it and establishes alternative ways of handling its current hard-currency problems. It is highly unlikely that any such decisions will dramatically affect trade volumes over the next few years.

How Much Technology Does the Communist World Buy From the West, and How Important Is That Technology to the Economies of the Importing Countries?

Communist imports of technology, including technology-intensive products, constitute in value terms a minor share of total purchases from the West, but the value to the East of the technological component of Western trade is high relative to other imports. Some contend that this is due to Western underpricing on technology sales to the East. Be that as it may, it is certainly true that Eastern purchasers carefully choose only those processes and products with the highest possibility of productivity gains. Estimates of the macroeconomic impact of Western technology imports on the Soviet economy vary, but it is clear that impacts in discrete areas of the economy have been significant. Thus, while aggregate Western resource inflows into the U.S.S.R. have relatively little impact on overall growth, there is no doubt that imports of certain commodities—capital equipment and associated technology in particular—have played a large role in the expansion of key sectors, and have thus made a significant contribution to total economic growth. This is particularly manifest in the chemical and motor vehicle industries.

The U.S.S.R.'s most productive domestic use of imported technological developments has come in industries that were based on well-established technologies. But in most industries in which significant technology transfer from the West has occurred, the technology gap between the U.S.S.R. and the West has not diminished substantially over the past 15 years. This may be due to the fact that imported technology substitutes for the development of domestic capabilities and therefore actually impedes the ongoing domestic innovation necessary to close technological gaps. In these sectors, Western technology has been extremely important, but it has never acted as a panacea for Soviet economic difficulties.

The Soviets will experience a sharp decline in the growth of the labor force in the near future. Technological improvement, aided by imports from the West, is to be the basis of planned increases in Soviet labor productivity. However, the rigidities of central planning inhibit the diffusion of imported technology in Communist nations. This is particularly true in the U.S.S.R., where lack of communication between producer and user, and lack of effective cost criteria hamper the Soviet ability to effectively assimilate and diffuse imported Western technology throughout the economy.

What Are the Prospects for Future Eastern Purchases of Western Technology?

In the U.S.S.R., the allocation of convertible currency is the most important single element in import planning. Decisions to allocate currency among purchase options in the West are made either within the framework of regular 1- and 5-year plans governing the entire economy or through irregular (ad hoc) decrees that concern single branches of industry or individual enterprises.

The Soviets are careful customers. Each ministry, nearly all R&D organizations, and many large enterprises systematically collect and process available Western scientific and technical data. The Ministry of Foreign Trade collects technological and marketing data abroad. Nevertheless, import priorities change over time and the U.S.S.R. has developed no consistent and universally applicable criteria for selecting Western imports. The decisionmaking process is time-consuming, complex, and often inconsistent.

The hard-currency debt of the East, although small from the standpoint of worldwide borrowing, has risen dramatically in the last 8 years. In spite of debt increases the credit ratings of nonmarket economies in the West are good, and their debt in the United States is relatively small. But because the short-term prospects of greatly increasing exports to the West are dim, the cost of hard-currency capital may be expected to rise. As the accumulation of hard-currency debt in the East increases, further borrowing will become more expensive. Should this occur, there are three alternatives open to Communist nations. They can allow more direct Western involvement in their enterprises in the form of joint ownership; they can resort to internal financing; or they can expand and diversify their hard-currency earnings from exports. On examination, only the latter option seems viable in the long term. It will require the import of Western manufacturing and marketing technology.

In the short and medium term, credit will constrain further technology purchases only insofar as it becomes difficult to obtain financing for large and costly projects (as in the U.S.S.R.). As the cost of capital increases, nonmarket economies may begin to place a higher priority on purchasing technology and technology-intensive commodities since these products promise the highest returns. The purchases of Western commodities as short-term means of achieving 5-year-plan targets will probably diminish as planned technology purchases increase in relative importance.

The situation of the PRC is quite different from that of the U.S.S.R. Imports of Western products and technology have traditionally been regarded by the Chinese leadership with ambivalence. The history of Chinese technological interaction with the West is punctuated with attempts to ignore or suppress the cultural consequences engendered by the transplant of Western productive techniques. But while Chinese imports of industrial plant and technology have (until recently) been relatively small, their importance in providing a cumulative qualitative improvement in key industries has been substantial.

Foreign technology provides the cutting edge of the general program of economic modernization announced by Premier Hua Guofeng in February 1978. The current leadership has manifestly committed itself to generating policies that will control Western cultural influence during the process of technological development, but it will nevertheless aggressively push for technological modernization. While the overly optimistic plans announced at that time have since been scaled down, major purchases of Western plant and technology will still occur, financed in part through increased levels of borrowing in the West.

The Chinese will experience no difficulty in obtaining credits. Their borrowing in the West, until now, has been extremely modest and their credit rating is high. They also would appear to have significant potential for hard-currency earning exports such as oil and labor-intensive handicrafts and consumer products.

The PRC has shown increasing sophistication in the search for and acquisition of foreign technology. There has been a coordinated national effort to accumulate as much published technical and scientific data as possible from Western sources. The last 2 years have seen an enormous growth in Chinese technical delegations traveling abroad to Western Europe, Japan, and the United States. The Chinese also seek to extract the maximum amount of information from technology negotiations as well as to utilize contractual arrangements that will yield as much experience to China as possible. Productivity gains resulting from Western capital and technology inflows will be most marked in centrally controlled urban industries, which stand to benefit greatly from the import of modern process equipment and complete plants.

Chinese import selection and hard-currency allocation procedures are relatively decentralized in comparison with the Soviet Union. Accordingly, it is possible that recent orders of foreign plant and equipment were permitted to outstrip China's ability to generate the hard currency needed to pay for them. Even though the decentralization of decisions regarding allocation of currency for foreign trade may be expected to continue, decisions regarding the purchase of whole plant and high-technology items are likely to remain centralized and the central government will exert closer control over foreign exchange.

Could the Sale of American Technology to the Communist World Produce Effects Detrimental to Sectors of the U.S. Economy?

Some sectors of the U.S. economy are more vulnerable to the repercussions of technology transfer than others. Perhaps the most important of these is the chemical industry, where American plants have been sold to the U.S.S.R. and are to be paid for by products produced in them. As a result of this "buy-back" transaction, U.S. firms producing anhydrous ammonia have experienced domestic plant closures and significant declines in prices over the last 2 years.

Photo credit: TASS from SOVFOTO

American equipment at the construction of the second Shatyk—Khiva pipeline, U.S.S.R.

The U.S.-U.S.S.R. contract has a life of 20 years and it will probably result in serious market disruption for domestic producers of ammonia. Other problems in the same industry have arisen in Western Europe where product buy-back provisions in contracts for turnkey plants have required the import of large quantities of chemicals, to the detriment of domestic producers.

Despite this growing threat to the West European chemical industry, existing legal mechanisms have proven ineffective in dealing with the glut. "Dumping," i.e., selling goods cheaply in overseas markets at below domestic production costs, is also a problem. Chemical firms have found it difficult to demonstrate dumping because the required evidence includes the exporters' prices in the home market or actual costs (prices in Communist nations are administered and are therefore unusable for comparisons). Although rulings in the United States have found dumping of some Eastern goods, often it takes Western firms at least a year to assemble a case based on the exporter's internal costs. By this time the damage has already been done.

It is highly unlikely in the near or medium term that any Eastern economy could offer serious competition to the United States in a product area involving advanced design and manufacturing technology. But Eastern nations are anxious to increase their export potential and any significant increase in East-West trade in the long term must include more Eastern exports.

In the long term, the proliferation of industrial technology in the East might weaken the competitive position of U.S. firms as suppliers of technology to newly industrializing nations. Nonmarket economies are increasingly attempting to break into this market.

On the other hand, there may be significant advantages to sectors of the U.S. economy actively engaged in East-West trade. Although the total volume of trade in technology between the United States and the Communist world has been relatively small in dollar terms, some firms contend that such technology sales play an important role in their corporate strategy and are linked both to potential sales of much greater size— in third markets as well as in the East—and to decisions regarding innovation and extension of product lifecycles. Should trade volumes increase, so too may these indirect effects.

Another potential benefit may lie in reverse technology transfer—from East to West—which is at present miniscule. The failure of Western firms to search for technology in the East and the inability of the centrally planned economies to market effectively in the West has resulted in significant opportunities for Western technology purchases being missed.

POLITICAL

What Basic Positions Have Been Taken Regarding the Use of Trade Leverage, i.e., Using Trade To Achieve Political Objectives of Foreign Policy?

The question of the political uses of trade has generated considerable controversy and at least three schools of thought. The first rests on a judgment that trade is not an effective instrument to achieve political objectives. This is the official view of the Soviet Union and is held by a number of OECD governments which contend that history has shown that efforts to obtain political concessions from the nonmarket economies through policies of economic pressure or in-

ducement have been unsuccessful. Consequently, each trade and credit transaction should be judged on its economic merits alone.

The second perspective is associated with detente. It rests on the proposition that trade can have a moderating effect on international politics by enmeshing national economies in a web of interdependence. The Soviet economy's acute need for imports of technology and capital equipment from the West provides the opportunity to deliberately bind the U.S.S.R. in such a web. Additional benefits could include a strengthening of the Soviet consumer economy as a claimant on domestic resources and a moderating factor in national policymaking; and increased opportunities for the penetration of Soviet society with Western products, culture, and values. This perspective adopts a limitationist view of American power and is skeptical of the extent to which the United States can coerce Soviet policy. Washington is seen to have little real choice but seek a stable cooperative relationship with Moscow. Through a combination of economic inducement and benign political subversion, trade offers one means of drawing the Soviet Union into such a relationship.

The third perspective accepts the proposition that technology transfers can be harnessed to political purposes, but is profoundly skeptical of the hypothesized connection between such trade and political moderation. Proponents of this view contend that the basic relationship between the Soviet Union and the West is, and will remain, one of conflict due to deep-seated differences in ideology, social and political systems, and foreign policy objectives and interests. Consequently, Western transfers of technology may have the net effect of strengthening an adversary—particularly if they are financed by credits at low rates. Such transactions can only be justified if the West obtains concessions in Soviet domestic or foreign policy in return. Thus the need for Western technology can and should be exploited as a source of leverage on Soviet policy.

Is There Evidence to Either Confirm or Deny the Utility of Trade Leverage in East-West Relations?

To date the effort to use trade for political leverage has focused on establishing the freedom of Soviet Jews to emigrate. An analysis of Jewish emigration in the context of the Jackson-Vanik amendment, which links U.S. trade concessions to such emigration, provides no conclusive evidence that the amendment either has or has not had a significant impact on Soviet emigration policy.

Much disagreement surrounds the question of whether there exist technologies critical to the U.S.S.R. in which the United States has a clear worldwide monopoly. To be effective for this purpose a technology must be highly valued by the Communist countries and must be unavailable from alternative sources. Few technologies meet these tests. Attempts to identify technologies with the greatest promise as instruments of U.S. leverage have focused on advanced oil and gas exploration and extraction equipment and on certain types of computers. Even in these cases, careful analysis suggests that while some leverage may indeed be possible, it will be of limited potency and duration as supplies from alternative foreign sources and domestic production become available. Leverage in other technologies would depend on cooperative efforts in the Western alliance.

Efforts to use trade to moderate Soviet policy as part of a broader detente policy have led to inconclusive results. During recent years there has, in fact, been a substantial growth in Soviet trade with the United States but there is little evidence that such trade has so far had the desired effect on Soviet foreign policy or domestic politics. Whether it will do so in the future is open to debate.

MILITARY AND STRATEGIC

Can the Military Risk Entailed in the Proposed Sale of a Dual-Use Technology Be Determined?

A conclusive determination is probably impossible. Assessment of the military contribution of a product or process entails consideration of the following: the capabilities of the technology itself; the nature of the transfer mechanism; the character of the recipient environment, including infrastructure capabilities; the relative technological capabilities of the seller and the recipient; the available deterrents to diversion of end use; the priorities and intentions of the recipient; and the character and volume of related purchases in the past. Much of this information is necessarily based on informed speculation. Determinations of the motives and probable behavior of potential adversaries, for instance, are judgmental and can never wholly account for the impact of unforeseen events on priorities and decisions. The sale of any dual-use technology—and this means virtually all high technology—therefore necessarily entails some degree of security risk, which end-use guarantees or monitoring arrangements cannot eliminate.

Will the Compilation of a List of Militarily Critical Technologies, Embargoed to the Communist World, Substantially Reduce This Risk?

The critical technology approach currently under examination in the Department of Defense (DOD) is far from implementation. One present difficulty is the degree of confusion in Government and business community alike over its intention and probable consequences. For instance, it is hailed both by those who believe it may reduce the number of items currently controlled and by those who feel that it will make export controls more extensive. The recent DOD reorganization may well add momentum and a renewed sense of purpose to the critical technology exercise, but it is unwise to regard it as a panacea to the difficult problems inherent in administering export controls. A list of embargoed technologies cannot simply replace the existing licensing system. No export control system can be effective unless it makes provision for case-by-case reviews of export applications. This case method approach may be combined with procedures such as the critical technologies list, designed to screen the number of applications subjected to detailed analysis, but it cannot be wholly eliminated.

Have American Technology Exports Contributed Significantly to the Military Capabilities of the Soviet Union?

Most observers of the export-licensing process would agree that U.S. and other Western technology has contributed to Soviet military capabilities. There is no agreement, however, on the degree or significance of any such contributions. This is partly due to lack of explicit policy guidance in the present export administration system on the specific military objectives and desired relative force capabilities of the United States; but disagreements also stem from divergent perception of Soviet capabilities and basic intentions, and even from different assessments of the technological capabilities of exported items. In this connection, it is relevant to note that no export license has ever been granted over the objection of the Secretary of Defense.

It is unlikely that these contributions could have been totally avoided without a complete economic embargo of the Communist world by the entire West.

AMERICA'S ALLIES

How Effective Is CoCom?

CoCom (Coordinating Committee for Multilateral Export Controls), the multinational organization that attempts to implement a uniform export control system throughout the Western bloc, remains a viable, albeit imperfect, organization despite its informal nature, the lack of sanctions or adequate policing mechanisms, and the equivocal attitude of several of its members towards the continuation of present levels of export control. There are frequent charges, both in the business community and Government circles, that firms in other CoCom nations have evaded or ignored CoCom restrictions. There is at least convincing anecdotal evidence to support such charges, but the extent of foreign government connivance at such practices is open to question.

How Do Other Members of the Western Alliance View the Problems Raised by East-West Trade and Technology Transfer?

East-West trade has always been economically more important for Western Europe and Japan than for the United States. Germany and Japan lead the United States in exports of "high technology" products to the East and consider such sales desirable elements of their normal foreign trade. West Germany, for instance, is the leading Western overall supplier of machinery and equipment to the U.S.S.R., providing nearly one-third of such Soviet imports. Japan supplies approximately 20 percent and the United States less than 10 percent. Japan is the U.S.S.R.'s leading Western supplier of oil-refining equipment.

America's allies do not deny the basic necessity of withholding items of direct military relevance from the Communist world. But although there does not appear to be much enthusiasm for disbanding CoCom, its European and Japanese members would grant it a narrower role in export control than would the United States.

There is little, if any, debate similar to that in the United States over the political, military, and strategic implications of transferring technology to the East in Western Europe and Japan. It appears that Japan, West Germany, France, and Great Britain all consider the sale of technology a primarily economic issue and are content to rely on the self-interest of the companies affected to protect domestic industry. Any use of export controls for political purposes is largely eschewed. West Germany, in fact, considers trade with East Germany a part of its domestic commerce and not "foreign trade," although it does observe CoCom restrictions in its sales of dual-use and military items to East Berlin.

With the exception of the small number of cases subject to delays in CoCom, or held up in the U.S. reexport licensing system, Japanese and West German export controls work quickly and efficiently and appear satisfactory to their business communities. Unlike U.S. firms, companies in these countries usually know how their cases will be resolved before a license application is submitted.

What Is the Likely Future of CoCom and How Much Influence Can the United States Expect to Exercise Over Its Policies?

Because of its position of leadership in a number of technologies of critical military significance, the United States feels it has a special responsibility to ensure their safekeeping. If it can play this role with intelligence and integrity, the United States may be able to initiate and maintain a strong and unified Western bloc position on the transfer of technology. Policymakers, however, must be cognizant of the fact that the attitudes and behavior on the issue of technology transfers to the East of at least four major CoCom partners (West Germany, France, Britain, and Japan) differ from those of the United States. Without major changes in the international climate and U.S. policy and behavior, attempts to strengthen the organization or impose formal sanctions on its members are likely to be resisted. Meanwhile, there is no immediate reason to expect any fundamental changes in the operation of the organization or the behavior of its members.

POLICY OPTIONS

There are three basic sets of options for future East-West trade and technology transfer policy. Each rests on a basic orientation toward the Communist world and set of beliefs and expectations regarding America's future relations with it. These orientations were discussed in the introduction to this chapter.

Present U.S. export control policy is the result of a decision to forego attempts at economic warfare against the Communist world and to further the dual aims of encouraging trade with the Eastern bloc and protecting U.S. national security. Legislation has attempted both to eliminate procedural barriers to trade and to strengthen national security safeguards. At times provisions of the law have pulled in opposite directions, but the trend over the past 10 years has been toward liberalization of export controls.

There are three broad categories of policy which can impact on East-West trade and technology transfer. Suggestions in each category are listed here and discussed at length in the body of the report.

1. Actions in Keeping With the Existing Policy But Designed to Make Current Procedures More Efficient.

The vast majority (90 to 95 percent) of U.S. exports are shipped under a general license that requires no formal application procedure. Similarly, only a minority of exports to Communist destinations require validated licenses. However, these must enter an export-licensing system which is complex and which has come under severe criticism for the delays it occasions. In fact, given the volume of applications handled by the system, it works reasonably efficiently and only a small number of cases are actually subject to excessive delays. Delayed cases assume a disproportionate importance, however, because they are often large and highly visible and concern areas of high trade-growth potential.

Suggestions for increasing the efficiency and reducing delays in the licensing process include the following: increasing appropriations for export-licensing administration; establishing a new form of export license; instituting timetables to curtail excessive delays; ensuring that application rejections are undertaken at the recommendation of all agencies in the review system; improving the data base for East-West trade; and enhancing the foreign availability assessment capabilities of the Office of Export Administration.

Procedures can and should be instituted to streamline the system without tampering with its basic structures or effectiveness, thus eliminating unwarranted, costly delays. Most promising among this family of suggestions is finding a means of systematically monitoring the availability of technologies desired by the Communist world from sources outside the United States. This is a crucial task of the export administration system. The establishment of a continuing capacity for undertaking such "foreign availability assessments" would be an important resource for administrators and policymakers alike.

2. Actions That Could Increase Restrictions on East-West Trade or Strengthen the Use of Trade as a Foreign Policy Lever.

Suggestions that aim at shifting the balance of U.S. policy in the direction of restricting increases in trade with the East include the following: enhancing the role of the Secretary of Defense in the licensing process; compiling a list of embargoed critical technologies; exercising trade leverage through foreign policy controls, MFN, and official credit restrictions; strengthening CoCom; and curtailing academic and scientific exchange programs.

The effectiveness of several of these suggestions is problematic. First, the critical technology exercise currently underway in DOD has made slow progress in the past 3 years and is the subject of widespread misconceptions. Even the compilation of a critical technologies list will not allow easy or comprehensive solutions to the problems posed by East-West technology transfer. Second, the United States is restricted in the degree of potential trade leverage it can exercise. As discussed above, evidence of the past effectiveness of such leverage has been ambiguous. Finally, the United States at present has a limited ability to persuade its allies to strengthen CoCom. Such changes might be possible only if the United States itself embarked on a new and clearly confrontational policy vis-a-vis the Communist world.

3. Actions Designed to Expand East-West Trade.

The third group of suggestions for shifting the balance of U.S. policy in the direction of increasing trade with the East includes the following: expansion of official export financing; granting MFN to Communist nations not presently enjoying it; limiting Presidential discretion in imposing export controls; reducing and/or indexing the Commodity Control List; bringing U.S. export control procedures into closer conformity with those of other CoCom nations; and a family of measures designed at export promotion in general.

Here, providing access to official export financing is probably the Government policy with the highest potential for increasing the volume of U.S. trade with the East. The impact of granting MFN varies greatly among individual recipients. From a purely economic perspective, MFN to selected Eastern European nations might have greater impact than granting it to the U.S.S.R. and PRC. And while the removal of items from the Commodity Control List might affect U.S. market shares in certain industrial sectors, this would have less overall impact on U.S. trade with the East than would changes in credit and tariff policies.

CHAPTER II

Future Policy Governing East-West Trade and Technology Transfer

CONTENTS

Future Policy Governing East-West Trade and Technology Transfer

INTRODUCTION

The continuing objective of U.S. regulation of East-West trade has been to balance both the commercial benefits of trade and the objectives of detente against the need to safeguard U.S. security interests. Continuing controversy about the proper balance is inevitable: there is no objective test of whether such a balance has been achieved, the economic and political circumstances affecting East-West relationships are in constant flux, and the United States has no comprehensive East-West trade policy. Alternatives for reforming existing policy can be broadly divided into three categories:

- **Policy options premised on an assumption that existing policy has, on the whole, achieved an acceptable balance between trade and security interests.** Such recommendations are designed primarily to make existing procedures more efficient and less costly.
- **Policy options designed to increase restrictions on East-West trade in order to decrease the risk that such trade could enhance the military interests of the nonmarket nations; and to use the threat of trade curtailments to exact political concessions from the East.**
- **Policy options designed to move U.S. policy closer to that of our European and Japanese allies by relaxing some of the current restrictions on East-West commerce, making licensing procedures less onerous, and providing trade incentives through the Export-Import Bank (Eximbank) and other mechanisms.**

The disagreements that have characterized the debate on U.S. trade policy as it applies to nonmarket nations result primarily from differences in the interpretation of the broader direction of East-West relationships. These are discussed at length in chapters I and IV. This chapter describes policy measures (some already incorporated in the Export Administration Act of 1979) that can affect the implementation or direction of U.S. trading policy with the East; analyzes whether these policies would have the desired effect; and reviews any inadvertent or unintended consequences that might result from their implementation.

REFORMS AIMED AT EFFICIENCY

The administration of U.S. export control policy has come under repeated attack by U.S. businessmen who charge that it is cumbersome, expensive, and slow. While statistics in chapter VII indicate that the situation has recently improved (i.e., the licensing procedure is being speeded), few would argue with the desirability of streamlining some of the more mechanical parts of the decision-making process. Suggestions to accomplish this have included the following:

Increase appropriations for export-licensing administration.

One way to accelerate the licensing process would simply be to increase the funding and staff available for processing applications; most offices in the Departments of Commerce and Defense dealing with these areas appear to be overworked. This solution, of course, would not resolve any flaws which might exist in the basic structure of the system. Moreover, the Office of Export Administration (OEA) claims that, at present, lack of qualified applicants has made it unable to fill all the positions for which funds have already been authorized.

Institute a new form of export license.

Recent legislation has created a new category of export license, a "qualified general license." These authorize multiple exports to the same end user for items that have been routinely approved in the past, and will hopefully reduce the volume of validated license applications as well as the cost and delay associated with the present need to apply for a separate license for each transaction. The strength of controls should not be diminished and their implementation should be made more efficient through the consequent reduction of the stress placed on the Federal administration of licensing. However, one potential ramification should be noted. At present, the U.S. Government keeps records only of those technology transfers that proceed under validated license. Unless data is also kept on trade conducted under qualified general licenses, there is the possibility that information on sales volumes, which could be valuable in inferring the intentions of the purchasers or the impact of sale on the U.S. economy, could be lost.

Improve efficiency and accountability

- Establish detailed timetables and deadlines for review of applications for validated licenses;
- Establish procedures by which applicants could take legal action against the Government if undue delays occur; and
- Improve reporting to applicants on the reasons for denial of applications.

Each of these measures was incorporated in the Export Administration Act of 1979. Taken together, they could significantly improve the relationship between applicant and Government.

Deadlines, for instance, would provide greater predictability for applicants and eliminate cases dragging on for months, even years, without resolution. Some industry critics have charged that present licensing procedures are overly complex and involve too many layers of consultation, and that important policy decisions are being made by midlevel bureaucrats. Administration goals for processing claims have already had the effect of speeding up the licensing process, and the recent reorganization of the Department of Defense's (DOD) export-licensing activities may also help in this regard (see chapter VII). But deadlines can only accelerate licensing to the extent that delays result from overworked, unresponsive, or sluggish bureaucracies. This is plainly not always the case; major delays in controversial cases may be occasioned by internal disagreement or uncertainty and extensive analysis on the part of the agencies involved. The speediest approach to resolving such cases would be simply to summarily deny controversial licenses. A forced deadline, therefore, could conceivably have the unintended consequence of reducing the number of approvals granted.

Until now, OEA has been neither always prompt nor explicit in informing applicants of the grounds for the rejection of their applications. The reason given has been as vague as simply "national security objections." Prompt disclosure will improve the atmosphere of business-Government relations; fuller details (security classifications permitting) will help industry prepare future cases and might forestall applications subject to the same objections.

Require all license denials to occur at the level of a Deputy Assistant Secretary.

This reform alone would have little bearing on the de facto policymaking occasioned by delay. Indeed, it might even increase it. The requirement could, however, have some impact on the process by creating a situation in which midlevel officials could approve, but not deny, license applications. Chapter VII demonstrates that, as the system now stands, the vast majority of validated license applications are eventually approved in any case; this change would affect those relatively few rejections that occur at midlevels by requiring the participation of all the agencies involved in the review process.

Improve assessments of foreign availability.

The determination of whether products or processes equivalent to those in the United States are available to controlled destinations from sources outside the United States is an important factor in decisions to grant validated export licenses. Determining "availability" requires either establishing that the controlled country already possesses the technology or product in question or that another Western nation has a technology functionally equivalent to the one proposed for sale and is itself prepared to sell it.

Problems of assessing foreign availability begin with gathering information; but this is only a prelude to the difficult task of developing policy guidelines for deciding claims of equivalency and comparability for a wide variety of complex technologies.

Until now, the assessment of foreign availability has proceeded in an ad hoc manner—carried out on a case-by-case basis by various agencies involved in the technical assessments of applications. A new plan to manage this activity from a single office within OEA can improve the continuity of foreign availability assessment in the Government and help create institutional expertise. The data base established by such a coordinating organization could become a valuable resource for ensuring uniform and equitable treatment of foreign availability issues. If properly staffed by engineers and technicians familiar with a large spectrum of technologies, it will also give the Government an independent resource for verifying and interpreting the assertions of comparability made by the industrial Technical Advisory Committees (see chapter VII). Because no one body in the licensing system has been responsible for foreign availability assessments, the U.S. Government has been criticized by some industry observers for failing to adequately monitor activities such as trade fairs or to take advantage of opportunities to inspect products and processes in the East. This situation could be remedied by hiring individuals specifically assigned to travel abroad, especially in the East, to make such inspections.

It must be pointed out, however, that the establishment of such a capability will not eliminate difficult decisions. Just as the determination of strategic significance is partly judgmental, so too the determination of the degree of technical equivalency necessary before items are deemed comparable can never be automatic.

Improve the monitoring of trade in technology.

Chapters III and VI discuss the difficulties encountered by analysts in locating statistics that accurately characterize the size and character of U.S. trade in technology. Better data would undoubtedly improve the analysis of trade policy both in the Administration and in Congress. But steps to ensure that the development of techniques for improving monitoring of technology and translating this technique into a workable data-gathering program entail agreement on the following difficult definitional problems:

1. obtaining a clear definition of what represents technology, including embodied technology and high-technology products;
2. refining the definition of "high-technology products;"
3. arriving at some measure for the level of activity conducted under cooperation agreements between U.S. firms and Communist nations; and
4. arriving at a way to acquire accurate, up-to-date, and easily accessible information on the number of turnkey facilities constructed by U.S. firms in controlled-market economies.

Data gathering could be improved in the near term by:

1. obtaining interagency agreement on the format for collecting information from U.S. firms trading in the East; and
2. revising the manner in which data on sales of patents and licenses is kept in accordance with the General Accounting Office (GAO) recommendations (see chapter III).

More explicit reporting requirements have now been applied to the details of any agreements between U.S. companies and Communist nations. This will almost certainly be strenuously opposed in the business community. Similarly, more comprehensive data might be collected on the activities of Communist nationals engaged in academic exchange programs. This would elicit objections from those concerned with the issue of academic freedom.

Clarify the Coordinating Committee for Multilateral Export Controls (CoCom) procedures.

As chapters VIII and IX indicate, the United States has little freedom to change CoCom procedures. Many of our allies are extremely sensitive on the whole issue of CoCom. Some are willing to cooperate in the organization but not willing to publicly acknowledge that they are doing so for fear either of alienating Communist nations or political groups within their own countries. With some caution, however, it may be possible to open the proceedings and deliberations of CoCom to public scrutiny without embarrassing any member nation. For example, it may be possible to be more explicit about the precise nature of the CoCom list reviews and about the internal procedures that lead to alterations. At present, it is difficult to be overly optimistic about the chances for persuading some of our allies to take even such simple steps, for any increase in public information about CoCom could make it more difficult for them to maintain the fiction that they are not active members.

POLICIES DESIGNED TO RESTRICT
EAST-WEST TRADE AND TECHNOLOGY TRANSFER

Recommendations designed to move the balance of current policy in the direction of increased concern about security (and possibly a corresponding relative diminution of interest in trade) would reverse the direction of the evolution of U.S. trade policy since the major export control reforms of 1969 (see chapter VII). Since 1969, the United States has moved in the direction of incrementally normalizing trade with the East. While few argue that we should attempt to restore the strict export controls which were in force immediately following the Second World War, there has been repeated concern that U.S. policy has drifted too far in the direction of promoting short-term corporate profits at the expense of fundamental security concerns. A number of measures have been proposed for reversing this drift:

Enhance the role of the Secretary of Defense in licensing.

This proposal would probably have a greater symbolic than operational impact. DOD now plays a major role in all cases involving questions of military relevance. If the Secretary of Defense is concerned about a license application, the Secretary is in a position to make these views known and, in disputed cases, to demand a ruling from the President. In fact, no item has ever been exported to the East over the objection of the Secretary of Defense. The proposal's symbolic value could be significant, however. It might be a first step in an overall program to persuade allies that the United States is making a serious attempt to increase the importance of security interests in overall trade policy. In the U.S. business community it would certainly be taken as a signal that Government policy was moving in the direction of restricting trade with the East.

Move rapidly to compile a list of critical technologies, and embargo their export.

The preparation of a comprehensive "critical technologies" list was advocated in a report of the Defense Science Board Task Force early in 1976 (the Bucy report). At that time, work was initiated in DOD with a view to implementing the report's recommendations. Little substantial progress has yet been made, however. The chief difficulty is probably not a lack of resources (see chapter V) so much as the inherent difficulty of the task and lack of consensus over the object of the exercise. There has been uncertainty, for instance, about whether the completion of the exercise will have the effect of increasing the restrictiveness of U.S. trade controls (by placing more severe tests on any product or technology qualified for export and increasing the number of such items), or of weakening the controls by limiting the categories of products that would require export licenses. If the former, the establishment of such a list provides an opportunity for strengthening trade restrictions and reducing the flow of products and know-how to the East.

It is difficult to be overly sanguine either about the imminence of results or the degree to which the appearance of a list of embargoed critical technologies will solve the difficult national security problems posed by the existence of dual-use technologies. As chapter V points out, determinations of the degree of risk posed by any particular sale cannot be made in an automatic way. Case-by-case analysis will always be necessary and subjective judgments and policy considerations inevitably enter the decisionmaking process. The critical technology exercise may, at the very least, be a useful in-house exercise for DOD, but it is unlikely to be a panacea for the complex problems posed by East-West technology transfer.

Retain the existing restraints on East-West trade exercised through foreign policy controls, tariffs, and official credits and work to strengthen the leverage available through these mechanisms.

Although the utility of attempting to use trade to extract political concessions is the subject of considerable debate, the tactic is plainly a current part of U.S. East-West trade policy. Indeed, the effect of the Trade Act of 1974, which prohibits the United States from extending most-favored-nation (MFN) tariff treatment and Eximbank credits and guarantees to nations which, in the view of the President, are violating human rights, has been to institute the maximum restrictions possible in this area. The Export Administration Act, moreover, gives the President the authority to deny certain classes of export licenses for reasons of foreign policy (see chapter VII). European and Japanese trading policy contain no such provisions (see chapter IX).

An examination of the effectiveness of using trade leverage as an element of foreign policy must address the following issues:

* Would denial of U.S. technology have a significant impact on the economy of the target nation?
* Would the tactic of using trade to extract concessions have the desired effect?
* What technique could best be employed to curtail trade when such curtailments are required by the leverage policy?

Opportunities for Exercising Leverage

There are only two areas in which the United States is widely thought to have a significant unilateral lead in a technology critical to the Soviet Union: computers and equipment for discovering and producing oil

and gas. But even in these cases, equipment available from other Organization for Economic Cooperation and Development (OECD) nations could provide a workable substitute for U.S. technology. Leverage with respect to the People's Republic of China (PRC) and nations in Eastern Europe may be somewhat greater, but in all cases the flexibility of unilateral U.S. actions is severely constrained by foreign availability. It is clear that a coordinated OECD effort would be needed to significantly increase areas where a threat of an embargo of technology would have significant impact on Eastern economies.

The Utility of Attempting to Exercise Leverage

How will the Soviets or any other Eastern nation react to the exercise, or the threat to exercise trade leverage? There can be no unambiguous answer to this question. Several criteria must be considered.

* How central to the maintenance of the entire system of the target nation and how integral to the country's aims abroad is the object of the leverage?
* To what degree can the issue be framed so that the country can comply without losing face, either at home or abroad?
* Has the country yielded before on this or a similar issue?
* How will compliance be viewed within the country, in the United States, by the rest of the world?
* What stake does the United States have in the issue, i.e., how important are the activities in question of the target nation to the United States?

Soviet response to previous attempts to exercise trade leverage is discussed in chapter IV. The record is ambiguous at best, but certainly when leverage fails, trade is constricted. It must be recognized, of course, that some observers feel that regardless of the practical impact on Eastern economies, the practice of withholding trade from nations whose internal policies are offensive to the United States is amply justified by ethical considerations and the need to focus international attention on those aspects of the Communist system that violate human rights.

The reaction of an Eastern nation to a U.S. attempt to exercise leverage cannot be easily anticipated. In some cases these nations may prefer to accept economic damage rather than make political concessions under pressure. Moreover, even in cases where the United States could inhibit an economic activity clearly critical to the target economy, the nation could take actions that might adversely affect U.S. foreign policy interests in unanticipated areas. A case in point is the embargo of oil and gas extraction equipment to the U.S.S.R. Here,

- The U.S.S.R. might turn to Japanese and Western European firms for massive orders of equipment and technology, many of which would otherwise have come to U.S. industries. The United States would thereby be deprived of revenues and the Soviet Union would receive the help it needs in any case.
- The U.S.S.R. might become unable to export enough oil to Eastern Europe to fulfill its needs. This would increase the economic distress of many Eastern European countries and contribute to domestic unrest, place stresses on the relationships between Warsaw Pact nations, and perhaps force countries of Eastern Europe increasingly on world markets for oil.
- The U.S.S.R. might itself become a net oil importer. This would cause a serious disruption in the present market situation. These potential liabilities would need to be weighed against the potential foreign policy gains that could be realized from an attempt to exercise leverage.

Policy Mechanisms

The United States has restricted trade for political purposes in three ways: Presidential intervention in the granting of export licenses, tariff policies, and denial of official credits and guarantees.

Executive Discretion in Granting Export Licenses

Existing law gives the President the authority to withhold export licenses in cases where the denial would serve broad foreign policy interests. This power, and the flexibility it implies, can be used in a timely way to influence fast-moving events: licenses are always pending in OEA. The need to make a rapid political point, however, may lead to a somewhat arbitrary selection of the license to be denied, the most convenient licensing issue at hand being seized for the purpose.

Presidential use of the power to deny exports for reasons of foreign policy has been the subject of a number of recent controversies. Use of such controls may be effective as statements of principle on the part of the United States, but there is no doubt that the use or threatened use of such power has introduced an element of unpredictability into the export-licensing system. This may have adversely affected the bargaining position of U.S. corporations in the East. A good example is the recent denial of a license for a Sperry-Univac computer ordered by the U.S.S.R. This action has had greater symbolic than practical value. The computer itself was identical to models already installed in the U.S.S.R. and its capabilities inferior to those of previously licensed computers. Moreover, the Soviets have now purchased an equivalent computer from France (see chapter IX).

It is possible for the President to be given greater latitude in controlling trade as an element in an overall program to indicate U.S. interest in strengthening its determination to confront the Soviet Union and other Communist nations. But, as the above example would suggest, the value of such Presidential latitude as a foreign policy tool in the absence of complete cooperation from our allies is open to question. This is the subject of more complete discussion in chapter IV.

Most-Favored-Nation Status.—The economic value of MFN varies among individual Communist nations, depending on their export mix. A detailed discussion of the economic aspects of granting MFN appears later in this chapter. For leverage purposes, however, its symbolic importance to the PRC and U.S.S.R. is perhaps even more significant. Granting MFN to China while still withholding it from the Soviet Union will almost certainly in itself be construed as an attempt to apply leverage in foreign policy.

Official Credits.—The availability of cheap Government credits and loan guarantees is a matter of substantial importance in the Communist world where hard-currency shortages restrict the ability to purchase Western goods and technologies (see below and chapter III). The high level of Soviet borrowing before the cutoff of Eximbank credit in 1974 (see chapter VII) suggests that the impact of this policy was felt strongly in the U.S.S.R. China's modernization plans will require substantial use of Western credit (see chapter XI) and the PRC is obviously anxious to take advantage of credits in the United States. Again, extension of Eximbank and other official credits to the PRC and not to the U.S.S.R. could have serious foreign policy repercussions. As noted above, restrictions on official credits to most Communist nations are already as stringent as possible. Further contraction would require legislation covering the extension of U.S. private commercial credits to the Communist world.

Attempt to limit as far as possible the foreign availability of technologies that appear only on the U.S. export control lists.

One way of accomplishing this is through present U.S. policy requiring reexport licenses for technologies originally developed in the United States, sold to a CoCom member, and subsequently shipped to a controlled country. Under this rule, West European and Japanese businessmen obtain both U.S. and CoCom licenses for such shipments. The United States is the only CoCom member to impose this kind of control, and its existence conveys an impression that the United States lacks confidence in the CoCom mechanism. But such a policy obviously has no impact on the availability of technologies not of U.S. origin. Efforts to limit such sales through diplomatic efforts outside of CoCom have met with limited success (see chapters VIII and IX). It may be possible to undertake bilateral or multilateral agreements with OECD nations with conservative governments if these nations can be persuaded to reverse the trend of European and Japanese trade policy and entertain more restrictive export programs.

Attempt to strengthen CoCom by recognizing the organization through treaty; increasing policing of CoCom decisions, and/or formalizing sanctions to be used against transgressors in member nations; or expanding the CoCom list to more closely conform to the U.S. Commodity Control List (CCL).

There is every indication that, at present, suggestions for a longer list, more stringent policing or the imposition of sanctions in CoCom would be strenuously resisted by some members. Attempts to strengthen the organization would probably better prosper through the continuation of quiet, informal, high-level negotiations. Only given a different international climate and a broad change in U.S. foreign policy, might the United States be able to persuade its allies to alter their policies on East-West trade. If, for instance, the United States entirely abandoned detente and adopted a clearly confrontational policy with respect to the Soviet Union, parties and individuals in other Western governments more sympathetic to a strengthened CoCom might be expected to grow in influence.

Curtail the transfer of information through academic and scientific exchange programs by controlling the subject matter and/or facilities to which visiting scientists and scholars are admitted.

It has frequently been charged that academic and scientific exchanges are an important and relatively unmonitored and uncontrolled source of technology transfer. A consistent attempt to restrict the flow of potentially strategic information to the East would have to include careful supervision of this channel, at least as it allows visits from high-level technicians in strategic areas. This might amount to the determination that the potential danger to the national security of the United States of such visits outweighs the political, scientific, and cultural benefits which accrue from exchange programs. It must be pointed out, however, that many regard these passive mechanisms of transfer as far less likely to result in the ability to absorb, diffuse, and improve on a technology than are other commercial channels. Moreover, even those who see such exchanges as important channels of technology transfer are often reluctant to impose any requirements on the institutions involved for fear of violating academic freedom. In this sense, there may be something of a double standard in regard to the constraints placed on academic versus commercial exchange activities. Proposals to require company reporting of commercial exchange agreements have received more widespread support.

POLICIES DESIGNED TO ENCOURAGE EAST-WEST TRADE AND TECHNOLOGY TRANSFER

The special role of technology in East-West trading relationships is difficult to define. It is theoretically possible to increase trade volumes without increasing trade in high technology, but this opportunity is ultimately limited by the fact that without some increase in technological sophistication, the Eastern trading partners will be unable to improve their capacity for earning hard currency. Moreover, it is often impossible to decouple sales of technology from sales of products in the corporate strategies of many U.S. firms (see chapter III).

Policies oriented toward expanding and encouraging U.S. trade with the East fall into two categories: 1) measures designed to change the trade/security balance in the direction of increased trade in technology by relaxing some of the current restrictions inherent in the licensing regulations, and 2) measures designed to increase the overall level of U.S. exports to the East. Most of the latter are indistinguishable from the family of suggestions that have been made for generally improving the U.S. export enterprise.

Many elements of the U.S. business community have pressed anxiously for a relaxation of those regulations restricting U.S. technology trade with the East which, in their view, put them in a position of disadvantage vis-a-vis European and Japanese competitors. Further steps to encourage the transfer of U.S. technology would inevitably risk repercussions on U.S. national security, while more general trade policies aimed at products rather than technology entail lesser risks.

It must be recognized, however, that there is little likelihood that even complete removal of political barriers would lead to vastly increased trade between East and West. Such trade is now limited primarily by a shortage of hard currency in the Communist world. While increasing the availability of debt financing can provide short-term gains, the only long-term gains likely to be achieved must be the result of an increased ability of Eastern nations to export. Promoting trade, therefore, must ultimately lead to a promotion of the strength of the trading partner.

This imposes commercial risks in any trading relationship, since there is a danger that a strengthened trading partner will begin to replace U.S. exports with domestic production and possibly begin exporting into third nations that currently purchase U.S. goods. In the case of Communist nations, there is the additional risk that a strengthened economy will inevitably lead to increased military prowess.

The following policies aimed at the expansion of U.S. trade with the Communist world must be considered in light of these caveats, and the potential benefits accruing in the U.S. economy must be weighed against them.

Expand export financing.

Given the chronic hard-currency shortages in the East, access to Eximbank and Commodity Credit Corporation guarantees and credits is probably the single most important factor in significant expansion of overall levels of U.S. trade with the East.

The United States ranks fifth behind West Germany, France, Japan, and the United Kingdom as a supplier of official credits to the Communist world. In 1977, West Germany extended nearly eight times

Photo credit: U.S. Department of Commerce

Former Secretary of Commerce Juanita Kreps in China for U.S.-PRC trade negotiations

more export credit to the East than did the United States, France over seven times, and Japan about five times. While no quantitative estimates of the impact of this situation on overall trade volume are available, many U.S. exporters contend that it puts them at a serious competitive disadvantage vis-a-vis other OECD nations. There are indications, for instance, that the availability of credit is an important reason for multinational corporations' preferring to handle their Eastern transactions through European subsidiaries.

Although purchases of U.S. technology would almost certainly increase with the availability of credits, a program to expand credit in the East might be targeted at financing for "nonstrategic" commodities such as grain. Credit expansion would require both raising the ceilings on available credit, and also eliminating barriers to financing exports to the East posed by the Trade Act of 1974. Congress might, for instance, make financing equally available to all countries with which it is U.S. policy to encourage trade, subjecting the policy to periodic review for individual nations. The availability of Government credits would be attractive to those Eastern nations in which borrowing has risen dramatically in recent years, but which are still considered good credit risks in the West (see chapter III).

Grant MFN to countries not presently enjoying it.

Amending the Trade Act to allow nondiscriminatory trade treatment for Communist countries not presently receiving it would also have an impact on levels of U.S. trade with the East. However, at least in the short run, these increases would not materially affect the U.S. balance of trade—either in terms of flooding the United States with Communist imports or of significantly increasing the prospects for U.S. exports. In order to purchase goods and technologies from the West, Eastern nations must earn hard currency through their own exports (see chapter III). Nondiscriminatory tariff levels, by encouraging import of Eastern goods in the West, can thus indirectly affect the volume of Eastern purchases.

A number of factors determine the extent to which levels of trade with individual nations can be increased through the granting of MFN. Foremost among these is the commodity composition of the exports. The extension of MFN does not result in a uniform reduction of tariffs. Under the existing Hawley-Smoot tariff scheme, tariffs on some items are more severe than on others; further, negotiations over the years have resulted in differing rates of relaxation, and the granting of MFN leads to considerable variation in tariff reductions. For example, the Soviet Union is predominantly an exporter of raw materials which at present have relatively low tariffs. Czechoslovakia, on the other hand, is an exporter of light manufactured products on which high tariffs are levied. From a purely commercial point of view, extension of MFN would affect Czechoslovakian exports relatively more, although MFN retains a great symbolic value to the U.S.S.R.

In 1977, the U.S. Department of Commerce estimated that the extension of MFN to the two largest Communist economies, the U.S.S.R. and the PRC, would result in a very modest increase in U.S. imports from those countries—together in the $30 million to $40 million range. This may be contrasted with the expected $50 million increase resulting from recent MFN extension to Hungary. Extension of MFN to Czechoslovakia and East Germany if accompanied by concurrent normalization of commercial relations could increase imports from those countries by as much as $200 million in the absence of other, nontariff barriers. And MFN extension to all other Communist nations could mean an increase in imports from those countries of between $200 million and $225 million. This figure would represent less than 0.2 percent of total U.S. merchandise imports in 1977.

These figures are based on the assumption that volume of U.S. trade with the East and

its commodity composition remain static. However, long-term effects could be significantly greater. To the extent that the removal of political barriers fosters familiarity both with U.S. markets and with U.S. producers, U.S. imports from the East and consequent U.S. export potential may be improved in the future. Eastern nations will not only earn hard currency through their sales; MFN agreements create legal and financial structures under which commercial interaction can be carried out more efficiently and with more certainty for U.S. entrepreneurs. In this regard U.S. business operates at a disadvantage vis-a-vis West European and Japanese firms in marketing products in the East.

Owing to relatively late normalization of trading relations with the East, however, U.S. business may be at a permanent disadvantage with its OECD competitors in Eastern markets, even if MFN is extended in the near future. Moreover, a note of caution is warranted: while such background agreements are a necessary condition for greater U.S. exports, they are not sufficient. Expectations regarding significant increases in Romanian purchases in the United States since extension of MFN in 1975, for example, have not materialized. The entire trade climate, including availability of credit and level of export controls, must be taken into account when projecting the possible impacts of MFN extension.

Restrict the President's ability to impose trade restrictions for reasons of foreign policy.

A proposal recently adopted by Congress requires that the President stipulate that reasonable efforts had been made to achieve foreign policy objects through other instruments than trade before the leverage of trade restrictions could be exercised. In addition, some have argued that Congress also be empowered to overrule such prohibitions by concurrent resolution. The enactment of such a regulation would have some effect in reducing the uncertainties inherent in the potential use of trade for foreign policy purposes. The uncertainties would not be entirely removed, however, since in many cases Congress could be expected to concur with the President. Moreover, restrictions on the President's freedom to use trade in this way could, in some circumstances, dilute the admittedly weak leverage now available to the United States.

Bring the U.S. CCL into closer conformity with the CoCom list and/or "index" the list to allow for automatic removal of obsolete or out-of-date technologies.

It is argued that U.S. producers are uniquely disadvantaged by the fact that the U.S. CCL is more restrictive than those maintained in other CoCom nations. At the heart of this issue is the question of whether the United States because of its technical strengths has a special responsibility for restricting categories of products beyond the CoCom list. While the perception (and the reality) of broad U.S. technological leadership has changed during the last decade, the United States does retain some supremacy in certain military technologies and therefore a special responsibility for safeguarding them. Whether this special responsibility is fairly represented in the U.S. export list has become a matter of contention.

A less comprehensive means of scaling down the CCL is to index it, i.e., to require annual updates on the performance levels of goods and technologies and the automatic removal of items that fall below these levels. There is obviously a range of standards by which such levels might be set. Items removed from the list because they are obsolete in terms of the Western state-of-the-art might still significantly improve existing military capabilities in the East. Other standards, based perhaps on levels already sold to or developed in the East, might involve less chance of this, but any automatic alteration of CCL entails the danger of eliminating items of potential military significance.

Alter the present export-licensing system so that it more closely resembles those of other CoCom nations.

Descriptions of the export-licensing systems of West Germany, France, the United Kingdom, and Japan (see chapter IX) reveal significantly more informal consultation between industry and Government over license applications than in the United States. In many cases this includes prior consultation which permits firms to know the disposition of their cases before applications are actually submitted.

Unquestionably, U.S. export-licensing procedures are universally regarded as the most time-consuming, rigorous, and uncertain of all CoCom nations. To the extent that delays or denials of licenses result in loss of contracts or deter Eastern countries from seeking out U.S. suppliers or U.S. firms from actively pursuing business in the East, this may have an impact on U.S. market shares in existing East-West trade. It is unlikely, however, that the relaxation of these controls would have much effect on increasing overall trade volume without concurrent alteration of credit and tariff policy in the United States and increase in export capabilities in the East (see chapter III). Moreover, the licensing systems in Europe and Japan reflect judgments on East-West trade and national security that have not been shared by U.S. policymakers. They also reflect close and consensual business-Government relationships that are not typical of private and public sector relations in the United States. It is unlikely that U.S. institutions would readily lend themselves to procedures which are predicated on such consensual relationships.

Create a single Government agency charged with advancing U.S. trading interests.

Proposals have been made for establishing a Department of Trade incorporating a variety of trade-related Government activities. Alternatively, the existing Office of the

Special Trade Representative in the White House could be established as a permanent organization and enlarged to embrace trade policy coordination and trade negotiations with the East.

The United States has no concerted, coordinated policy on East-West trade. The development of such a policy would be welcomed in many quarters of the Government and business communities. Proponents of a Trade Department argue that it could help to counter activities of organizations in other nations where the interests of business and Government are more closely tied than is the case in the United States. Japan, where trade-related ministries work very closely with private industry, is an extreme example. Whether such a system is either appropriate to or even possible in the United States is open to question, and a complete analysis of the issue is beyond the scope of the present study. Objections may be expected to the creation of a new bureaucracy, especially if this is not accompanied by conscious formulation of coherent and internally consistent aims in East-West and other world trade.

Relax antitrust restrictions inhibiting consortia of U.S. industries organized for export.

Present antitrust law is vague and has sometimes been narrowly interpreted as it applies to the cooperative activities of U.S. firms abroad. Revision to allow various export trade associations, trading companies, etc., could help U.S. firms to compete with Japanese trading companies and European bidding consortia, without precluding the possibility of penalties for impeding fair competition.

Increase participation and improve performance of overseas Government personnel in fostering U.S. exports.

American businessmen have charged that, unlike their Western European and Japanese counterparts, U.S. embassies do little if anything to further U.S. commercial interests abroad. Redefinition of the responsibilities of commercial attaches or other State Department personnel to explicitly include active support for businessmen attempting to conclude foreign contracts (so long as such aid does not discriminate among U.S. firms) would help to eliminate some of the competitive edge presently enjoyed by firms in other nations. If it is felt that such activities are inappropriate for existing embassy staff, new export-related offices could be created in appropriate countries.

Bolster the U.S. R&D enterprise.

All other things being equal, trade with the East will benefit from the same kinds of measures that promote U.S. foreign trade in general. Moreover, regardless of whether East-West trade in technology expands or contracts, the best way to ensure continued U.S. superiority in technology is to maintain a vigorous program of Federal and private R&D projects. Attempts to control the export of technology can be effective only up to a point. It would be foolhardy to rely entirely on such controls to maintain a position of relative technical strength. Investigation of all the actions that could be taken to strengthen the R&D enterprise in the United States would require a much lengthier analysis than can be conducted here. Several proposals for mitigating certain perceived barriers to expansion of U.S. R&D and for providing incentives for accelerated R&D are listed in table 1.

Table 1.—Recommendations for Bolstering the U.S. R&D Enterprise

Amelioration of disincentives	Establishment of incentives
• Modification of antitrust regulations to permit easier pooling of research efforts for environmental improvements. • Institution of a uniform patent and licensing policy for Government-sponsored research. • Passage of legislation controlling third-party product liability litigation. • Partial stabilization of raw material costs by stockpiles, trade agreements, and long-term national planning. • Modification of Government regulations to make them less expensive and time-consuming, while still achieving the desired goals. • Continuation of the effort to strengthen the present patent system. • Development of better integration of antitrust laws and patent laws. • Extension of the life of a patent beyond 17 years, possibly for a period of 10 to 15 years after final Government approval, if long-term testing is required for Government approval, or other factors delay commercialization.	• Greater than 100-percent deductibility of research expenses from taxable income. • Grants-in-aid for cost of new research facilities and/or equipment. • Tax credits (possibly at the rate of 80 percent) for increases in industrial R&D over base-period expenditures. • Exclusions from taxable income or part of any royalty received from the export of technology. • Accelerated depreciation allowances for research facilities and equipment. • Long-term low-interest loans for high-risk R&D. • Cash grants repayable only from successful projects for high-risk R&D. • Inclusion of R&D expenses under the 10-percent investment tax credit provision. • Initiation of a technological depletion allowance program. • Deduction (or accelerated depreciation) of the cost of new technology or patents. • Special, low capital gains taxation for small businesses engaged in R&D. • Institution of an option for small businesses to capitalize their research expenses. • Direct deduction from Federal income taxes of all expenses incurred in the performance of research associated with Federal regulations. This deduction can be prorated, at 80 percent for example, so that the Federal Government and companies can share the expenditures roughly in proportion to the direct benefits obtained from the research. • Increase of Federal support of basic research in universities to compensate for decreased basic research in industries. • Encouragement of cooperative research between universities and private industries.

SOURCE: The American Chemical Society, U.S. Chemical and Engineering News, Apr. 30, 1979.

The Economic Implications of East-West Trade and Technology Transfer

CONTENTS

The Economic Implications of East-West Trade and Technology Transfer

Policy decisions regarding the future of U.S. trade, including transfer of technology, with the Communist world require weighing the economic and political benefits of such trade against the military risks it may incur. This is difficult, not only because it entails the comparison of unlike things, but because the economic merits, particularly of technology sales, are difficult to assess and the results of such assessments are controversial. For instance, the profits and other returns of technology exports to the East must be balanced against the possibility that unrestrained technology transfer by U.S. corporations could be detrimental to this country's long-term economic interests. In this connection, the economic dangers of technology sales currently lie primarily in transactions with our Western trading partners and not in trade with the East where major risks are military; the Communist nations at the moment have a relatively small export capacity and a systemic difficulty in rapidly assimilating and diffusing Western technology. This is not to say, however, that this situation could not change, especially with the help of Western management expertise and Eastern impetus to expand trade in technology.

The economic balance sheet which must be drawn up in technology trade includes the following considerations: On the positive side are the final gains resulting from the sale of patents, licenses, construction of turnkey plants, and the sale of items that may embody sophisticated technology. The balance of payments in such items has historically been decidedly in favor of the United States. Further, even where the direct income from technology transfers is small, broad agreements in other trading areas often depend critically on such transfers, and there may be indirect commercial benefits to U.S. firms operating in Eastern markets.

On the other hand, the possibility exists that transferred technology can be used to build industries in the purchasing nation which will eventually supplant U.S. export markets in that country or in other nations, perhaps eventually even in the United States itself. These situations would clearly threaten a loss of employment in the United States. The difficulties now being encountered by Occidental Petroleum in the United States and by other companies in Italy and West Germany as a result of buy-back agreements with Eastern Europe highlight these fears, as does the U.S.S.R.'s emergence as a competitor to Fiat in Europe and Canada with cars produced at the Italian-built Togliatti (Volga) auto plant.

In an attempt to evaluate the economic value of East-West trade and technology transfer to the United States, this chapter examines the following issues:

- the volume of East-West trade in general and trade in technology in particular,
- the potential for growth in East-West trade, including trade in technology, and

- the impact of sales of technology to the East on the U.S. economy.

Discussions of the value of Western technology to the Soviet and Chinese economies appear in chapters X and XI.

Photo credit: The National Council for U.S.-China Trade

Ammonia concentrator, La Madian #2 Multipurpose Pump Station, Taching, China

EAST-WEST TRADE AND U.S. MARKET SHARES

THE PRESENT VOLUME OF EAST-WEST TRADE

Trade with the Communist world has never constituted a large part of U.S. foreign trade. Despite the fact that the total turnover of American trade with the East grew by approximately 50 percent between 1977 and 1978, the volume of this business in absolute terms, including sales of agricultural commodities, is small. In 1978, the United States earned about $4.5 billion from exports to Communist nations, half of which came from the U.S.S.R. The net trade balance with these countries was $2.7 billion. This must be evaluated in the context of 1978 U.S. worldwide trade turnover of over $315 billion and overall deficit of $28 billion. The Communist world thus accounted for only 3.1 percent of U.S. exports and 1 percent of U.S. imports in 1978. Even the recent acceleration in trade with the People's Re-

public of China (PRC) has done very little to alter this overall trade picture (see table 2).

Part of the reason for these magnitudes lies in the fact that for both trade in general and trade in technology in particular, the United States has captured only a small share of the Eastern market relative to the other countries of the industrialized West. Since the end of World War II, the United States has never held more than 10 to 15 percent of the total Western trade with Communist nations (see table 3 and figure 1). There are a variety of reasons for this: because of its vast domestic market the United States has traditionally been relatively less active in foreign trade than Japan or Western Europe; Western and Eastern Europe are natural trade partners; and as chapters VII to IX argue, America's allies have been less restrictive in controlling trade with the Communist world.

Table 2.—U.S. Trade With the World and With Selected Nonmarket Economy Countries (in millions of U.S. dollars)

U.S./world trade

	1977	1978	January to June 1979
Exports	121,206	143,659	85,532
Imports	147,492	172,025	95,506
Balance	− 26,286	− 28,366	− 9,973
Trade turnover (exports plus imports)	4,077	6,303	3,972

U.S. trade with selected nonmarket economies

	Exports			Imports			Balance		
	1977	1978	Jan-Jun 1979	1977	1978	Jan-Jun 1979	1977	1978	Jan-Jun 1979
U.S.S.R.	1,623	2,249	1,457	422	254	243	1,201	1,995	1,214
People's Republic of China	171	818	704	197	324	245	− 26	593	459
Poland	437	677	275	327	439	212	110	238	63
Romania	259	317	260	231	347	167	28	29	93
Czechoslovakia	74	105	83	36	163	25	38	47	58
East Germany	36	170	138	17	205	19	19	135	119
Hungary	80	98	42	47	69	48	33	29	− 6
Bulgaria	24	48	31	26	19	23	− 2	29	8
Total	2,704	4,483	2,990	1,303	1,820	982	1,401	2,663	2,008

	1977	1978	January to June 1979
NME share of total U.S. trade			
Exports (percent)	2.2	3.1	3.5
Imports (percent)	.9	1.09	1.03

NOTE: Both imports and exports are valued on a free-along-side basis.
SOURCE: U.S. Department of Commerce.

Table 3.—The Trade of the Industrial Market Economies With Eastern Europe and the Soviet Union

	Imports, c.i.f.					Exports, f.o.b.					Trade balances		
	Value (million U.S. dollars)	Percentage share of country's total imports	Percentage change over the same period of the preceding year			Value (million U.S. dollars)	Percentage share of country's total exports	Percentage change over the same period of the preceding year			Exports minus imports[a] (million U.S. dollars)		
	1977	1977	1976	1977	Jan.-May 1978	1977	1977	1976	1977	Jan.-May 1978	1976	1977	Jan.-May 1978
West Germany . . .	$ 4,474	4	25	11	27	$ 6,649	6	− 3	6	14	$2,439	$2,327	$ 952
Italy	2,596	5	28	6	2	2,287	5	− 10	16	− 10	− 339	− 153	− 94
Yugoslavia	2,714	28	16	26	4	2,044	39	8	1	14	55	− 446	− 123
France.	2,216	3	18	11	22	2,781	4	5	2	− 12	832	663	119
United Kingdom. .	2,172	3	23	16	5	1,457	3	− 9	23	35	− 521	− 584	11
Finland	1,795	24	2	11	9	1,709	22	14	14	24	2	32	85
Austria	1,249	9	15	14	15	1,416	14	1	10	17	228	214	114
Sweden	1,141	6	4	0	− 22	945	5	− 6	− 8	− 1	− 19	− 116	56
Netherlands	1,040	2	18	11	15	816	2	− 4	7	5	− 116	− 171	− 57
Belgium-Luxem. . .	718	2	− 5	22	28	760	2	− 8	− 4	− 2	222	77	1
Denmark	593	4	10	4	9	286	3	− 9	2	1	− 263	− 275	− 124
Switzerland.	598	3	45	20	78	878	5	7	11	34	309	298	102
Spain.	345	2	− 7	− 20	− 11	284	3	18	− 6	15	− 91	− 32	17
Greece.	385	6	47	− 5	54	334	12	9	16	8	− 70	− 10	− 1
Norway	395	3	38	15	− 3	276	3	8	0	11	− 48	− 107	− 71
Turkey	344	6	31	7	1	172	10	35	4	35	− 121	− 135	1
Ireland.	119	2	0	30	1	29	1	− 43	46	51	− 67	− 85	− 30
Iceland	75	12	7	15	12	62	12	1	51	− 51	− 14	− 6	− 9
Portugal.	166	3	97	1	14	81	4	96	− 3	− 37	− 66	− 70	− 41
Total Western Europe.	$23,135	5	18	12	13	$23,266	5	0	7	9	$2,352	$1,421	$ 908
United States	977	1	19	5	49	2,542	2	26	− 27	35	2,638	1,626	1,203
Canada	189	0	22	− 4	− 11	546	1	31	− 31	15	595	362	132
Japan.	1,627	2	− 1	19	0	2,669	3	27	− 5	2	1,590	1,225	625
Subtotal	$ 2,793	1	7	12	16	$ 5,757	2	27	− 18	19	4,823	3,213	1,960
Grand total	$25,928	3	17	12	13	$29,023	4	5	1	11	$7,175	$4,634	$2,868

[a]Exports f.o.b. minus imports f.o.b. The latter have been adjusted according to data taken from IMF, *International Financial Statistics*, for each industry.

SOURCES: OECD, *Statistics of Foreign Trade, Series A*, Paris; IMF, *Direction of Trade* and *International Financial Statistics*, Washington, D.C.; national statistics; and U.N. Economic Bulletin for Europe, vol. 30, No. 1, New York, 1978.

But equally important is the fact that overall volumes of East-West trade are artificially low. Foreign trade has played a relatively minor role in the Communist world and within this already circumscribed arena, the volume of East-West trade is particularly small. China until very recently virtually excluded itself from world markets. The Soviet Union has been far more active in world trade, but in 1977 imported only $150 in goods per capita, as compared to $700 for the United States.

U.S. POLICIES AFFECTING TRADE VOLUMES

American policies on trading with the Communist world probably influence U.S. market shares in existing trade more than they do the volume of East-West trade overall. These shares are determined generally by U.S. foreign trade and export promotion strategies and in particular by credit, tariff, and export control regulations directed at Communist countries.

Figure 1.—East-West Trade and U.S. Share

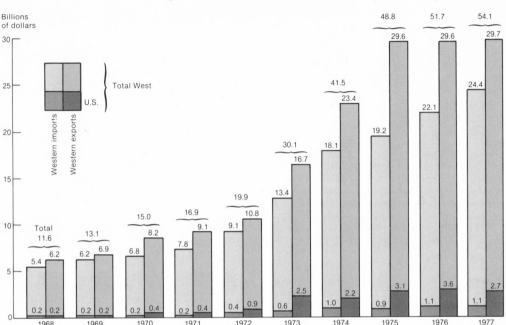

East = Bulgaria. Czechoslovakia. East Germany. Hungary. Poland. Romania. U.S.S.R.. and PRC.
West = Austria. Belgium. Canada. Denmark. West Germany. France. Italy. Japan. Luxembourg. The
Netherlands. Norway. Sweden. Switzerland. United Kingdom. and the United States.

SOURCE: *Selected Trade and Economics Data of the Centrally Planned Economies,* U.S. Department of Commerce, Industry and Trade Administration, Bureau of East-West Trade, 1979.

Import Barriers

Eastern exports have elicited strong protectionist sentiment among some American producers, and the tendency has been for commercial import policies to remain restrictive even in the face of stimulative export strategies. In a number of Western countries, the United States among them, both tariffs and quantitative restrictions, and voluntary restraints inhibit the quantities of Eastern goods that are imported. The aim of nontariff barriers is to help the balance of payments and in particular to assist import competitive labor-intensive industries such as woodworking, textiles, and shoe manufacturing. There is now even discussion of extending protection to such technology-intensive products as electronics and chemicals, in which Western countries enjoy or have enjoyed a comparative advantage.

U.S. action on tariffs—notably the denial of most-favored-nation (MFN) status to most Communist nations—has been politically rather than economically motivated (see chapter VII). It is virtually impossible to link the lack of MFN status directly to existing levels of trade, although it is unlikely that the extension of MFN over recent years would have led to dramatic increases in Eastern imports. As chapter II has pointed out, however, the removal of the political barriers to trade symbolized by the U.S.'s withholding of MFN status might, over the long run, contribute to more regular and expanded trade relations with the East.

Credit

Chapter VII documents the history of the U.S. restrictions on the amount of subsidized official export credits available to the Communist world. The availability of such financing is often an important factor in the choice of a Western supplier. The curtailed role of the United States in this area can be seen by comparing it to other Western nations (see table 4). The Chase World Information Company has estimated that at the end of 1977 outstanding commitments on export credits extended to Eastern Europe and the U.S.S.R. by Western governments totaled nearly $32 billion. The U.S.S.R. and Poland, which together are responsible for nearly 60 percent of the Communist bloc's total hard-currency debt, received the bulk of these —$14.2 billion and $8.3 billion respectively. West Germany was the greatest lender, providing official credits of $7.5 billion; France ranked second at $7 billion, followed by Japan at $5 billion. The United States, with the activities of the Export-Import Bank (Eximbank) severely curtailed by Congress, ranks fifth, after the United Kingdom, with $945 million.

This fact may support the contention of some U.S. exporters that the availability of cheap official credits in other nations puts them at a competitive disadvantage. It also highlights the limited role of the U.S. Government in promoting exports to the East. The impact of such credit policies on U.S. trade cannot be assessed with any precision, but the general effect seems to have far outweighed any positive actions to encourage trade with the Communist world.

U.S. Export Controls

A third important factor in the maintenance of low levels of trade with the Communist world is the restrictions imposed on technology sales. These, as well as the attitudes of U.S. businessmen toward them, are described in detail in chapter VII. There is a widespread perception among businessmen that U.S. export control policies are a significant, if not the most important, barrier to expansion of U.S. trade with the East.

It is impossible to estimate the amount of business lost to American companies because of the stringency or inefficiency of export controls and licensing procedures, but it is probably safe to assume that—the perceptions of some businessmen apart—it is by no means the predominate factor. An as yet unpublished report being prepared by the U.S.

Table 4.—Official Export Credit Commitments to CMEA Countries, as of End-1977
(in millions of U.S. dollars)

	Bulgaria	Czech.	East Germany	Hungary	Poland	Romania	U.S.S.R.	Total
Commitments on signed contracts offered by:[a]								
Austria	$ 183	$ 85	$ 455	$ 395	$ 600	36	$ 260	$ 2,014
Britain	30	50	45	40	960	100	720	1,945
Canada	0	3	0	0	454	9	173	639
France	540	350	480	110	1,800	390	3,400	7,070
West Germany	140	450	1,200[b]	65	1,900	430	3,300	7,485
Italy	80	70	530	70	800	200	1,950	3,700
Japan[c]	280	0	400	200	450	500	3,150	4,980
United States	0	0	0	0	408[d]	74	463	945
Other	265	195	465	95	950	215	750	2,935
Total	$1,518	$1,203	$3,575	$ 965	$8,322	$1,954	$14,166	$31,713
Estimated drawings on official credits[e]	798	841	2,455	460	5,775	1,256	10,730	22,315
Undrawn balance	720	362	1,120	515	2,547	698	3,436	9,398

[a]Refers to active commitments of official credit. Figures take into account maturing credits and are adjusted for repayments.
[b]Intra-German trade swing credits.
[c]Includes supplier credits that are provided jointly by Japan's Eximbank and commercial banks.
[d]Includes $220 million in U.S. Eximbank commitments and $188 million in CCC credits.
[e]Approximate disbursements.

SOURCE: Adapted from a review of CMEA debt by Miriam Karr in *East-West Markets*, Chase World Information Co., May 15, 1978, p. 3, and May 29, 1978, p. 3.

International Trade Commission (ITC), for instance, investigated cases involving the loss by U.S. firms of 85 separate contracts with the U.S.S.R. between 1972 and 1977. Noncompetitive price was cited by the firms involved more than twice as often as any other reason for the failure. Inability to obtain Government credits, guarantees, and insurance; and competition from firms with a better foothold in the Soviet market were next more frequently mentioned. Export controls and license delays appeared far down the list.

EASTERN POLICIES AFFECTING TRADE VOLUMES

While part of the reason for the low volume of U.S. trade with the East may be attributed to American failure to capture high market shares, decisions on the other side of the Iron Curtain have had greater impact on the nature and extent of East-West commercial relations.

The great majority of Soviet and Eastern European trade is conducted within the Council for Mutual Economic Assistance (CMEA or COMECON). The members of CMEA are the U.S.S.R., Poland, East Germany, Czechoslovakia, Romania, Hungary, Bulgaria, Cuba, Mongolia, North Korea, and Vietnam; associate agreements have been concluded with Yugoslavia and Finland. The CMEA, founded in 1949 as the Soviet response to the Marshall plan, was intended to give the Communist bloc economic as well as political and military cohesion. It provides for the exchange of economic and technical information among Socialist countries, and approximately 70 percent of all Eastern-bloc trade takes place within it. Potential trade with the West is circumscribed by the politically motivated controls imposed on CMEA members. These are both direct and indirect. For example, Eastern European dependency on Soviet raw materials diminishes opportunities for Western raw material exports. Further, Eastern European manufactures are frequently of such design and quality that they can be marketed only in the Soviet Union. The effect of CMEA, together with

the barriers to complete economic interdependence posed by the structural differences between market and nonmarket economies, and the lack of hard currency in the East (discussed below) work against the possibility that East-West trade will ever rise to levels comparable to those between Western nations.

THE GROWTH OF EAST-WEST TRADE

Barriers to increased trade in both the East and the West have eroded steadily since the onset of the era of detente. As figure 2 demonstrates, although absolute

Figure 2.—U.S.-Eastern* Trade, 1972-78**

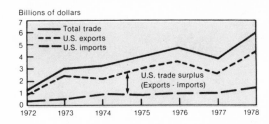

Billions of dollars

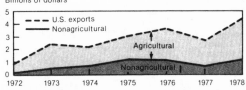

U.S. exports:
Billions of dollars

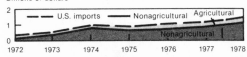

U.S. imports:
Billions of dollars

*Bulgaria, Czechoslovakia, East Germany, Hungary, Poland, Romania, U.S.S.R., and PRC.
**1978 trade estimated imports do not include U.S. imports of nonmonetary gold from U.S.S.R.

SOURCE: *Selected Trade and Economic Data for the Centrally Planned Economies,* U.S. Department of Commerce. 1979.

levels of East-West trade have been small, it has grown rapidly in recent years. Table 5 shows this growth in absolute terms. Table 6 demonstrates the fact that the rate of growth in trade between the industrialized West and the East has consistently outrun world trade as a whole from 1955 to the present. This trend has been particularly manifest in Eastern exports of raw materials and labor-intensive commodities and imports of manufactured goods.

On the Eastern side, an important set of reasons for this growth lies in three interrelated decisions made at some point in the development of each Communist nation. First, the policy of detente involved a political decision to expand contacts with the West in all fields. An attempt was made to exploit the advantages offered by trade with Western States, but to avoid if possible the social and political liabilities inherent in East-West communication. A second decision involved a shift in development strategy, which required the use of advanced Western capital and techniques to increase productivity in specific sectors. Finally, purchases in the West began to be utilized on a wider basis to compensate for shortfalls in annual plans. This has been especially true for agricultural products and, to some extent, consumer goods. The result has been that increases in Eastern imports from the West have occurred at a greater rate than has expansion of exports.

Table 5.—Trade With the Developed West
(in millions of U.S. dollars)

	1972	1973	1974	1975	1976	1977
Bulgaria						
Exports	$ 310	$ 403	$ 403	$ 363	$ 420	$ 392
Imports	349	480	928	1,204	940	821
Balance	− 39	− 77	− 525	− 841	− 520	− 429
Czechoslovakia						
Exports	921	1,266	1,639	1,600	1,600	1,698
Imports	1,056	1,513	2,031	2,178	2,178	1,443
Balance	− 135	− 247	− 392	− 578	− 578	− 245
East Germany						
Exports	1,406	1,915	2,646	2,586	2,850	2,695
Imports	1,929	2,735	3,540	3,630	4,050	2,906
Balance	− 523	− 820	− 894	− 1,044	− 1,200	− 211
Hungary						
Exports	739	1,085	1,221	1,096	1,290	1,562
Imports	851	1,135	1,862	1,843	1,860	2,195
Balance	− 112	− 50	− 641	− 747	− 570	− 633
Poland						
Exports	1,397	2,063	2,865	3,026	3,330	3,495
Imports	1,772	3,431	5,233	6,076	6,660	4,570
Balance	− 375	− 1,368	− 2,368	− 3,050	− 3,330	− 1,075
Romania						
Exports	826	1,203	1,402	1,653	1,450	1,682
Imports	1,043	1,451	2,436	2,164	2,150	2,152
Balance	− 217	− 248	− 534	− 511	− 200	− 470
U.S.S.R.						
Exports	2,570	4,121	6,341	6,750	8,773	10,079
Imports	3,317	4,957	6,250	10,714	11,653	11,412
Balance	− 747	− 836	91	− 3,964	− 2,880	− 1,333
PRC						
Exports	1,085	1,825	2,415	2,620	2,695	2,939
Imports	1,670	3,525	5,305	5,480	4,110	3,585
Balance	− 585	− 1,700	− 2,890	− 2,860	− 1,415	− 646

SOURCE: U.N. Trade Data from the U.S. Department of Commerce, East-West Trade Center; CIA, "PRC-International Trade Handbook," 1976.

Table 6.—Average Annual Rates of Change of East-West Trade[a] and World Trade by
Commodity Category, 1955-76
(percentages computed on the basis of current prices)

Period	Food and beverages	Raw materials	Fuels	Raw material- and labor-intensive manufacturers	Capital- and skill-intensive manufacturers	Total exports
Western exports to the East						
1955-60....	3.9	8.7	18.3	27.3	20.0	17.0
1960-65....	27.2	5.4	30.3	2.4	14.1	11.0
1965-70....	2.9	1.7	42.5	18.1	13.4	10.9
1970-76....	31.0	20.3	16.8	25.7	20.0	26.1
1955-76....	14.6	9.3	15.4	18.3	18.6	16.5
Eastern exports to the West						
1955-60....	10.9	9.3	9.7	15.9	11.0	10.5
1960-65....	6.9	8.6	8.7	18.0	10.7	9.1
1965-70....	6.5	6.3	10.5	12.3	11.3	12.2
1970-76....	15.7	27.7	53.8	24.0	25.7	30.2
1955-76....	10.2	13.3	20.8	17.8	15.0	15.9
World exports						
1955-60....	4.1	4.2	4.3	7.7	10.0	6.6
1960-65....	6.8	3.1	7.2	8.5	10.5	7.9
1965-70....	5.9	5.9	9.6	11.8	14.0	10.9
1970-75....	20.4	15.1	41.9	18.7	22.3	22.8
1955-76....	9.1	7.0	15.0	11.6	14.1	11.9

[a]Excluding inter-German trade.
NOTE: Figures are rounded.

SOURCES: United Nations, *Monthly Bulletin of Statistics*, 1955-76; Organization of Economic Cooperation and Development, *Trade Commodities: Country Summaries*, series B (Paris: OECD, 1955-76).

THE CONSTRAINTS ON GROWTH OF EAST-WEST TRADE

Some U.S. corporations point to this growing volume of Eastern imports from the West as evidence of the fact that U.S. policies that inhibit trade in general and trade in technology in particular with the Communist world exclude the United States from the economic benefits of lucrative and growing markets. This claim assumes that the patterns and growth rates of recent years will continue. This assumption, however, must be evaluated against the economic forces at work to inhibit the continued growth of East-West trade and to ensure changes in the structure of that trade. By far the greatest of these forces is the chronic shortage in Communist nations of the hard currency with which to pay for imports.

Trade between the nations of the industrialized West is denominated in "hard" or "convertible" currencies, i.e., currencies whose value is determined by market forces

outside the complete control of individual countries. CMEA and the PRC have chosen not to participate in this system. To do otherwise would be to allow outside forces to make de facto decisions with significant impact on domestic economies. Such an alternative is unacceptable to Communist nations which desire to concentrate economic decisionmaking in hands of central planners and which wish to be as insulated as possible from world market forces.

The decision not to have a convertible currency, however, entails drawbacks in trading with the Western nations that accept cash payment only in hard currency. An Eastern country therefore has three choices: it can earn currency by selling to the West; it can arrange countertrade agreements (i.e., transactions in which the seller delivers technology, finished products, and/or machinery and equipment and at the same time, contractually agrees to purchase goods from the buyer equal to an agreed percentage of the original value of the contract); or it can go into debt. The policy decisions that have resulted in expanded overall levels of East-West trade have greatly increased demand for Western goods in the East, but they have not been accompanied by a corresponding growth in demand for Eastern goods in the West. So far, Eastern nations still lag in their capacities to produce salable manufactured goods for export. Moreover, although countertrade is important, it involves complex transactions that Western firms enter into with reluctance. It is the latter choice, therefore, which has most often been made. This means that Eastern nations have had increasingly to be willing to resort to borrowing to finance their trade, and Western nations have had to be willing to supply the necessary credits.

The rapid growth in East-West trade turnover has therefore been accompanied by a rise in Eastern balance of payments deficits. In 1976, this deficit for the Communist world as a whole was $7.3 billion, and the only country that managed to achieve surplus in trade with the developed West was the U.S.S.R. in 1974, the direct result of the increase in the price of oil in world markets.[1] The paradox of the current chronic Western export surpluses vis-a-vis the East is that, desirable as these balances may be in the near term, they are financed largely through debt and cannot continue indefinitely. The greatest single curb to the continued expansion of East-West trade has become the limitations posed by this debt.

It is important to note, however, that nothing sets the East apart in this connection from other nations, such as less devel-

oped countries (LDCs), plagued with hard-currency shortages. The size and composition of the East's hard-currency debt has become a matter of controversy. Allegations are sometimes made that Communist nations borrow huge and disproportionate amounts from the West, that these sums are virtually "given away" both because they are provided in the form of cheap Government credits and because the debts go unpaid, and that for nonmarket economies, there is no incentive to restrict borrowing. None of these contentions hold up under examination.

THE SIZE OF THE EASTERN DEBT

Estimates of the net amount of Communist debt in the West in 1977 range from between $37 billion to $40 billion (U.N. Economic Commission for Europe Secretariat), to $42 billion, excluding the PRC (Bankers Trust), to $47 billion to $58 billion (U.S. Department of Commerce).[2] This variation is probably due to different methods of accounting. Until 1979, the PRC had not made extensive use of Western credit facilities; its hard-currency debt in 1978 had yet to exceed $1.6 billion. Table 7 demonstrates the expansion of CMEA debt since 1970.

A recent Department of Commerce study compared the magnitude of CMEA external debt to that of other nations, and found that Eastern debt is relatively small compared with the aggregate external debt of many Western borrowers.[3] Table 8 shows Eastern external debt as compared with other States with large loan commitments. Here, Eastern

---

[1]In 1976, the PRC had a favorable trade balance in total world trade. This was due to its large trade surplus with the less developed countries. Deficits have been partially offset by Eastern revenues from shipping, tourism, and sales of arms and gold, in all of which the East has a positive balance of payments. But the ability of individual countries to utilize this method of financing varies greatly, and only in the case of the U.S.S.R. is it a major means of significantly redressing trade balances.

---

[2]In 1977, the debt was distributed as follows:

	% of total hard-currency debt
Bulgaria	5.3
Czechoslovakia	5.3
East Germany	11.9
Hungary	6.7
Poland	25.4
Romania	7.5
U.S.S.R.	31.7
PRC	2.6
CMEA banks	3.4

[3]L. Theriot, "Communist Country Hard Currency Debt in Perspective" in Joint Economic Committee, *Issues in East-West Commercial Relations*, 1979.

Table 7.— Estimated Net Hard-Currency Debt
of Eastern Europe, U.S.S.R., and CMEA Banks,
End of Year, 1970, 1974-77
(in billions of U.S. dollars)

	1970	1974	1975	1976	1977
Bulgaria	$0.7	$ 1.2	$ 1.8	$ 2.3	$ 2.7
Czechoslovakia	0.3	1.1	1.5	2.1	2.7
East Germany	1.0	2.8	3.8	6.0	5.9
Hungary	0.6	1.5	2.1	2.8	3.4
Poland	0.8	3.9	6.9	10.2	13.0
Romania	1.2	2.6	3.0	3.3	4.0
Total Eastern Europe	4.6	13.1	19.1	25.7	31.7
U.S.S.R.	1.9	5.0	10.0	14.0	16.0
CMEA banks	0	0.1	0.5	1.1	1.7
Grand total	$6.5	$18.2	$29.6	$40.8	$49.4

SOURCE: Paul Marer, statement in *U.S. Policy Toward Eastern Europe* (hearings before the Subcommittee on Europe and the Middle East, Committee on International Relations, U.S. House of Representatives, 95th Cong., 2d sess., Sept. 7 and 12, 1978) (Washington, D.C.: U.S. Government Printing Office, 1979), p. 100.

borrow to an excessive degree compared to other nations.

Care must be taken in drawing these kinds of comparisons. First, comparisons of the U.S.S.R. with even the largest developing nations are distorted to the extent that they fail to take into account the size and sophistication of the Soviet economy. Second, in comparing Communist to capitalist nations, the criteria of relative debt size or level of debt servicing must be modified to reflect the points on which State-controlled and market economies differ.[4] Even with these caveats, however, it is clear that the levels of Eastern debt are by no means alarming or unusual in the context of the world economy.

nations compare favorably to countries with similar gross national product (GNP). Another method for measuring the economic burden of the debt—relating its size to exports (in the Eastern case, hard-currency exports)—is shown in table 8. From this perspective too the Communist world does not

---

[4]For instance, Communist nations have no recourse to risk capital (i.e., the sale of stocks). Second, much East-West trade is conducted under "self-liquidating" countertrade agreements, i.e., the creditor accepts as payment the goods produced by the facility for which credit was given. Third, the great legal and social powers of a centrally planned economy give Eastern Governments much greater flexibility in meeting international financial obligations than is possible in the West.

Table 8.— Hard-Currency Debt and Foreign Trade, 1977 of CMEA and Selected Western Countries

Country	Net debt ($ billions) 1977	Exports ($ millions) 1977	Imports ($ millions) 1977	Balance ($ millions) 1977	Balance ($ millions) 1976	Ratio of debt/hard-currency exports 1977	Ratio of debt/hard-currency exports 1976
Bulgaria	$ 2.7	$ 608	$ 997	$ − 389	$ − 422	4.4	4.7
Cuba	2.2	784	1,565	− 872	− 711	2.8	1.7
Czechoslovakia	2.7	1,903	2,639	− 736	− 758	1.4	1.2
East Germany	6.0	2,900	4,070	− 1,140	− 1,456	2.1	1.7
Hungary	3.7	1,712	2,441	− 729	− 474	2.2	2.3
Poland	12.8	3,852	6,374	− 2,522	− 3,235	3.3	3.1
Romania	3.8	2,270	2,660	− 390	− 4	1.4	1.7
U.S.S.R.	11.3	11,666	14,747	− 3,081	− 5,516	0.97	1.4
Vietnam	0.2	128	434	− 306	− 183	1.6	1.8
Total CMEA	$49.6	$25,823	$35,927	− $10,104	− $12,759	1.9	1.8
Other developing countries							
Argentina	$ .5	$ 5,800	$ 4,400	+ $1,400	− $883	0.84	1.7
Brazil	19.3	12,139	13,229	− 1,090	− 2,200	1.6	2.6
Colombia	2.6	1,900	2,000	− 100	− 125	1.38	1.4
Mexico	20.9	4,166	5,489	− 1,323	− 2,732	5.0	6.5
South Korea	8.5	10,047	10,814	− 767	− 1,059	0.84	0.96
Spain	7.0	10,223	17,835	− 7,612	− 8,732	0.7	1.2
Venezuela	4.5	9,487	9,269	+ 218	+ 2,844	0.47	0.3
Yugoslavia	6.5	3,600	7,400	− 3,800	− 2,515	1.8	1.2

SOURCE: Lawrence H. Theriot, "Communist Country Hard-Currency Debt in Perspective," Department of Commerce Project D-66-74.

THE COMPOSITION OF THE EASTERN DEBT

There are three major sources of financing available to Communist nations—Western Government financing in the form of guarantees, insurance, and direct credits; regular private commercial credits, including Eurocurrency financing; and supplier credits. The mix of these varies among countries. In the U.S.S.R. approximately 60 percent of the gross foreign debt is financed by official credits; Western commercial banks hold 25 percent, and supplier credits constitute the remainder. Eastern Europe, however, relies much more heavily on commercial bank loans, although again the mix varies among individual countries. PRC debt still consists almost exclusively of supplier credits.

It is impossible to generalize about the degree to which the Communist world as a whole relies on "cheap" Government credits and guarantees, but it is clear that the shortage of negotiable currency in the East means that Western official and private credits can have a significant impact on the growth of East-West trade.

Not all this borrowing is subsidized, however. At year end 1977, Western commercial banks held approximately $25 billion in net claims on CMEA nations (see table 9). Again, the value of comparison between market and nonmarket economies is limited, but some perspective on this figure may be gained by considering that for the same period non-OPEC LDCs owed Western banks approximately $30 billion.

Furthermore, the share of CMEA debt in public and private facilities in the United States is relatively modest. As of June 1977, U.S. private bank claims totaled $4.9 billion or about 10 percent of the net debt of the U.S.S.R. and Eastern Europe. In contrast, U.S. banks hold 41 percent of Brazilian debt and 44 percent of Mexico's. In addition, although U.S. banks hold about 10 percent of CMEA external debt, their claims represent only a relatively small commitment of the total equity capital of the banks. The shares

Table 9.—Estimated Composition of Net Hard-Currency Debt of
Eastern Europe, U.S.S.R., and CMEA Banks, Dec. 31, 1977
(in millions of U.S. dollars)

	Drawings on official credits	Supplier credits[a]	Net liabilities to Western banks[b]	Outstanding bonds & notes	IMF and IBRD[c] drawings	Total
Bulgaria	$ 798	$ 100	$ 2,065	$ 0	$ 0	$ 2,963
Czechoslovakia	841	200	884	0	0	1,925
East Germany	2,455	400	3,729[c]	0	0	6,584
Hungary	460	0	3,630	180	0	4,270
Poland	5,775	1,200	6,890	82	0	13,947
Romania	1,256	200	1,073	0	670	3,199
Total Eastern Europe	$11,585	$2,100	$18,271	$262	$670	$32,888
U.S.S.R.	10,730	2,200	3,411	0	0	16,341
CMEA banks	0	0	3,500	0	0	3,500
Grand total	$22,315	$4,300	$25,182	$262	$670	$52,729

[a]Including outstanding a forfait obligations.
[b]Banks in Group of Ten countries, Switzerland, and foreign branches of U.S. banks in the Caribbean and Far East.
[c]International Bank for Reconstruction and Development.
[d]Excluding net liabilities of East Germany to banks in West Germany.

SOURCE: East-West Markets, May 15, 1978, pp. 3 and 10.

of capital accounted for by outstanding loans range from 0.9 percent for Czechoslovakia to 5.6 percent for the U.S.S.R..

THE CREDIT-WORTHINESS OF THE COMMUNIST WORLD

The borrowings of Communist nations are, therefore, not alarmingly large in absolute terms, and the private market's evaluation of the risk entailed in lending to them has been generally favorable. Commercial banks reflect their evaluation of this risk through the terms at which individual Eastern nations are granted loans; i.e., the interest rate spreads between their rates and the London Interbank Borrowing Rate (LIBOR), the risk-free rate utilized in the Eurocurrency market.[5] The interest spread on commercial loans to the East, therefore, reflects the private market's objective and carefully weighted evaluation of credit worthiness.

Poland, with the highest debt-export ratio of the countries under consideration and the highest interest rate spread, is the least credit-worthy of the Eastern nations. Nevertheless in April 1979, Poland received $550 million, the largest syndicated loan it had ever obtained in the Euromarket, and the 1¼ point spread over LIBOR was identical to that granted on a similar loan to Egypt. This loan was oversubscribed, a fact interpreted in Warsaw as a relatively positive market evaluation of Poland's credit-worthiness, although there are growing indications that the Poles may be increasingly hard-pressed to begin hard-currency repayments to the West.

A syndicated loan of similar magnitude ($500 million) was recently granted to the PRC. As might be expected from the very

low debt/export ratio as well as conservative Chinese borrowing practice in the past, the Chinese were granted an extremely low rate —one-half percent over LIBOR. The size and interest spread of this loan indicate a positive evaluation not only of Chinese ability to repay, but also of political stability in the near term.

During 1977, Communist countries arranged for approximately $3.4 billion in publicized Eurocurrency credits. While considerable, this borrowing accounted for only about 8 percent of total borrowing on the Euromarket during 1977. Borrowing by all Eastern nations in that year was roughly equal to that of Canada, and in general, with the exception of Poland and the CMEA investment bank, Eastern use of the Eurocurrency markets has been relatively modest compared to many developing countries. Furthermore, international bankers have not only been willing to increase the debt, but have rendered relatively favorable interest rate judgments on the Eastern economies.

THE GROWTH OF THE EASTERN DEBT

Discussion thus far has centered on the level of the Communist world's external debt and concluded that its volume and structure are unexceptional in the context of world trade as a whole. This does not mean, however, that this debt can continue to accrue at its present rate.

Between 1974 and 1977, the debts of Bulgaria, Czechoslovakia, East Germany, and Hungary roughly doubled. Growth in Polish and Soviet debts was proportionately even higher, 230 and 220 percent, respectively. Only the debts of Romania and China grew somewhat more slowly (see table 10), but as chapter XI discusses, the hard-currency debt of the PRC can be expected to grow rapidly over the next several years.

In addition to debt incurred by individual nations, CMEA's two international banks, the International Bank for Economic Cooperation and the International Investment

---

[5]These spreads presently are as follows:

Poland	1 1/4
Romania	3/4
Hungary	5/8
Czechoslovakia	5/8
U.S.S.R.	5/8
PRC	1/2

(LIBOR = 11% for 6 months)

Table 10.—Growth in Debt, Selected Communist Countries, 1974-77

Country	% growth in debt, 1974-77
Bulgaria	125
Czechoslovakia	145
East Germany	115
Hungary	150
Poland	230
Romania	48
U.S.S.R.	220
PRC	30

SOURCE: Office of Technology Assessment.

Bank, have been active borrowers in Western private credit facilities. The Eurocurrency obligations of these two banks rose from $100 million in 1974 to $1.7 billion in 1977.[6]

These enormous growth rates reflect the expansion of East-West trade. But while additional loan capital seems to be available in varying degrees to all the Eastern nations, in the long run continued growth of East-West trade cannot be financed through borrowing, even should the East wish to do so. According to Department of Commerce estimates, East European nations would have to sustain growth rates of between 6 and 9 percent and cut import growth to zero to stabilize their debt levels by 1985.[7] As both these possibilities are highly unlikely, all other things being equal, the accumulation of debt will probably increase. If this happens, the financial risk component of interest rate spreads on East European loans will increase until borrowing becomes uneconomical.

ALTERNATIVES TO BORROWING

Should further borrowing become prohibitively expensive to the East or should Western lending be restricted for noneconomic reasons, three alternatives are open to the

---

[6]Morris Bornstein, "Issues in East-West Economic Relations," unpublished paper for Research Conference in East-West Relations in the Eighties, Rockefeller Foundation Study and Conference Center, Bellagio, Italy, 1979.
[7]Allen Lenz, "Potential Hard-Currency Debt of the U.S.S.R. and East Europe Under Selected Hypotheses" in Joint Economic Committee, op. cit.

centrally planned economies. They can allow more direct Western involvement in their enterprises; they can resort to internal financing; or they can expand and diversify their hard-currency earnings from exports.

Western Involvement in Eastern Enterprises

There are at least two ways of increasing Western involvement. A country can obtain risk capital by establishing joint enterprises which enable foreign firms to invest directly in its economy. Such entities are permitted in Hungary, Romania and, to a lesser extent, in Poland. In the PRC the possibility is under discussion. The current contribution of these enterprises is small, however. Alternatively, greater use may be made of leasing, although the existence of foreign-owned property in a Socialist country raises ideological problems. This has not prevented the Soviets from leasing containers from the West, but there are no prospects for rapid or widespread basic policy changes in this regard.

Internal Financing

Internal financing requires the allocation of a larger share of the national income to investments. Because standards of living inevitably suffer as a result of this tactic, it is subject to political constraints. Poland, for example, has found it extremely difficult to raise internal consumer prices without immediate and violent reaction from the populace. While this is an extreme example, the increases in savings necessitated by internal financing make this alternative unattractive.

Increased Exports to the West

Given the limitations inherent in both these approaches, it seems inevitable that the hard currency necessary to finance trade and economic growth in the Communist world over the long term can be obtained in sufficient quantities only through the sale of goods in the industrialized West and the LDCs. Western imports must ultimately be paid for through Eastern exports, and presumably through the reduction or elimination of present Western trade surpluses with the Communist nations. The Western technology sold to the East will help to accomplish this to the extent that it is aimed at capacity expansion or long-term productivity increases in potential export sectors.

Undoubtedly, the attempts of many Communist nations to transform themselves into net exporters of manufactured goods have already been aided by technology imported from the West, much of which is specifically directed into export industries. At the same time that Eastern markets for technology-intensive goods have been expanding, Eastern exports to the West have become increasingly capital intensive. Structural changes in Eastern exports in favor of capital-intensive products do not, however, adequately reflect the progress in industrialization or capital accumulation and technological expertise which has been achieved by Eastern countries. In particular, relative to the level of economic development in the East, too few technologically advanced and sophisticated capital-intensive products of too low a quality are produced for sale to Western countries. This is largely due to the nonmarket economies' inherent systemic difficulty in developing products suitable to Western demands and effectively marketing them. Comparisons of Eastern export development to that of Japan, Taiwan, or Korea are therefore invalid.

CONCLUSIONS

In the last analysis, deliberate policies in both the East and West may be hostage to larger economic conditions. CMEA behavior during the 1974-75 recession provides an example of the problems many planned economies have experienced in controlling their trade balances with the West. Except in the U.S.S.R., which is the sole oil exporter in CMEA, growth in East-West trade and resulting trade imbalances became a particularly acute problem to CMEA members after 1974, when the slow pace of world economic recovery hindered the growth of Eastern export earnings at a time of greatly expanded imports of food and other items. In Eastern Europe, restrictive action directed at import-elastic sectors such as industrial investment was instituted. As a result, industrial expansion in the region declined from 8 to 8.5 percent in 1974, to 5.5 percent in 1978. This decline in the growth rate of domestic output seems to have affected the expansion of exports more strongly than that of imports, with the result of a further widening of the deficit in 1978. In other words, attempts to reduce the deficit indirectly have only increased it.

In contrast, when the PRC was faced with lagging demand for its goods in Western markets in 1975, it simply slashed its agricultural imports by $1 billion. Its ability to take such incisive action was predicated on the low absolute value of its trade with the West and its consequent lack of dependence on Western imports, a situation which, at least in some sectors, no longer exists in many CMEA countries.

It may be, therefore, that world energy prices and Western economic growth levels (and their effect on import demand) ultimately have as much direct and indirect impact on the level of East-West trade as any policy decisions taken in either East or West.

In any case, it is clear that U.S. (and other Western States') willingness and ability to purchase more Eastern exports are vital conditions for the long-term expansion of East-West trade. How large could this trade ultimately grow? One optimistic assessment has been made by Michael Forrestal, President of the U.S.-U.S.S.R. Trade and Economic Council, who estimated recently that "over a relatively tranquil five-year period ahead with no remedial U.S. tariff or credit legislation, U.S.-U.S.S.R. trade could reach 20 billion; 15 billion in U.S. exports and 5 billion in Russian sales to the U.S. If tariff and credit limitations were removed, the total would be substantially higher."[8]

---

[8]*Industry Week,* Mar. 5, 1979.

While significantly expanded East-West trade rates may be possible in the long run, Forrestal's estimates are spectacularly optimistic for the near future. Given the size and diversity of its economy, the U.S.S.R. retains an extremely low level of foreign imports per dollar of gross domestic product, and there is no reason to expect this policy to change in the near future. Moreover, Soviet imports from the United States could increase fivefold and still be only $7 billion to $10 billion annually. And these figures fail to take into account the limitations posed by present Eastern export potential and the limitations of demand for Eastern exports in the West. Nor do they allow for the fact that the United States has never captured a large fraction of Eastern markets. Large increases in East-West trade as a whole would benefit other Western countries proportionally more, especially in the absence of vigorous U.S. export promotion campaigns, favorable financing terms, relaxation of export controls, and other policies aimed at foreign trade expansion in general.

THE ROLE OF TECHNOLOGY IN EAST-WEST TRADE

FUTURE PROSPECTS

Trade in technology has remained a relatively stable and relatively small component of East-West trade as a whole. There is reason to believe, however, that Eastern imports of technology may rise, and that this will occur regardless of whether East-West trade expands or whether world economic conditions, U.S. commercial or political policies, or the pressure of increasing hard-currency debt cause it to stagnate or contract. Indeed, the very structure of East-West trade is creating a situation in which Eastern importers will have higher incentives to acquire foreign technology.

In the future, the Communist world is likely to place a greater emphasis on obtaining technology than on pure capital inflows.[9] Presently technology-intensive products constitute only a minor share of the total resource inflows from the West. These, however, have a disproportionate importance to the economies of the Eastern nations (see chapters X and XI). This is not only because of the need to expand exporting sectors of the economy, but also because the increase in productivity resulting from the use of the new technology usually more than offsets the cost of the credit needed to obtain it. So long as this remains the case, Eastern nations will be increasingly eager to borrow in order to purchase Western technology. By the same token, if as Eastern debt continues to grow, interest rate spreads increase, other types of imports will become even less economical; high-productivity technology imports will thus begin to constitute a larger relative share of Eastern imports. This suggests caution in concluding that debt constraint will inhibit technology purchases. On the contrary, it may create incentives for purchasing more technology at the expense of other imports. Demand for technologically intensive products in Communist nations is, therefore, unlikely to abate in the future. In the absence of foreign production and marketing know-how, however, long-term ability to market usable products cannot be created without major structural changes that such countries are unwilling to make. The demand for Western management technology is therefore expected to grow enormously. The medium-term result of this is that Western technology-intensive industries and firms providing management expertise will benefit most from expanded East-West trade over the next several years, while import-sensitive capital and labor-intensive industries may be injured by increased competition.

[9]Padma Desai, "The Productivity of Foreign Resource Inflows to the Soviet Economy," in *American Economic Review*, LXII (2), p. 74.

PRESENT U.S.
MARKET SHARES

The implications of this for the U.S. economy must be understood in the context of the U.S. share in Eastern technology purchases. This is impossible to determine with any precision. A rough picture of the value of U.S. sales of technology to the East relative to those of America's major Western trading partners may be constructed, but this is possible only through categories of technology transfer for which data exists—trade in high-technology products and industrial cooperation agreements (see chapter VI).

Sales of High-Technology Products

As table 11 demonstrates, in 1977 U.S. aggregate sales of high-technology products to the U.S.S.R., Eastern Europe, and the PRC amounted to less than $300 million, and in no case did these products constitute a major share of U.S. exports to individual Communist countries. High-technology sales

thus ranged from a high of 19.7 percent of U.S. exports to Bulgaria to 3.3 percent of total exports to East Germany.

Nor is the United States a leading source of high-technology products among Western sellers. In the case of the U.S.S.R., West Germany is by far the largest single exporter of high-technology products, followed by Japan and France. In 1977 those three countries accounted for more than 62 percent of total Soviet imports of such items from the West. The U.S. share in high-technology products in that year amounted to only 9.1 percent. Nearly a third of Western high-technology exports to the PRC originate in Japan. West Germany and France account for another 29 percent, and the United States ranked fifth in this category with a 6-percent share.

Table 11 has also demonstrated that Eastern purchases of high technology from the United States have, if anything, occurred at a slightly lower rate than purchases from the

Table 11.—Comparison of High-Technology Exports With Manufactured Goods and Total Exports—15 Industrialized World (I.W.) Countries to the Communist Countries and to the World (in millions of U.S. dollars)

Destination	1977	High-tech. exports as % of	1976	High-tech. exports as % of	1974	High-tech. exports as % of	1972	High-tech. export as % of
U.S.S.R.								
High-technology I.W. exports	$ 2,003	—	$ 1,627	—	$ 1,036	—	$ 582	—
Manufactured goods I.W. exports	9,537	21.0	9,169	17.7	5,546	18.7	2,430	24.0
Total I.W. exports	11,412	17.6	11,653	14.0	6,250	16.6	3,317	17.5
Eastern Europe								
High-technology I.W. exports	1,741	—	1,525	—	1,223	—	619	—
Manufactured goods I.W. exports	11,769	14.8	11,438	13.3	10,432	11.7	4,738	13.1
Total I.W. exports	12,866	13.5	12,757	12.0	11,322	10.8	5,098	12.1
PRC								
High-technology I.W. exports	248	—	342	—	410	—	64	—
Manufactured goods I.W. exports	2,986	8.3	3,094	11.1	3,166	13.1	1,090	5.9
Total I.W. exports	3,585	6.9	3,423	10.0	4,369	9.5	1,445	4.4
Total all Communist countries								
High-technology I.W. exports	4,886	—	4,140	—	3,197	—	1,562	—
Manufactured goods I.W. exports	29,991	16.3	27,955	14.8	23,714	13.5	10,266	15.2
Total I.W. exports	34,263	14.3	32,808	12.6	27,261	11.7	12,234	12.8
World								
High-technology I.W. exports	71,576	—	64,366	—	49,314	—	29,092	—
Manufactured goods I.W. exports	523,890	13.7	459,351	14.0	381,983	12.9	214,182	13.6
Total I.W. exports	669,393	10.7	590,833	10.9	498,470	9.9	273,045	10.7

SOURCE: *Quantification of Western Exports of High Technology Products to Communist Countries*, prepared by John Young, Industry and Trade Administration, Office of East-West Policy and Planning, U.S. Department of Commerce, Project No. D-41.

industrialized world as a whole. In 1977, the fraction of high-technology, as a percentage of total Soviet imports from the United States was 11.3 percent as opposed to 17.6 percent for the total industrialized world. The comparable figures for Eastern Europe were 10.6 percent for the United States versus 13.5 percent for the industrialized world. Only in the PRC did America garner a higher than world share of high-technology sales—8.8 percent as opposed to 6.9 percent.

As is evident in table 12, Communist world (including Yugoslavia and Cuba) shares of total high-technology exports from the United States are slightly higher than overall world averages (14.3 versus 10.7 percent). Of the Eastern countries, the U.S.S.R. purchased the highest proportion of high technology (17.6 percent). In the cases of both the U.S.S.R. and Eastern Europe, however, these shares have not risen notably over the past 5 years, despite large increases in the total volume of East-West trade.

Industrial Cooperation Agreements

The paucity of information available on the value of coproduction agreements, licenses and patents, and turnkey ventures can be seen from a brief survey of the best existing data. In 1975, U.S. firms participated in 424 agreements. Nearly four-fifths of these (79.3 percent) were in the manufacturing sector. Within this sector machine building and chemicals predominated, each with approximately one-fifth of total agreements. Electrical machinery and petroleum processing industries were also important. The U.S.S.R. and Poland signed the largest number of agreements with U.S. firms.

Care must be taken in interpreting this information however. Although the U.S.S.R. ranks third, after Hungary and Poland, in the number of substantive arguments concluded, it has been estimated that the total value of the Soviet agreements exceeds that of all Eastern European cooperation agreements combined.[10] Thus, the number of

[10]Paul Marer and Joseph C. Miller, "U.S. Participation in East-West Industrial Cooperation Agreements," *Journal of International Business Studies*, fall-winter 1977. p. 21.

Table 12.—U.S. High-Technology Exports to the Communist Countries and to the World, 1977

Exports to:	Millions of dollars	High tech. as % of
Bulgaria		
High technology	$ 4.7	—
Manufactured	20.1	23.4
Total exports	23.9	19.7
Czechoslovakia		
High technology	7.1	—
Manufactured	18.4	38.5
Total exports	74.0	9.6
East Germany		
High technology	1.2	—
Manufactured	4.1	29.1
Total exports	36.1	3.3
Hungary		
High technology	12.9	—
Manufactured	44.8	28.7
Total exports	79.7	16.2
Poland		
High technology	37.0	—
Manufactured	114.2	32.4
Total exports	436.5	8.5
Romania		
High technology	23.6	—
Manufactured	61.0	38.6
Total exports	259.4	9.1
Total Eastern Europe		
High technology	86.5	—
Manufactured	262.6	32.9
Total exports	909.6	9.5
U.S.S.R.		
High technology	182.7	—
Manufactured	547.4	33.4
Total exports	1,623.5	11.3
Total Eastern Europe & U.S.S.R.		
High technology	269.2	—
Manufactured	810.0	33.2
Total exports	2,533.1	10.6
PRC		
High technology	15.1	—
Manufactured	86.9	17.4
Total exports	171.3	8.8

SOURCE: *Quantification of Western Exports of High Technology Products to Communist Countries.* prepared by John Young. Industry and Trade Administration. Office of East-West Policy and Planning. U.S. Department of Commerce. Project No. D-41.

agreements tells nothing of their magnitude or technological significance. Unfortunately, no comprehensive data exists to fill these gaps. This is a reflection not simply of the complexity of the deals, many of which involve other countries as well as U.S. foreign subsidiaries, but also of the reluctance of firms to divulge details of their transactions. There is at present, therefore, no way of accurately estimating the amount earned by U.S. firms in cooperation agreements.

The little information that is available for the United States is in data for the value of license and patent sales collected by the Department of Commerce. Unfortunately, this is presented in a form that makes interpretation difficult. Although for the past 3 years, the General Accounting Office has suggested that the Department of Commerce disaggregate this data, Commerce continues to report only cumulative revenues from royalties, not payments collected annually.

Among Western countries, West Germany is the leading licensor to the East. It is followed by the United Kingdom, the United States, France, Japan, Italy, Sweden, Switzerland, the Netherlands, and Belgium. But while certainly a common mode of technology transfer, licensing is by no means a major money-earner for any nation. While the number of transactions involving the sale of licenses by Western firms is not accurately known, a 1976 estimate placed the figure at less than 2,400. Again, this in itself is deceptive. The U.S.S.R. has sold more licenses to the West than it has bought, but the price paid for Western licenses has been estimated by Licensintorg, the Soviet licensing agency, as an average of 10 times greater than the price paid by Western firms for Soviet licenses. It has been estimated that in the mid-1970's annual proceeds in the West from Eastern license purchases were in the order of $300 million. Much of this, however, was paid for in the goods produced by the license under countertrade agreements. There is no official estimate of the share of this revenue accruing to the United States— in cash or in goods.

Moreover, although there are persistent rumors about patent infringements by the Soviet Union, no reliable estimates exist on the magnitude of this problem. A recent study by the National Research Council reported that "conversations with several experts on international patent law have led the panel to believe that Western companies tend not to take legal action even when they believe their rights have been infringed upon by the U.S.S.R. simply because 'it is too great a hassle.' "[11]

Conclusions

The only reliable information for measuring the value of U.S. technology sales to the East is in data for high-technology product exports. Even this must be treated with extreme caution since many subjective judgments are made in preparing quantitative estimates. The information is valuable primarily for indicating changes in overall trade volumes and for making crude estimates of rates of change. The gross outcome of this analysis suggests that U.S. trade in technology and technology-related products with the East is relatively small (less than $300 million in 1977) and has been growing at roughly the same rate as overall East-West trade. The data does not clearly support the thesis that the nonmarket nations have made a concerted effort to extract technologies from the United States on a massive scale to support economic or military interests. Nor can it be taken as a certain rebuttal of the thesis. The Soviet Union and Eastern Europe did import relatively more technology as a fraction of total imports than the world average, but the PRC imported considerably less. The differences may be due primarily to relative degrees of industrial development and deliberate Chinese policy which, as chapter XI demonstrates, is changing. Beyond this, it is safe to conclude that sales of technology constitute only a small fraction of U.S. trade with the Communist world, trade which itself has been very circumscribed. If, as is likely, technology purchases from the West accelerate, U.S. policies designed to capture larger market shares would be necessary for American firms to benefit as much as would firms in allied nations.

---

[11]National Academy of Sciences, "Review of U.S.-U.S.S.R. Interacademy Exchange and Relations," National Research Council, May 1977.

THE IMPACT OF TRADE WITH AND TECHNOLOGY SALES TO THE EAST ON THE U.S. ECONOMY

Any evaluation of the merits of expanding U.S. trade as a whole or trade in technology with the Communist world must take into account the net effects of such trade on the U.S. economy. Many attempts have been made to approach this conceptually complex question, but qualitative generalizations have addressed themselves to relatively narrow segments of the issue. Satisfactory quantitative assessments, except for single sectors or commodities, simply do not exist. The reasons for this paucity of analysis are manifest. Technology is notoriously difficult to measure empirically, either for particular commodities or through macromodels of entire economies. At present, for instance, there is no universally accepted model for assessment of aggregate technical change in the U.S. economy. Furthermore, any satisfactory model of the macroeconomic effects of technology transfer in the United States would entail an accurate assessment of technology not only as a factor in U.S. growth, but in the nonmarket and third-country economies as well. This is not only beyond current capabilities; it is unlikely that a sufficient data base for such an attempt will ever be assembled. In light of this, assessment of the impact of East-West technology transfer on the U.S. economy must be limited to narrowly defined generalizations.

In the United States, those with a stake in commercial technology transfer to nonmarket economies may be divided into four categories: the vendors of U.S. technology; industries that must compete with Communist exports both in the United States and in third markets; purchasers of Eastern technology; and the U.S. consumer. Policymakers must aggregate and balance the interests of all four.

TECHNOLOGY VENDORS: U.S. CORPORATE STRATEGY

The primary motive for American firms' sales of technology to the East is profit. Export income is generated by the sale of "high-technology" commodities and know-how, and also by the sale of associated plant, equipment, and services. In addition, indirect results of the transfer transaction may bear fruit in the medium or long term in the form of future sales. Highway construction equipment, for example, may be purchased in the future as a result of the transfer of automotive manufacturing technology.

Gains to individual firms obviously increase the aggregate income of the United States as a whole. Moreover, sales resulting from growing demands for exports lead to increased employment, not only in research, design, and engineering services directly associated with technology sales, but in associated industries which benefit from the derived demand. In addition, there is an important sense in which overall trade levels, analyses of numbers of plants and licenses, or of the revenues received from these sales, may not provide a useful estimate of the full role of technology sales to the East in international commerce. In most cases, technology transfers are only a part of complex transactions which include barter, two-way technology transfers, coproduction agreements, buy-back agreements, and other arrangements often involving third countries. The participation of a U.S. firm in such relationships can become an integral part of its corporate strategy and therefore assumes an importance disproportionate to the dollar value of the transactions.

Companies engaged in East-West trade, therefore, often contend that its value to U.S. corporations cannot be measured solely by its present volume or profitability. The continued ability to compete in the sale of technology to the East is important to a variety of other corporate activities. OTA sought to explore this argument by conducting interviews with high officials in 10 firms,[12] all of which have clearly articulated positions on the importance of trade with Communist nations to their corporate strategy. No attempt was made to assemble a representative or statistically relevant sample, and the following discussion should therefore be regarded merely as an attempt to synthesize as a cohesive argument the views of an identifiable segment of U.S. industry. Interviewees by no means agreed on every point, but the case presented here is faithful to the opinions of many of those interested in fostering trade with East.

U.S. firms seem rarely to enter Communist markets as part of a deliberate global strategy. More often their initial involvement is the result of an isolated opportunity which comes about either as the result of other international activities, or of approaches by representatives from the East. In some cases, however, the contacts developed in an initial venture result in the establishment of closer forms of cooperation with the Eastern nation.

Control Data Corporation (CDC) provides a good example of the way in which such an opportunity can grow into a larger relationship. In 1968 it sold a CDC 1604 computer to the Soviets. This model was being phased out in the United States; nor did it represent

a major technological innovation for the Soviets. The sale resulted in additional contacts which led CDC to evaluate the Soviet market potential and eventually resulted in a protocol agreement between the company and the Soviet State Committee on Science and Technology. CDC is now actively involved in marketing products and technology in the U.S.S.R.

Once an initial transaction is successful and longer term contacts are established, U.S. firms evaluate their involvement in the East in the context of their worldwide activities, and begin to examine broader forms of cooperation, e.g., coproduction agreements. This sequence of events is not unique to dealings with the Communist world. Early transactions generally have little or no impact on corporate strategy. Similarly, the complexity of the issues associated with closer cooperative relationships necessitates a building of trust that may only be obtained through extended personal or corporate contact. But as the involvement in the East increases, there is a tendency to consider these markets as a concrete part of the new product planning and development process.

The development and introduction of new products is a large, complex, and costly process. Because initial activities in the Communist world are usually based on exploitation of isolated opportunities, there is no indication that companies explicitly consider these markets in their early new product decisions. This situation may now be changing, and some large firms consider Eastern markets in the evaluation of worldwide market potential for new products. This tendency is particularly marked in companies with coproduction and joint venture agreements with Eastern-bloc partners. Movement towards explicit consideration of the Communist world market potential appears to be less a reflection of corporate philosophy than

---

[12]The firms included an international chemicals company and an international consumer industrial manufacturer (both of which declined to be identified); Control Data Corporation; Corning Glass Works, Inc; Hewlett-Parkard Corporation; International Harvester; Herman Corporation; Levi-Strauss; Texas Instruments; and Satra Corporation.

of the broadened geographical perspective of corporate decisionmakers.

Those executives who argue that East-West transactions are part of new product decisions also contend that increases in this trade will stimulate the U.S. economy by increasing innovation, creating new jobs, and improving productivity.

Other firms contend that, for the first time, they are considering Eastern market potential as part of the R&D justification on new product technology because East-West trade is now sufficiently institutionalized to become part of the global marketing plan. This may include ongoing discussions with Eastern trading companies which involve cooperation in design, development, testing, and production of new product models.

The past profitability of East-West transactions has been mixed. Some companies openly admit that business with the Communist world has not been as profitable as was originally expected. In fact, although no firm would provide concrete examples, several stated that in retrospect their expectations had been unrealistic. Throughout history examples abound of companies' continuing to fail to realize profits in their dealings with the Communist world. This is particularly true of the Soviet Union, where it is almost impossible to document cases of American corporations' showing direct profits. Despite this disappointing record, many firms continue to believe that it simply takes time before profits begin to accrue from trade with the East. There are several reasons for this:

- It takes time to develop enough insight into centrally planned economies, their institutions, and people, to know what business opportunities are possible and where to look for them.
- The authorization and security procedures within such countries are rigid and complex. It may take many years before a firm's counterparts in the East feel secure enough to propose meaningful deals.

- The difficulty in getting access to end users and to research institutions makes it difficult to collect the information often essential to transactions.

It may be that some companies are not getting an adequate return because they are not working hard at developing closer relationships in the East. Those most willing to discuss complex joint ventures are the most likely to identify meaningful areas for future transactions. Moreover, some returns in East-West trade do not involve direct profits, but rather may involve the acquisition of design, engineering, and technical development capabilities of Eastern counterparts.

A firm may also benefit from the sale of technology that is no longer competitive in Western markets, but is appropriate to Eastern technical sophistication. This is a way of partially recouping R&D costs, and is likely to be a factor in industries with a particularly high rate of technical innovation (e.g., computers or integrated circuits).

At the same time, however, U.S. firms have made very extensive investments in the East. Anecdotes of negotiations that have taken place over several years, costing millions of dollars, abound. To these costs must be added those of participation in activities viewed as necessary to the development of successful East-West ventures—e.g., membership in trade councils or maintaining foreign offices. These expenses are of sufficient magnitude to warrant continued efforts, even in the absence of short-run profits. They represent a vested interest in the health and continuity of East-West commercial relations.

Perhaps most importantly U.S. companies fear that difficulties in dealing with the Communist world will have the long-run effect of shutting them out of other markets. World trade relationships have implications that go far beyond the contact between two companies. Often transactions are initiated because they meet the needs of worldwide marketing strategies. For example, a U.S. company may enter into a coproduction agreement with an Eastern European counterpart to produce products that are no longer cost-effective within the United States, but which can be sold in LDCs as part of the company's larger strategy.

As a company's experience and expertise in foreign commercial relations increase, it begins to evaluate the role of these relations in terms of global market needs. Any relationship with an individual country tends, therefore, to be regarded as a potential lead to new markets. In this area, involvement with some Eastern European countries is viewed as an entree elsewhere. Two quotations from OTA's interviews with selected businessmen indicate this:

> We started negotiating with the Chinese in their Embassy in Bucharest many years before there were serious thoughts about regularizing relations with the PRC.

> We have been contacted by trade representatives from one of the Eastern European countries regarding the possibility of joint ventures to address the needs of LDCs, particularly Africa. We are actively following up on these possibilities since they are a logical extension of our total marketing interests.

Thus, both China and Eastern Europe are looked towards for potential assistance in dealing with third countries. Trade with China is seen by some firms as an entree into parts of the Far East. Similarly, Eastern European ventures can become part of a strategy to address markets throughout the CMEA, in Western Europe, and in LDCs.

Because of these interrelationships, there are fears that the dimunition of East-West trade will have effects beyond immediate bilateral relationships, including isolation from other markets. The problem is exacerbated by the prevalence of barter and countertrade in East-West transactions. This form of trade is often new to American firms, but once involved, the need to market the items purchased in the transaction perforce involves breaking into new markets.

In sum, it would appear that both direct and indirect benefits accrue to those export-oriented industries that engage in technology transfer. Moreover, some of these benefits can be diffused throughout the economy, although it is usually impossible to disaggregate the effects of technology transfers from other sales. However, there can be negative effects stemming from such transfers, and policymakers must decide whether individual firms can be depended on to prevent transactions in which long-run harmful effects will make themselves felt in their own industries. There is some evidence, for example, that U.S. firms are encountering increasing difficulties in adjusting to technical change and are considering the marketing of their technology as an alternative to aggressively engineering for competitive production in the high-wage U.S. economy.[13] The nonmarket economies encourage this trend through providing a market for technology no longer competitive in the West. While such transactions may indeed improve the cash flow of an individual firm, the long-term effects on an entire industry can be devastating.

Further, the proliferation of industrial technology in the Socialist economies may be weakening the bargaining position of U.S. firms as suppliers of technology to newly industrializing nations. At present, U.S. firms reap the greatest return on technology through sales to LDCs. Communist nations are attempting to break into this market, notably in order to procure raw materials. When the long-term interests of a given industry are considered, the immediate short-term gains resulting from the sale of industrial technology may be more than offset both by the possibilities of future inability to compete in Western markets, and by increased competition in technology sales in third world markets.

---

[13]Jack Baranson, *International Transfers of Industrial Technology by U.S. Firms and Their Implications for the U.S. Economy,* U.S. Department of Labor, 1976, p. 35.

IMPORT COMPETITIVE INDUSTRIES

The negative effects of technology transfer rebound most acutely in the second major group interested in the process—those industries that compete with Communist exports both in domestic and foreign markets. Owing to the centrally planned economies' desire to increase their exports, their technology purchases are often in export oriented sectors. Indeed, compensation agreements by their very nature involve U.S. imports of the commodity produced as a result of the technology transfer transaction; other types of countertrade involve Eastern exports of unrelated goods which may also affect U.S. markets.

The negative effects of Communist imports as a whole are rare but relatively easy to document. Victims of Eastern imports may initiate import restraint petitions with the ITC charging market disruption. Problems arise however in connecting specific exports of technology not only to export capability in the same sectors in the East, but also in identifying sectors which may become problems in the future.

One clear example of a U.S. transfer of technology to the East that resulted in a direct and significant increase in imports occurred in 1976, when a U.S. firm signed a $3.2 million contract with Hungary for the sale of equipment, designs, and know-how to manufacture women's shoes. The direct result of this transaction was a fivefold increase in Hungarian shoe exports to the United States between 1977 and 1978. In 1978, women's footwear became the largest single Hungarian export to the United States, and the value of U.S. imports in that year alone was nearly double the value of the original contract. It is relatively unusual, however, for cases of this kind to occur in the consumer goods sector.

There are sectors of the economy that are more vulnerable to the repercussions of technology transfer. Perhaps the most important of these is the chemical industry. In April 1973, Occidental Petroleum Corporation agreed to purchase from the U.S.S.R. 33.3 million metric tons of ammonia and 18.5 million metric tons of urea, most to be marketed in the United States. The Soviets in return agreed to make comparable purchases of U.S. goods, including 18.5 million tons of superphosphoric acid. The deal also involved the construction of several ammonia plants in the Togliatti area of the U.S.S.R., although the technology transfer involved in these plant sales was handled largely by another U.S. firm, Chemica.

In 1977, the U.S.S.R. exported no ammonia to the United States. As a result of this single transaction, 1 year later the Soviet Union became this country's second largest foreign supplier. Meanwhile, over the last 2 years the United States has experienced domestic plant closures and significant declines in ammonia prices. The U.S.-U.S.S.R. contract has a life of 20 years and ITC has already judged that it has led to serious disruption in the domestic anhydrous ammonia market.

Other problems in the same industry have arisen in Western Europe. After a crash program of expansion in the chemical industry greatly aided by technology sales by Western European firms, CMEA production in plastics, ammonia, fertilizers, urea, and soda ash has more than doubled since 1970. Now CMEA's growing self-sufficiency in chemicals has eroded one of the West European chemical industry's largest export markets. In 1975, CMEA purchases from Western firms amounted to over $2.5 billion; since then they have declined to less than $2 billion. In addition to the loss of export markets, CMEA producers have begun to challenge West European firms in their own markets for the sale of many petrochemicals and plastics. In 1976, CMEA accounted for 20 percent of world production of basic chemicals, compared with Western Europe's 30 percent and 25 percent for the United States. Forecasts predict that CMEA will overtake the U.S. share by 1986.

The major West European chemical firms are thus experiencing the results of a boom in technology sales to CMEA for which they negotiated countertrade deals and accepted payment in kind. So long as this payment was largely in the form of raw materials, there was little problem in utilizing it profitably. CMEA payments in intermediate chemical products were also welcomed. Some firms, in fact, came to rely on CMEA for quantities of bulk chemicals that they could not themselves supply without expensive capacity additions. Now, however, not only are compensation agreements becoming more common, they are involving more sophisticated chemicals. Once these are sold to user industries or placed in the spot market in Rotterdam, the Western companies lose control of the market. By now it may be impossible for chemical companies to stop this flow. Most large European producers are committed to long-term compensation arrangements, deals that proliferated because of a depressed market for chemical plants in the West.

Despite the growing menace to the West European chemical industry, existing legal mechanisms have not been able to deal with the glut effectively. Chemical firms also find it difficult to prove dumping under European Economic Community (EEC) procedures, where (as in the United States) relevant criteria are the exporter's prices in the home market or actual costs. CMEA prices are administered and are therefore unusable for price comparisons. Moreover, it usually takes Western firms at least a year to assemble a case based on internal CMEA costs. By this time the damage has already been done.

The case of chemicals illustrates the development of a novel export strategy in those nonmarket States whose exports have been largely composed of primary products. This strategy is to increase the degree of fabrication of primary exports in order to gain hard currency from the increased value added. This is a particularly attractive option because the resulting semifabricate can also be used in domestic industry, thus eliminating the danger of excess supply in times of lagging world demand.

This strategy is now being used in the PRC. In 1974, an American firm, SOHIO, licensed a process to the PRC for producing acrylonitrile, a chemical used in acrylic fibers. This process was to be used to produce 50,000 metric tons annually. Engineering and construction services were provided by two Japanese firms. The synthetic fiber produced in this scheme could be absorbed by the domestic market in the PRC. It is possible, however, that the Chinese may choose to export and use this product as a major foreign exchange earner. Already, synthetic fabric from China is being sold to Hong Kong and Macao where it is made into clothing and then exported to the United States under a favorable (MFN) tariff structure. While in the near future it is unlikely that China's production of synthetic fibers will compete directly in the U.S. market against domestic producers, the PRC is already breaking into U.S. export markets in the Far East.

The market disruption caused by technology transfer in the chemical industry is clear, as is the lesson it provides for the United States. But this case may not be generalizable to other industries. The chemical market is more open to CMEA assaults than other sectors because purchase decisions on chemical suppliers are made almost entirely on the basis of price. Soviet ammonia, in other words, is identical in quality to that produced anywhere else. It is likely that price elasticity of more sophisticated CMEA manufacturers (i.e., automobiles, tractors, etc.) will depend more significantly on nonprice factors—quality, design, availability of service—and aggressive marketing. For example, in spite of heavy infusions of Western technology and highly competitive prices abroad, Soviet exports of passenger automobiles constitute only 1.2 percent of Soviet exports (in value terms) to the industrialized West and have not significantly increased their share of world markets in recent years.

Photo credit: TASS from SOVFOTO

"Lada" cars, produced at the Italian-built Togliatti (Volga) plant. Similar cars are being exported to the West

Threats of disruption in American markets are therefore more likely to appear in categories of semifabricates. Most U.S. dumping actions against nonmarket commodities have, in fact, occurred in these areas. The threat of disruption is also great in the area of metals such as nickel and aluminum, where capacity increases have occurred as a direct result of infusions of Western technology to the U.S.S.R.: Finland has provided nickel-refining technology to the Soviet Union and a French consortium built a 1-million-ton-per-year alumina plant on the Black Sea. It is apparently as common for West European and Japanese firms as it is for their U.S. counterparts to sell technology to nonmarket economies, conscious that in doing so they may ultimately decrease the market share of their capitalist competitors.

This is an important point, for only rarely is U.S. industry the sole contributor of technology necessary to increase nonmarket export potential. In the ammonia case, for example, both Japanese and French firms contributed heavily in terms of equipment and know-how to Soviet productive and delivery capability, and the American supplier of the technology did not possess unique or otherwise unobtainable technology. It is safe to assume, therefore, that had limitations been placed on U.S. sales of plant and technology, the Soviets would have obtained them elsewhere.

A U.S. Department of Labor study of the effects of industrial technology transfer has concluded that in most cases restrictions on technology transactions made by U.S. firms could not have eliminated the negative effects in terms of market disruption either in the U.S. or third-country markets. U.S. firms in most cases do not possess monopolies of the required technologies, and limiting sales only deprives the economy of additional income. The long-term negative effects on sales and market shares will still manifest themselves.

Obviously, this argument does not hold in those areas where U.S. firms hold a monopoly in a given technology, at least for the term of that monopoly. But there are very few of these areas and it is possible that embargoes in these instances may accelerate the development of the technology both in the East and the West. It has been asserted by Hungarian trade representatives, for instance, that when U.S. export controls denied them access to advanced computer-controlled machine tools, they were driven to develop their own models. These now compete with U.S. products in other markets.

In sum, the threat of net losses in U.S. development through technology transfer is most significant in those sectors where market disruption is likely. In these cases, loss in sales by U.S. firms in both domestic and foreign markets may be greater than the value of the transfer contract, and net loss in income will translate into a loss of employment. Moreover, the cost of the resulting loss of jobs in other sectors may be compounded by labor market adjustment, relocation, and retraining. In terms of the aggregate economy, however, these cases may be partially offset by instances where the increased sales of technology and technology-intensive products result in a gain in employment in the selling industries.

IMPORTERS OF EASTERN TECHNOLOGY

By any standard, Eastern sales of technology to the West have been small. As of October 1977, for instance, the total value of all license fees and royalties paid by the United States to the U.S.S.R. was less than $14 million. It has been asserted that there is considerable potential for increasing the amount of technology transfer from Eastern Europe and the U.S.S.R.,[14] but barriers to such expansion exist on both sides. In the Soviet Union, and to a lesser extent in Eastern Europe, inadequate organization of sales efforts and poor marketing inhibit the growth of such trade; in the United States, the widespread perception that no technology of interest to American firms is to be found in the East, and the resources required to learn and evaluate the market may preclude U.S. firms from taking advantage of such opportunities as do exist. The continued dearth of Eastern technology in the West thus becomes a self-fulfilling prophecy.

Despite these handicaps, however, a few U.S. firms have aggressively marketed products produced as a result of Eastern technologies. Notable instances from the U.S.S.R. include excavation machines and surgical stapling devices. Technologies in several

---

[14]See John Kiser, "Report on the Potential for Technology Transfer from the Soviet Union to the U.S.," prepared for the U.S. Department of State, October 1977.

other areas, including energy, are presently under examination by U.S. firms. U.S. industry and consumers have also begun to receive concrete benefits from a wide array of Czech patents and licenses.[15] Since 1961, U.S. firms have purchased 46 licenses from Czechoslovakia. These represent a total expenditure of less than $5 million.

It is clear that any impact of such purchases on the U.S. economy as a whole has been miniscule. This not to say, however, that the potential of such "reverse" technology transfer should be ignored. Soviet advances in peat, coal gasification, oil shale processing, and permafrost technologies all hold promise for U.S. firms interested in exploiting new energy resources. In Eastern Europe, a potential vehicle of technology transfer is U.S. utilization of Eastern R&D facilities. At a time of growing concern with declining outlays in the United States for R&D, access to Eastern European facilities through cooperative research agreements and contracting can provide new and relatively low-cost avenues for R&D investment.

[15]See John Kiser, "Briefing Paper on Czechoslovak Technology Transfer to the United States," prepared for the Czechoslovak-United States Economic Council of the U.S. Chamber of Commerce, Aug. 23, 1978.

Photo credit: Battelle Columbus Laboratories

Testing Soviet stainless steel. A section of a 750-lb high-nitrogen-steel ingot, sent to the United States under the Science and Technology Agreement, is hot-rolled at Battelle Laboratories to evaluate its workability. The Soviets made the ingot by plasma arc melting, an inexpensive way to produce good quality stainless steel

It is unlikely that hitherto unsuspected major technological breakthroughs will come from access to Eastern technologies. The real potential lies in the possibility of marginal improvements in products and processes. In some segments of mass-production industries these can be significant. Furthermore, access to new products or processes may be an entry to large bodies of associated technical information. Soviet construction equipment for permafrost conditions, for example, has evolved from a volume of basic and applied research on arctic conditions. This would be useful both in Alaska and in the development of Antarctica.

The poor showing of Eastern nations in selling their technology abroad is related less to the availability of useful technology than to systemic factors such as the lack of incentive to sell abroad, lack of personnel trained in marketing, and bureaucratic structures poorly suited to facilitating foreign sales. Technological performance in the East, especially in the U.S.S.R., is erratic. Generalizations concerning poor performance may often cloak formidable accomplishments in priority sectors.

Thus, the potential for increased technology transfer from the East is heavily dependent on the ability of the Communist nations to organize themselves efficiently and to make buying less difficult. On the other side, U.S. firms must actively seek these technologies. Systematic monitoring of technological developments of salable Soviet and East European technology in the civilian sector would greatly enhance the ability of U.S. firms to identify opportunities. Without such an effort, purchases of Eastern technology in magnitudes large enough to affect the U.S. economy as a whole, or even significant sectors of it, are unlikely.

THE CONSUMER

The effects of increased competition, even that induced by sales of technology abroad, are not always negative. Such competition may result in increased initiative in product design, manufacture, and marketing. It may also be argued that, given protection against predatory trade practices, inefficient producers should be eliminated if they cannot compete. In this way consumers benefit, and through them, the entire economy. Disposable income that was previously used to purchase more expensive consumer goods can be used elsewhere. Factor costs are lowered, raising profit ratios. These gains may be slight but are well distributed throughout the economy.

SUMMARY AND CONCLUSIONS

The volume of U.S. trade with the Communist world has been low, and the sale of U.S. technology to the East has as yet made little impact, either positive or negative, on the U.S. economy as a whole. A number of economic benefits have accrued to the United States through technology transfer to nonmarket economies. These are primarily in increased sales and employment in those industrial sectors that conclude the sales. In other sectors these benefits may be outweighed by potential negative effects such as decreasing market shares for U.S. firms both at home and abroad. It is unlikely, however, that deregulation of technology transfer can ameliorate these adverse economic effects except in a few cases where the U.S. completely controls the relevant technology. Furthermore, given the present magnitude of East-West trade, any aggregate effects on the U.S. economy have been minimal. Should this trade grow significantly, Eastern exports to the United States will certainly increase. This may necessitate balancing the negative impact of the exports on individual industrial sectors with benefits in other parts of the economy.

Western technologies, while no panacea for Eastern economic problems, appear to benefit the economy of the purchaser to a much larger degree than that of the seller (see chapters X and XI). Barriers to expanding this trade exist on both sides, but the importance of restraints on the U.S. side—tariff and credit restrictions and export controls—may be outweighed by the problem of Eastern export potential. Overall volumes of East-West trade are unlikely to expand significantly in the absence of improved manufacturing and marketing capabilities in the East, although demand for Western technology in these and other areas is unlikely to abate. U.S. policies on the extension of MFN and credits and export controls may affect the market share of American firms in the Western technology sold to the East, but will have less long-run impact on overall trade volumes than will improved Eastern capacity to export.

CHAPTER IV

The Foreign Policy Implications of East-West Trade and Technology Transfer

CONTENTS

The Foreign Policy Implications of East-West Trade and Technology Transfer

One of the basic issues of technology transfer to the Soviet Union concerns the opportunities for and utility of using trade to achieve political objectives of U.S. foreign policy. Such efforts are distinct from controls on trade designed to restrict the transfer of military-relevant equipment and technologies. The question of the political uses of trade has generated considerable controversy and at least three major schools of thought. They divide according to whether they view trade as, in fact, usable for achieving the political objectives of U.S. foreign policy toward the Communist world and, if so, whether the strategy should be one of using trade to build a constraining web of interdependence within an overall framework of "detente" or as leverage to obtain specific policy concessions in the context of superpower conflict.

The following analysis sets forth the logical assumptions and policy implications of three different perspectives on the political utility of trade. These perspectives are not intended to reproduce the views of any particular individual. Nor does the analysis seek to capture all the manifold detail and nuance of the policy debate on this question. Finally, there is no effort to pass judgment concerning the relative merits of the different perspectives and the policy recommendations that flow from them. It is hoped, however, that an identification of the major ways of viewing the question and the logical assumptions and implications of those perspectives will help clarify what has become a highly complex and emotional debate.

PERSPECTIVE I

THESIS

The first perspective rests on a judgment that, for good or ill, trade is not an effective instrument to achieve political objectives. This is the official view of the Soviet Union and is held by a number of America's allies, who contend that history has shown that efforts to obtain political concessions from the nonmarket economies through policies of economic pressure or inducement will be unsuccessful. Consequently, countries like Japan and France have largely decoupled trading policy from other aspects of their foreign policy toward the Communist countries. Each trade and credit transaction is judged on its economic merits alone. What distinguishes this perspective from the two that follow is the belief that attempts to extract a political price for trade—however desirable the objective—will be ineffective and counterproductive in practice. The logic and implication of this approach is discussed in considerable detail in the chapters reviewing

European and Japanese trade policies (see chapters VIII and IX).

CRITIQUE

In the most general terms, this perspective has the advantage of allowing economic transactions to provide economic benefits unfettered by extraneous requirements. The vexing policy dilemmas that inevitably follow any attempt to attach political conditions to trade are thereby avoided.

On the other hand, any opportunities to use trade to further other, noneconomic, State interests are forgone. Moreover, there is an important asymmetry in economic transactions between pluralist and centralized economic systems. The former tends to judge the merits of each transaction in microeconomic terms, i.e., whether it is to the advantage of the particular corporation(s) involved. The latter judges the same transaction in macroeconomic terms—its net benefit to the economic system (i.e., the State) as a whole. A particular business deal may well be to the net advantage of a specific American company, but to the net disadvantage of the United States relative to the U.S.S.R. Finally, for a nonsuperpower with relatively circumscribed interests vis-a-vis the U.S.S.R., any effort to use trade as a means of altering Soviet policy may make little sense. But for the United States, whose interests engage those of the U.S.S.R. along a very wide front, trade controls may be one instrument among many in an ongoing and unavoidable effort to influence Soviet policy.

PERSPECTIVE II

THESIS

The second perspective rests on four propositions:

1. trade can and often does have political consequences and utility;
2. a stable cooperative relationship between the United States and the U.S.S.R. is achievable and the United States has no real choice but to try to build such a relationship;
3. trade tends to have a moderating effect on international relationships due in part to the interdependencies it fosters; and
4. the Soviet need for Western imports of technology provides one of the most effective means for inducing moderation in Soviet policy.

These propositions tend to be associated, in turn, with a series of assumptions and observations about the Soviet Union. **First,** there are powerful forces tending toward a lessening of the ideological fervor and revolutionary commitment of the regime. Among them are the gradual emergence of a consumer economy; the transformation over time of the ideology from revolutionary guide to ritual incantation; the aging of the Soviet leadership and Soviet society generally; the status of the U.S.S.R. as a "have" nation relative to most of the less developed countries of Afro-Asia; recognition of the dangers of pursuing a radical foreign policy in the nuclear era; the potential for a complementary economic relationship involving the exchange of Soviet raw materials for Western capital and technology; and the existence of several areas of common interest with the West, including arms control and such areas of common concern as the prevention of nuclear proliferation and the protection of the global environment.

Second, the Soviet Union is viewed as being in the early stages of a deepening systemic crisis manifested initially by economic stagnation and a failure to close the gap relative to the West in the civilian applications of science and technology. The U.S.S.R. is burdened with a chronic shortage of hard currency resulting from a seeming inability to develop a range of manufactured products that are competitive on world markets—this despite the advantages of a rich and varied resource base and a very large pool of trained scientific and technical manpower. These shortcomings are rooted in the inappropriateness of a rigidly authoritarian political and social structure for a complex, advanced industrial economy. The Soviet system does not permit and foster the flows of information, the innovation, the experimentation, and the general flexibility and adaptability such an economy requires. As a consequence, the Soviet economy has been unable with its own resources to provide for the broad modernization of Soviet life. The problem is greatly exacerbated by the heavy burden of military expenditures.

In this situation the Soviet regime has three broad options: 1) maintain or even tighten political-ideological controls in the name of Marxist orthodoxy, but at the price of economic inferiority vis-a-vis the West, 2) ease controls to stimulate economic growth, but at the risk of changing the political character of the system, and 3) retain controls, but try to escape the economic consequences by obtaining needed technological and managerial innovations and know-how from the West. A similar situation and set of choices confront the Soviet-occupied nations of Eastern Europe.

A **third** assumption posits the existence of conflicting views within the Soviet leadership concerning whether to adopt a general posture of negotiation or confrontation vis-a-vis the West. This division assumes particular importance given the imminent passing of the aged Brezhnev leadership and the uncertainty concerning the identity and policies of his successor. In short, the near future may witness a policy decision of historic dimensions by the Soviet Government regarding its relations with the West.

Based on these propositions and assumptions, proponents of Perspective II have advocated a broad U.S. strategy for dealing with the U.S.S.R.—an approach identified with the term "detente." The basic idea is to take advantage of the Soviet need for Western technology and capital and of other opportunities for interchange (e.g., tourism,

cultural and scholarly exchanges, sports, etc.) to build what Henry Kissinger called "a web of constructive relationships." This should have a number of beneficial effects:

1. It should give Moscow a greater stake in the existing world order and the attendant "disciplines of international life" by integrating the Soviet economy into the international economic system.
2. It should strengthen the Soviet consumer economy as a claimant on Soviet resources and as a generally moderating factor in national policymaking.
3. It should strengthen the hand of moderates in the Soviet leadership by demonstrating the opportunities for useful cooperation with the West.
4. It should provide increased opportunities for the penetration of Soviet society with Western products, culture, and perspectives, i.e., the greater number of peaceful interactions the U.S.S.R. has with the non-Communist world, the more likely that it will become responsive to Western canons of international and domestic behavior.
5. It should, with time, make Soviet policies and behavior increasingly susceptible to foreign pressures.

For example, scientists who have participated in exchange programs with Soviet counterparts argue that these contacts have been effective in achieving a better integration of Soviet scientists with the world scientific community. This has made it much more difficult for the Soviet Government to repress individuals without attracting world attention to the fact. It has allowed Westerners an opportunity to assist scientists officially denied the opportunity to receive literature in their fields or to communicate with other scientists working in related fields.

CRITIQUE

There is no agreed systematic formulation of the detente perspective that can be used to evaluate the concept. In large part the viability of detente as a strategy will depend on what is expected of it. It can be viewed as a comprehensive framework for integrating and managing the strategic, political, and economic dimensions of relations between the United States and the Soviet Union. Alternatively, it may simply be viewed as a series of specific agreements between the superpowers designed to ease tensions and build a network of mutually beneficial interactions, particularly in the economic area. The criteria for evaluating detente and the judgment reached concerning its viability as a strategy will clearly vary according to the definition used.

For the critics of detente, the key question is to what extent detente has served to restrain Soviet actions. In the economic sphere the value of U.S. exports to the U.S.S.R. rose from $105.5 million in 1969 to $546.7 million (1972), $1.2 billion (1974), $1.6 billion (1977), and $2.2 billion (1978). Although trade is still very low relative to the total Soviet economy and compared to other industrial nations (see chapters III and X), there has been a clear growth in Soviet involvement in international commerce. As Helmut Sonnenfeldt notes:

> Trade with the outside world has long been used to fill gaps that the Soviet economy itself could not fill. But the volume and diversity of this trade have steadily increased in recent years; the methods have evolved from barter or straight cash deals to more complex commercial arrangements, including considerable reliance on foreign credits. These latter have now risen to some $40 billion for the Soviet bloc COMECON countries as a whole; Soviet hard-currency indebtedness is in the neighborhood of ten billion dollars. A substantial volume of economic activity in the U.S.S.R. and other Eastern countries must now be devoted to earning hard currency to finance imports and to service mounting indebtedness.[1]

Other forms of interaction, e.g., scholarly exchanges and tourism, have shown a similar pattern of growth. However, the political implications of these trends are uncertain. It is difficult to demonstrate that Soviet international behavior in the 1969-79 decade has been significantly more considerate of Western interests than it was in the preceding decades. The same is true of the hypothesis concerning the moderating effect of international transactions on domestic Soviet policies. In the short term, the characteristic reaction of the Soviet internal security apparatus is to tighten controls during periods of international relaxation.

The most explicit agreement between the two superpowers concerning the political content of detente was embodied in the declarations of basic principles signed at the 1972 Moscow summit conference and the 1975 Helsinki Conference on Security and Cooperation in Europe. The intent was to develop a broad code of conduct or "rules of the game" to which the United States and U.S.S.R. agree to adhere in their interactions. Such codes are designed to impose restraints on the scope of acceptable behavior and thereby manage the competitive aspects of U.S.-Soviet relations. From the U.S. perspective, however, there are serious grounds for questioning whether subsequent Soviet policy, notably concerning Africa and the human rights of political dissidents within the U.S.S.R., has been faithful to these agreements. Other events to which critics point as evidence of the failure of detente to exert restraint on Moscow include Soviet aid to Arab forces during the Yom Kippur War, Soviet approval of North Vietnam's disregard of the Paris Accord, Soviet support for the Portuguese Communist Party's attempted *putsch*, and most importantly, the dramatic buildup of Soviet military capabilities.

Not surprisingly, Soviet spokesmen have taken pains to disabuse proponents of Perspective II of their belief that the U.S.S.R. can, in effect, be co-opted through trade or other economic arrangements. For example,

---

[1] Marshall D. Shulman, Special Adviser to the Secretary of State on Soviet Affairs, testimony before the Subcommittee on Europe and the Middle East of the Committee on International Relations, House of Representatives, Sept. 26, 1978. Committee print, p. 164.

D. Gvishiani, a Deputy Chairman of the Soviet State Committee of Science and Technology, has objected strongly to the proposition that joint production arrangements are a way to overcome political or ideological differences:

> We have different socio-economic systems and different ideologies—that is an existing reality to be reckoned with. Ideological differences between us exist and will continue to exist and we should not count on eliminating them by way of developing industrial cooperation or by some other way.[2]

In addition to ambiguities concerning the scope of detente and doubts concerning the lessons of history, it is not always clear how detente relates to efforts to exert pressure on Moscow. There is a theoretical tension be-

---

[2]D. Gvishiani, statement at 8th "Dartmouth Conference" held in Tbilisi, Republic of Georgia, 1974. Cited in National Academy of Sciences, *Review of the U.S./U.S.S.R., Agreement on Cooperation in the Fields of Science and Technology,* 1977, p. 34.

tween the notion of building a web of cooperative interaction and using those same transactions to coerce Soviet policy. This is particularly true if pressure is exerted before such transactions have become well-established. Thus, the opportunities for using detente to leverage Soviet policy will tend to be greater after the relationship has matured when, paradoxically, such pressures may become unnecessary.

Finally, criticism of detente focuses on the role of science and technology in U.S.-Soviet relations. Proponents of detente argue that these are the most promising areas and instruments for achieving cooperative solutions to common problems. Critics contend that the Soviet leadership views science and technology as the key determinants of national power in the modern era and hence a decisive arena of competition between East and West. By facilitating the transfer of scientific and technological knowledge to the U.S.S.R., detente plays into the hands of Soviet global ambitions.

PERSPECTIVE III

THESIS

The third major school of thought accepts the basic proposition that transfers of technology can have political consequences, but is profoundly skeptical about the linkage between simple economic interaction and policy moderation. It argues that the Soviet Union's behavior has not visibly moderated since it began to import Western technology on a substantial scale and that history is full of examples of wars and confrontations between major trading partners. Proponents of this view contend that the Soviet-U.S. relationship is, and will remain indefinitely, one of conflict. The Soviet Union needs imports of technology to compensate for the rigidities, low productivity, and lack of innovation in its economy, and, consequently, Western

transfers of technology may have the net effect of strengthening an adversary. This is particularly true if such imports are financed by credits at low rates; they then amount to a kind of foreign aid. Such assistance to the Soviets is only justified if we receive something in return in the form of more congenial Soviet policies.

Any real hope for a long-term easing of the confrontation between the U.S.S.R. and the West will require basic changes in the policies of the Soviet Government—changes that will not come voluntarily and can be induced only by outside pressure. Moscow's need for Western technology and the credits to pay for it offer one of the best, if not the best, instruments available to exert such pressure. For example, Senator Stevenson, in arguing for amendments to the Export-Import Bank Act of 1974 that would significantly curtail the extension of U.S. official credits and guarantees to the Soviet Union (see chapter VII), contended that:

> Credits and the withholding of credits can at times serve useful political purposes . . . . The $300 million limitation of credits to the Soviet Union for the next year will permit a tighter rein on Eximbank activities. Major project review by the Congress, whether it involves loans, guarantees, or insurance, would force a more careful assessment of the overall implications of Exim credit assistance and provide the Congress with a tool for exercising appropriate influence. The evolution of detente, peace in the Middle East, SALT, and human rights in the Soviet Union will all influence future Congressional decisions as to whether a particular large project should be financed or the availability of credits continued.[3]

Adherents of Perspective III agree with the proponents of Perspective II that the U.S.S.R. faces a deep-seated economic and political dilemma, but they are far less sanguine that the forces and trends shaping the Soviet system are pressing toward a moderation of Soviet policies. They see the regime as determined to avoid such an outcome and as using the importation of Western technology and capital as a key element in that effort. Perspective III agrees with Perspective II that divisions almost certainly exist within the senior Soviet leadership concerning policy toward the West and that it is in the interest of the West to strengthen the hand of those in the leadership that favor a more moderate, cooperative posture. Where Perspectives II and III diverge is on the best tactics for achieving such an outcome. Proponents of detente Perspective II argue that the overt use of trade pressures to alter unfavorable Soviet policies will tend to strengthen the hand of the Soviet hardliners whereas proponents of Perspective III main-

tain that Soviet hardliners are strengthened whenever the United States fails to vigorously defend its interests.

There is a variant of Perspective III that carries the assumption of the adversary nature of U.S.-Soviet relations to its logical conclusion. It is argued that if the Soviet Union is considered to be a direct and relatively immediate threat to Western security, the United States and its allies should respond with an embargo on all trade and capital flows that could strengthen Soviet capabilities. The objective here is not to leverage Soviet policy, but to limit Soviet power. Thus the U.S. embargo against Nazi Germany was not intended to modify Hitler's policies, but to deny that regime needed resources. To be effective such a strategy would require close cooperation among the major Western exporters to the U.S.S.R. Consequently, the feasibility of this approach hinges on calculations concerning the possibility of a hardening of attitudes toward the U.S.S.R.—born of disillusionment with detente—among the Coordinating Committee for Multilateral Export Controls (CoCom) governments. At the present time such a shift is not evident. Consequently, the predominant tendency with Perspective III is to look to trade as a lever on Soviet policy.

CRITIQUE

There are very few high-technology products that the Soviet Union wants to import in which the United States has an effective monopoly by virtue of the fact that the U.S.S.R. has no adequate alternative suppliers. It is generally agreed that the major plausible exceptions are computers and certain types of advanced oil- and gas-drilling equipment (see below). Wheat may also be in this category, but as a primary product export, it falls outside the purview of this report. When the United States does not enjoy a monopoly, leverage is feasible only if a coordinated approach can be negotiated among all the major suppliers. At present,

---

[3]Helmut Sonnenfeldt. "Russia, America, and Detente," *Foreign Affairs*, vol. 56, No. 2, January 1978, p. 286.

the other major exporters of technology to the East (West Germany, Great Britain, France and Japan) are generally opposed to the use of trade to achieve political leverage (see chapter IX), although a change in domestic political alinements might change that situation. Even with stronger coordinated actions, controls over trade in technologies tend to be effective for a relatively short period due to the emergence of alternate suppliers or the domestic development of the controlled technology in the embargoed State. The latter point is particularly applicable in the case of a large, advanced, and diversified economy like that of the U.S.S.R.

There can be no doubt that the Soviet Union's need for certain imported technologies is considerable and that the priority assigned to those imports is high. They are required to overcome serious bottlenecks in an economy that is showing unmistakable signs of stagnation. In all probability the need for technology imports from the West will increase over the foreseeable future. Nevertheless, it remains true that the U.S.S.R. can probably develop and produce domestically any technology that it wants badly enough and cannot obtain from abroad, although the opportunity costs for other sectors of the economy may be considerable as the required resources are diverted and committed. The consequences of forgoing imports are likely to appear most directly in the civilian economy, since it can be assumed that a concentrated effort has been made to avoid reliance on imported technologies in the military sector.

While the consequences of affecting the civilian Soviet economy are difficult to assess, there is little evidence that the regime need fear domestic political unrest as a consequence of such costs. The drive for economic advance probably comes less from a need to respond to public desires for increased living standards than from the regime's own ambition to match the economic and technological achievements of the West. Consequently, under present circumstances

it is unlikely that the importation of any particular technology will be viewed as sufficiently critical to justify altering policies that derive from the Soviet Union's basic objectives concerning national security, international influence, and preservation of the domestic political system and the dominant role of the Party. An important caveat may be necessary with regard to Eastern Europe. Events in recent years suggest strongly that domestic economic conditions can give rise to serious political unrest in at least some of the Eastern European countries.

The calculation of the cost of acquiescing to U.S. trade pressures is presumably affected, perhaps decisively, by the manner in which leverage is exerted. Critics of attempts to compel changes in Soviet policy by use of legislative sanctions argue that such public ultimatums inevitably raise the price of compliance to prohibitive levels by making national pride the overriding issue. They contend that quiet diplomacy outside the public spotlight is more likely to achieve results. A related argument questions whether it is advisable or even legitimate for the United States to demand changes in Soviet domestic policies—those being within the sovereign jurisdiction of the Soviet State.

Defenders of the legislative approach respond that proponents of quiet diplomacy have had ample opportunity to test that approach without notable success. As for domestic jurisdiction, it is contended that international practice has clearly established that such matters as human rights and emigration are the proper concern of the international community. Moreover, a long-term cooperative relationship between East and West will become possible only if some of the more totalitarian characteristics of the Soviet regime are substantially modified. Thus, Soviet domestic policy becomes an appropriate litmus test of the regime's intentions in its relations with the West.

To date, efforts to use technology transfer for political leverage on the U.S.S.R. have focused on the issue of the right of Soviet Jews to emigrate and to a lesser extent on the treatment of political dissidents in the U.S.S.R. and on Soviet activities in Africa, particularly the use of Cuban troops in that continent's conflicts. It is extremely difficult to rank order the policy priorities of the Soviet Union or any other State. Nevertheless, it seems clear that at least in the two instances of Soviet policy regarding domestic political dissent and foreign policy toward Africa, the cost of acquiescing to U.S. pressure would be high from a Soviet perspective. At stake in Africa is Moscow's claim to the rank of a great power with global interests on a par with the United States. It is most unlikely that any imported technology would be of such value to the Soviet Union that it would, in effect, back away from its claim to global great power status to obtain it. The same is true for the handling of political dissidence[4] where what is ultimately at stake is the basic authoritarian structure and ideological identity of the political system. It would take powerful pressures indeed to induce the Soviet leadership to undertake so profound a change—one that would in a real sense constitute a political revolution. This is not to say that no effective pressure can be brought to bear by the West in support of such objectives. It is to suggest that to accomplish the desired results, trade leverage would almost certainly have to be supplemented by other pressures and inducements in a coordinated allied strategy. Emigration policy, while also of considerable importance to the Soviet regime, probably ranks somewhat lower on the scale of priorities and is, therefore, presumably more susceptible to Western pressure. Whether enough leverage can be exerted through the trade sector to significantly alter emigration policies is a question that

has been put to an empirical test by the Jackson-Vanik amendment discussed below.

It is important to note that attempts to exercise leverage, even when successful, can have unanticipated and sometimes costly side effects. Marshall Shulman, Special Advisory to the Secretary of State on Soviet Affairs, has contended in congressional testimony that U.S. pressure on the U.S.S.R. concerning human rights, to which the administration is deeply committed, may adversely affect the prospect of achieving agreement with the Soviet Union in other areas:

> It is true ... that when we make an issue of human rights, and particularly, we speak of individual dissidents, who are being arrested, harassed, or tried, that is a factor of [sic] the political deterioration of the relationship, it affects other aspects, and perhaps it affects the degree of cooperation we can achieve in other matters.[5]

Another adverse side effect of imposing political conditions on trade may be to damage the competitive position of American exporters in world markets by jeopardizing the reputation of the United States as a reliable supplier.

Even if leverage is exerted successfully, the Soviet Government may vow "never again," and make extraordinary efforts to avoid any subsequent need for Western technology, thereby diminishing the prospects for successful leverage in the future. If the attempt to exert leverage fails, Soviet policy will not be adjusted to U.S. requirements and, consequently, the technology will not be transferred. In this event, the Soviets may again take steps to minimize future dependence while the United States incurs economic and possibly political costs from the embargo. Soviet officials have tried to give credibility to this argument by vigorously contesting the morality and utility of Western efforts to use trade as a lever on Soviet policy.

---

[4]A distinction must be made between "dissidence," and "dissidents." Whereas the prerogative to suppress dissidence as a political phenomenon is a vital interest to the Soviet Government, that same Government has shown some flexibility in its handling of individual dissidents, including allowing or compelling a number of the most prominent to emigrate.

[5]Adlai E. Stevenson III, "Views on Eximbank Credits to the U.S.S.R.," in *U.S. Financing of East-West Trade*, ed. by Paul Marer, International Development Research Center, Indiana University, Bloomington, Ind., 1975, pp. 253-4.

The most visible recent attempt to utilize a leverage strategy is the Jackson-Vanik amendment to the Trade Act of 1974 (Public Law 93-618), which explicitly links U.S. tariff policies and export credits to the willingness of other nations to allow free emigration of those desiring to leave. An unstated target of the amendment is Jewish emigration from the Soviet Union. The Jackson-Vanik amendment must be differentiated in one crucial respect from the use of technology exports for leverage as described above. The amendment is sweeping in its scope; by focusing on Export-Import Bank credits and most-favored nation (MFN) status it puts at risk the entire spectrum of trade between the Soviet bloc and the United States, not just the export of a particular technology or family of technologies. Consequently, the amendment represents a more forceful instrument than the manipulation of technology exports alone.

EMIGRATION POLICY

The question of the advisability of focusing leverage on Soviet emigration policy as opposed to other Soviet policy targets has been much discussed. In support of the former choice, the following arguments can be made:

1. For the Soviet Government emigration policy is important, but not so important as to render it immune to external pressures.
2. It is a relatively discrete policy, subject to monitoring by outside observers.
3. Emigration is not an academic question; it involves real costs to the U.S.S.R. in terms of the loss of valuable human resources and in terms of political symbolism as significant numbers of persons overtly and successfully reject the Soviet system.
4. Conversely, the decision by many of the emigres to come to the United States provides this country with a steady infusion of talent.
5. Support for freedom of emigration is a tangible manifestation of U.S. concern for human rights.

The two principal contrary arguments are:

1. Emigration is a domestic policy and therefore an inappropriate target for leverage.
2. Soviet foreign policy would be a more appropriate target both under international law and in terms of the potential benefits to the U.S. national interest.

In attempting to assess the impact of this legislation it must be remembered that U.S. attitudes are not the only factors influencing Soviet emigration policy. Additional considerations include the reaction of other non-Russian ethnic minority groups in the U.S.S.R., the advantage to Moscow in some instances of permitting or even compelling particularly troublesome individuals to leave, the strenuous objections of the Soviet Union's Arab allies to the continued flow of Soviet Jews to Israel, and the costs in terms of political symbolism and the loss of human resources noted above.

The first year of significant Jewish emigration from the U.S.S.R. was 1965, when 1,500 were granted exit permits. The number rose dramatically in 1971 to more than 14,000, and again in 1972 (31,500) and 1972 (almost 35,000). The number fell sharply to 20,700 in 1974 and to 13,300 in 1975. The Jackson-Vanik and Stevenson amendments became law in December 1974. Through 1976 and 1977 the outflow remained at low levels (14,300 and 16,700 respectively) before beginning a marked upturn in 1978 to nearly 30,000 with a projected increase to perhaps 50,000 this year (1979)—the highest total ever.

These numbers have been interpreted in a variety of ways. How they are interpreted speaks directly to the utility of Jackson-Vanik and by implication, the utility of leverage more generally. Two principal schools of thought can be identified. The first begins with the proposition that fluctuations in the number of emigrants reflect in large part deliberate Government policy, i.e., the Soviet regime can and does turn the emigration spigot on and off to serve its own domestic and foreign policy objectives. Instruments available for this purpose include:

1. interfering with the delivery by Soviet mail of the formal "invitations" from Israel to potential emigres;[6]
2. making the application process prolonged and difficult,
3. intimidating potential applicants with harsh reprisals; and
4. manipulating, as necessary, the rejection rate of applications.

From this perspective the rise in emigration in the early 1970's can be attributed to Moscow's desire to forestall Jackson-Vanik by adopting a relatively liberal policy on emigration. However, the actual passage of the amendment then put the Soviet Government under public pressure, which it defied by sharply reducing the number of exit permits. The recent upturn is seen as an effort to breathe new life into detente, and perhaps more specifically to win Senate approval of the SALT II agreement and forestall a further warming of U.S.-China relations.

Even among those who agree on this basic interpretation of recent events, the question is still open as to the impact of the amendment. Some argue that the amendment (both as a threat and then as an accomplished fact) had the net effect of forcing Moscow to permit more emigration than it would have otherwise tolerated—although the impact was delayed. Others contend that the amendment served a useful purpose as long as it was only a threat, but that its actual passage was counterproductive as indicated by the sharp fall in emigration following its enactment into law. Still others feel that it is as yet too early to reach a conclusion about the consequences of the legislation. Jewish

groups and Soviet specialists in this country are divided on the issue, but Jewish activists in the Soviet Union apparently favor the first interpretation.[7]

An alternative school of thought accepts the proposition that the sudden increases in emigration in 1965 and 1971 were a consequence of a liberalization of Soviet Government policy—due in part to the perception that such an action would facilitate U.S. support of detente. Since 1971, however, fluctuations in the level of emigration are attributed not to Soviet policy but to attitudes within the Soviet Jewish community concerning the desirability of emigrating.[8] That is, the number of actual emigres monthly and annually correlates closely with the number of applications submitted for exit visas. In support of this argument are the following contentions based on limited (and unconfirmed) informal surveys done within the Soviet Jewish community:

• Although the Government could in theory regulate the number of invitations delivered to Soviet citizens from Israel, it has not done so. Inquiries within the Soviet Jewish community indicate that all invitations that are requested are eventually received.
• A comparison of the number of invitations sent and the number of persons actually emigrating is not meaningful because: 1) it is common for an individual to solicit and receive more than one invitation and 2) many of those who solicit and receive invitations choose to defer or forgo application for an actual exit permit.
• The percentage of applications for emigration permits that are denied has re-

---

[6]Soviet law requires that any Soviet citizen desiring to emigrate must receive a written invitation from relatives in another country.

[7]Some supporters of the amendment feel that a larger principle is at stake: human rights (of which the treatment of Soviet Jews is a major example). Proponents argue that U.S. policy should insist that the principle remain central to the overall U.S.-U.S.S.R. relationship, and the importance of doing so transcends the question of the utility of economic leverage, i.e., the principle embodied in Jackson-Vanik should be upheld even at the risk of a counterproductive effect in terms of the amount of actual emigration.

[8]See Igor Birman, "Jewish Emigration from the U.S.S.R.," unpublished manuscript.

mained relatively low (probably less than 5 percent) and stable throughout the period.

- There is no evidence to attribute the 1974-75 decline in emigration to any effort by the Government to make the process of emigration more burdensome. In fact, after the Helsinki accords, the fee required for emigration was reduced and the procedures involved became increasingly routine making entry into the process psychologically easier for a potential emigrant.
- Although emigration to Israel fell off sharply from 1974 through 1977, this was not true of Soviet Jewish emigration to the United States, which increased throughout the period. Several possible explanations are suggested. The 1973 Middle East war indicated that Israel could be a dangerous place to live. Also, it became increasingly clear that the capacity of Israel's economy to absorb intellectual and professional emigres was quite limited. Emigration to the United States grew only slowly from 1971 because it constituted a leap into the comparative unknown and because many of the first emigrants reported difficulty in finding employment during the recession of the early 1970's. However, the number began to grow as conditions in the United States became better known— particularly the fact that there was a demand for doctors, scientists, and other professionals—and as a core emigre community became established that could assist new arrivals.

As far as the Jackson-Vanik amendment is concerned, the implication of this line of argument is that the legislation has had no significant impact on the level—up or down—of Soviet Jewish emigration.

There is a cautious middle position between the two extremes of asserting that fluctuations in Jewish emigration are due either entirely or not at all to Soviet Government manipulation. It is based on the proposition that a variety of factors probably affect the level of emigration, including both Soviet Government actions and attitudes within the Jewish community. If this is correct, the observed fluctuations in Soviet emigration cannot be used to support or rebut arguments concerning the utility of the Jackson-Vanik amendment.

At this time it is not possible to reach any definitive conclusion as to which of the perspectives on Jackson-Vanik is more accurate. A key to doing so would be to obtain official data concerning the number of applicants for exit visas to enable comparison with the number of exit permits actually granted. The latter number is known but the former is kept classified by the Soviet Government.

The State Department seems to be of two minds on the question of whether or not the fluctuations in the level of emigration reflect a deliberate policy on the part of the U.S.S.R. Mr. Shulman has noted, however, that most of the recent increase seems to have come from new applicants.

It is noteworthy that two Communist governments, Romania and Hungary, have successfully appealed for annual waivers of the ban on MFN and Government credits under the 1974 Trade Act. Both have assured the United States that their citizens are being accorded the right to emigrate and both have had their performance in this regard judged adequate by the administration and Congress. Whether either case has implications for U.S.-Soviet relations on this issue is speculative.

CHINA

Although policy questions regarding technology transfer have focused overwhelmingly on sales to the U.S.S.R., during the 1950's and 1960's similar issues arose in the context of U.S. and allied relations with China. Like the U.S.S.R., the PRC was viewed as an adversary State that posed a serious threat to U.S. security interests. Consequently, a U.S. trade embargo was imposed against the Mainland regime. With the end of the Vietnam War and the Nixon Administration's opening to China, official perceptions changed. China was viewed less as a threat and more as a potential weight in the balance of power against the U.S.S.R. Perspectives on technology transfer to China hinge heavily on which view of China is adopted. If Beijing is seen essentially as an adversary, then the issues surrounding sales of technology are similar to those involving transfers to the U.S.S.R.—although the specific technologies involved and the volume of transactions will differ significantly. Proponents of Perspective I on trade leverage have no difficulty applying it to China. Perspective II also has its China analogue, but whereas restrictions on transfers to the Soviet Union are usually advocated as a way to exert leverage on Soviet policy, controls on transfers to China would more likely be justified as forestalling the growth of Chinese power. The major potential exception might be an effort to influence Beijing's policy toward Taiwan. Perspective III would be rationalized and applied much as in the case of the Soviet Union, but with a different technological content because of China's much lower level of economic and technological development. If Beijing is viewed as

more of a threat to the U.S.S.R. than the West, then questions of technology transfer to China are posed largely in terms of their implications for U.S.-Soviet relations. Three broad policy options emerge in this circumstance:

1. Technology can be sold to China in an essentially uncontrolled manner with a view to strengthening Beijing vis-a-vis Moscow. Such a policy will almost inevitably antagonize the Soviets.

2. Sales to China can be controlled and regulated with a view to either pressuring or reassuring the U.S.S.R. Whether or not this approach successfully influences Moscow, it will have the effect of making U.S. policy regarding exports to China a hostage to U.S.-Soviet relations. Beijing is unlikely to view such an arrangement with enthusiasm.

3. U.S. trade policy can be made identical toward the two Communist powers. The advantage of apparent impartiality will be purchased at the price of forswearing any effort to use trade as an instrument for exploiting the Sino-Soviet rivalry to American advantage. Also, it is not clear that such an approach will be truly evenhanded because the same U.S. policy may have quite different consequences for Moscow and Beijing. On the other hand, any effort to manipulate the Sino-Soviet rivalry would be a hazardous and delicate enterprise and there is serious question whether trade control is a sufficiently refined instrument or whether the information needed to use it effectively is available.

COMPARISON OF PERSPECTIVES

In the debate over U.S. trade policy toward the East, proponents of Perspectives II and III share the belief that trade can be used to achieve desirable changes in Soviet domestic and foreign policy. They differ as to the most effective tactics for achieving the agreed end. They also tend to differ in their assumptions concerning the nature of the Soviet Union and U.S.-Soviet relations.

In terms of **assumptions,** proponents of Perspectives II and III see the basic character of U.S.-Soviet relations in somewhat different terms. Proponents of Perspective II tend to emphasize the areas of common interest and the potential for a stable cooperative relationship. Adherents of Perspective III are more impressed with the ingrained conflict between the two superpowers rooted in sharp differences in ideology, social and political systems, and foreign policy objectives and interests. A closely related assumption concerns the ease and extent to which the Soviet regime modifies its perceptions and policies in response to the actions of other countries. Perspective II tends to view the Soviets as capable of substantially moderating their approach to international relations if the actions of others, notably the United States, consistently suggest that it is in the Soviet Union's interest to do so. Perspective III is based on a more mechanistic model of Soviet policymaking. Seeing it as responsive to internal imperatives and largely unaffected by U.S. actions except to the extent those actions bring power to bear sufficient to compel Soviet policy to take them tactically into account. An analogous assumption relates to the conditions of long-term coexistence between the two countries. Perspective II sees detente becoming the basis for a long-term stable relationship as the Soviet system evolves in a moderate direction, i.e., the Soviet system is seen as dynamic and malleable over time. Proponents of Perspective III disagree and see the only real hope for coexistence in either of two strategies:

1. confronting the U.S.S.R. with durable countervailing power structures that force long-term restraint on Soviet policy or
2. depriving the U.S.S.R. of the economic benefits of detente in the hope of forcing the emerging Soviet economic crisis to a point of systemic transformation.

Although not necessarily following logically from the two perspectives, there tend to be contrasting judgments concerning the potential value to the U.S. economy of trade with the U.S.S.R. and Eastern Europe. Perspective II places a comparatively high value on such trade, citing the chronically adverse U.S. balance of payments, concern about a new recession, and the sales by European and Japanese firms to the Soviet Union. Perhaps more important, proponents of this perspective see a basic complementarity between the Soviet and American economies that should permit the large-scale exchange of Soviet raw materials for American capital and technology. Adherents of Perspective III are more impressed by the lack of hard currency available to the Soviet bloc for purchases in the West and the seeming inability of those countries to develop the product exports that could be used to earn such currency. Moreover, the Soviet bloc may soon be an importer rather than an exporter of the raw material most needed in the West—oil.

In terms of **policy strategies** and **instruments,** Perspective II tends to emphasize integration of the Soviet economy with the West through a range of trade opportunities and other economic inducements. Negative sanctions would be used reluctantly, particularly in the short term, because they would jeopardize the process of weaving the fabric of interdependence. In contrast, Perspective III stresses the utility of negative sanctions with a lesser emphasis on positive inducements for influencing Soviet policy in the present and near future. Perspective II tends to adopt a limitationist view of American power and to be skeptical of the extent to which the United States is capable of coercing Soviet policy. At the root of the tactical differences between the two perspectives is a sharply differing judgment concerning the consequences that will follow from a policy of facilitating economic transactions with the Soviet bloc. Whereas Perspective II sees the eventual outcome as a kind of benign subversion of the Soviet system, Perspective III argues the opposite—that Western credits and technology will provide the Soviet Union with the means of escaping the systemic crisis that currently confronts it. Without that timely infusion of external resources, the Soviet regime may be compelled to seriously consider a basic alteration of the economic and political structure of the U.S.S.R.

CASES

One means of assessing the utility of export controls as instruments of political leverage is to examine the most plausible candidate technologies in some detail. It is generally agreed that the two technologies with the most promise for this purpose are advanced oil- and gas-drilling equipment and computers.

The purpose of the following analysis is to assess the costs that export controls on the technology in question would impose on the Soviet economy. This is a necessary, but not sufficient, condition for any effort to predict the Soviet policy response to such controls. Costs are a function of such factors as the availability of alternative Western suppliers, the extent of the U.S. lead over the Soviet technology, the importance or value of the technology to the Soviet economy, the Soviet capability to develop the technology with domestic resources and the time required to do so, the inherent susceptibility of the technology to export controls, and the probable course that future innovation will take and what that implies for the other factors mentioned above. Clearly, these factors will weigh differently depending on the technology in question. Consequently, conclusions relevant to export policy must be based on a case-by-case analysis of the technologies to be controlled.

OIL AND GAS EQUIPMENT

There are a number of considerations that suggest oil- and gas-drilling technologies could be an effective source of political leverage. The United States retains a monopoly position in the West on a few items of petroleum equipment that the U.S.S.R. needs: electric centrifugal submersible pumps, high-pressure blowout preventers, and subsea blowout preventer systems. Because the U.S.S.R. has already purchased most of the important oil production equipment it needs through the mid-1980's (submersible pumps, a drill-bit plant, and gas lift equipment) the impact of a U.S. embargo on oil equipment would be negligible in the short term. However, although the longer term impact on production cannot be estimated with certainty, the effect on the Soviet Union of a complete cutoff of Western equipment and technology would be serious. There are several reasons for this. First, according to the Central Intelligence Agency (CIA) estimates, the Soviet oil industry is at a critical point in its development; i.e., production is likely to peak within the next few years. Output is now declining in all of the major oil-producing regions except West Siberia, and production gains there promise to be much more difficult to achieve now that the giant Samotlor oilfield is reaching its peak. Development of the smaller West Siberian fields is lagging. These are more remote and costly to develop and are proving to be less productive than other fields. This situation is exacerbated by the fact that previous Soviet extraction policies appear to have been shortsighted, stressing maximum current production at the expense of conservation and exploration.

There is reason to believe that the Soviet domestic oil equipment industry will have difficulty in meeting new demands; it lacks the physical capacity and technology to simultaneously sustain production, meet vast new drilling requirements in West Siberia, intensify exploration in remote areas, and accelerate offshore production. If it is to reverse current production projections and achieve productivity gains, the U.S.S.R. must convert from turbo to rotary drilling. This will require the import, not only of additional bit technology, but also of most of the associated drilling equipment—kelly bushings, drill pipe, drill collars, and tool joints. In these areas, therefore, Western equipment could ease critical constraints on oil development. This is especially important for the Soviet Union in view of the fact that oil accounts for approximately half of all Soviet hard-currency exports, and there is already a maximum emphasis on nonpetroleum energy sources with limited opportunities for additional energy conservation.

There are, however, reasons to question how much leverage the United States could in practice obtain by threatening to withhold exports of oil production equipment. Although U.S. firms presently dominate the world market in the kinds of equipment discussed above, their position could be seriously eroded in the next 2 or 3 years if the U.S. Government embargoes petroleum equipment and technology to the Soviet Union. Soviet oil production would be affected and Soviet development costs would rise, but substitutes for U.S. equipment would ultimately appear, and the substitution of equipment from other Western manufacturers would mitigate the impact of U.S. denials. Moreover, a likely effect of withholding exports will be to constrain Soviet oil production. It is at least debatable whether this outcome is desirable from the U.S. standpoint since it may hasten the emergence of the Soviet Union as an oil importer thereby putting more pressure on what is already a tight international supply situation.

COMPUTERS

Another logical candidate for exerting leverage is the computer industry. As the discussion in chapter X indicates, despite rapid progress in recent years, the Soviet Union still lags well behind the United States in such areas of computer technology as very large-scale integration, high-density magnetic disks and the precision manufacturing techniques required to produce them. Transfer of any of the more advanced computers and peripheral technologies will help the Soviet Union upgrade its own capability to produce and utilize computers—a capability that is a critical aspect of efforts to reverse the lagging productivity of the Soviet economy. Moreover, any such computer technology will have military capabilities, whatever the initial intended use. The latter point assumes particular significance given the difficulty of devising effective controls against the diversion of exported computer technologies to military use. Monitoring systems are not foolproof and sanctions involving the termination of maintenance services or actually reclaiming the equipment would have limited effectiveness in the one case and almost no credibility in the other. A computer can be modified ("lobotomized") to reduce its capabilities as a prerequisite for export to the U.S.S.R. However, such a procedure is expensive and clearly reduces the attractiveness of the sale to the purchaser. Ironically, the computer that was scheduled for sale to the Soviet news agency TASS had been altered in this way before the transaction was canceled to express U.S. displeasure over the treatment of Soviet dissidents.

While the potential value of U.S. computers to the Soviet Union would argue for the utility of that technology for political leverage, the existence of alternative non-Communist suppliers argues against it. Japan, in particular, has rapidly emerged as a potentially serious challenger to American supremacy in this field. As previously noted, Tokyo has resisted any efforts to attach political conditions to trade with the Communist countries.

Besides the size of the technological gap, the value of computers for the Soviet Union, and the availability of alternate suppliers, the potential for U.S. political leverage will be affected by the direction of technological innovation in the industry. For example, two areas that are currently attracting attention are mathematical modelling and new architectural configurations of the hardware and innovative ways of knitting hardware and software together in patterns and networks for specialized problem-solving. On a conceptual level each is essentially a task of applied mathematics—one of the strong suits of Soviet science. They are also types of innovation that will disseminate relatively easily through the scientific literature and the technological marketplace. Advances in mathematical programing allow certain types of calculations to be performed with less powerful computers. This suggests that export controls on computer hardware may have a declining utility. It is not to say that the Soviet Union will readily close the current gap in computer technology. There are, for example, very difficult institutional problems involved in networking that may be particularly troublesome for the Soviets.

These considerations, plus the growing capacity of the Soviet Union to acquire, imitate, and absorb Western technology, strongly suggest that the potential for political leverage inherent in computers is modest and probably declining. However, these considerations may be at least partly offset by the high level of technological sophistication required to actually produce and use software for managing the most advanced systems and by the recent tendency of computer companies to protect software as a proprietary corporate technology, something they have not done in the past. This, in turn, may be undermined by pressures to standardize software throughout the industry given the huge costs of developing and implementing advanced systems.

CONCLUSIONS

This complex situation does not easily lend itself to firm conclusions about the utility of export controls for political leverage. It does seem clear that the technological advantage in oil- and gas-extraction equipment and computers enjoyed by the United States relative to the Soviet Union will continue and that Moscow will place a high value on opportunities to obtain advanced technologies that will help bridge that gap. This should create some leverage potential for U.S. policy. That potential is circumscribed, however, by the emergence of alternate non-Communist suppliers and the increasing rapidity with which technological innovations in all fields disseminate internationally. It may be further limited in the case of oil-drilling technologies by doubts about the desirability of constraining Soviet petroleum output. In any case, U.S. policy must be formulated under the assumption that, at best, American technology can maintain a few years' lead over that of the Soviet Union and any opportunities for leverage will have to be exercised within that context.

On the basis of the above analysis it is not possible to state conclusively whether either technology can be used to generate enough leverage to move Soviet policy. In short, there is no "magic" technology as far as political leverage is concerned. Given the complexities involved it is very difficult to calculate with any precision the costs that could be imposed on the Soviet Union by denial of any particular technology. This element of unpredictability is exacerbated by the inherent limitations on our ability to predict the behavior of the Soviet decisionmaking apparatus and the reactions of Soviet policymakers in any specific situation.

CHAPTER V

The Military Implications of East-West Technology Transfer

CONTENTS

The Military Implications of East-West Technology Transfer

INTRODUCTION: THE CONCEPT OF MILITARY RISK

All trade, and trade in technology in particular, necessarily carries with it a risk that the trade will enhance the military capability or at least the military potential of the trading partner. In the case of U.S. trade with the Soviet Union, it can be argued that any accretion in Soviet military capacity weighs against the United States in an overall worldwide balance of power. Whether the political or economic benefits of trade with the U.S.S.R. offset the military costs is a matter of judgment. In order to consider such potential tradeoffs carefully, it is useful to distinguish among five categories of possible military risk:

1. Technologies which not only have a clear and direct military application that could have a substantial effect on relative force capabilities, but which also make possible the construction of weapons or the development of skills currently outside the realm of the recipient's technical competence;

2. Technologies whose immediate application would advance civilian industry, but which might also be applied to military purposes in a way that would give the recipient access to weapons or military skills that it does not now possess, including:

   A. Technologies that lend themselves to direct diversion, with or without modification. An example of the former might be the precision-grinding machines sold in 1972 to the Soviet Union. It has been alleged that these machines were instrumental in allowing the Soviets to produce precision ball-bearings needed for the guidance system in multiple independently targetable reentry vehicles (MIRVs), thereby providing them with a capability they would not otherwise have possessed at the time.[1] A hypothetical case of diversion with modification would arise if a large computer, sold to TASS or Aeroflot for a specific civilian end use, were reprogramed to perform military functions and/or actually moved from one site to another; and

   B. Technologies that lend themselves to indirect diversion, either by providing hands-on training that would be otherwise unavailable (again, the example of large computers applies), or by providing the opportunity for reverse engineering;

3. Technologies with clear and direct military application that could improve, simplify, or render cheaper or more efficient a military industrial activity already within the recipient's technical competence, or that would help to move

---

[1]See "Export Licensing of Advanced Technology: A Review," Hearing before the Subcommittee on International Trade and Commerce of the Committee on International Relations, U.S. House of Representatives, Apr. 12, 1976.

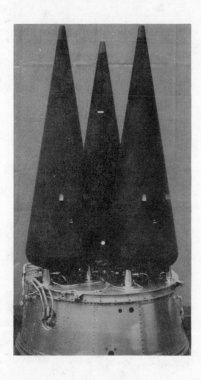

Photo credit: U.S. Department of the Air Force

Left: model of the MIRV warhead that is positioned in the nose cone (circled on right) of the U.S. Minuteman Missile

existing military development activities ahead on promising paths;

4. Technologies whose immediate application would advance a civilian industry, but which might be applied to military purposes in a military industrial activity already within the recipient's technical competence, or which would help to move existing military development activities ahead on promising paths. The same subgroups in #2 above apply here. An example of a civilian technology capable of direct diversion is semiconductor production technology; direct diversion after modification occurred when the equipment of the Kama River truck plant was altered to produce military vehicles. All computer sales carry with them the danger of indirect diver-

sion, as all provide important training opportunities for programers who may later be employed in the military sector; and finally

5. Technologies that would be applied to civilian industry, thereby releasing resources that might be used in the military sector. Any consumer good technology is an example, as are turnkey plants for the production of fertilizer. The latter not only make a significant contribution to agricultural productivity, but if the products of these plants are exported, also may generate the hard currency necessary for further purchases of other technologies, including those with direct military application.

These categories and the examples that illustrate them are summarized in table 13.

Photo credit: TASS from SOVFOTO

Civilian trucks on the main conveyor at the Soviet Kama River plant

Table 13.—Categories of Military Risk

Nature of equipment	Direct military application	Diversion from civilian use	
		Direct	With modification
Permits capabilities that would not otherwise exist in this time frame	1. Nuclear weapons design information in mid-1940's	2A. Precision machines	2B. Advanced computer hardware or software
Improves or makes more efficient an existing capability	3. Naval nuclear reactor production techniques	4A. Semiconductor production technology	4B. Turnkey truck plant
Frees resources for military use	N/A	5A. Fertilizer production technology, or technology expanding production of manufacturing goods for export, thereby contributing to hard-currency earnings	N/A

SOURCE: Office of Technology Assessment.

The likelihood that a given technology will markedly improve the recipient's military capacities decreases as one moves from category one to five. There is unanimity among the United States and its allies that technologies in the first category should be stringently protected, and there is little argument that the third category deserves protection as well. Similarly, most would oppose the blockage of items falling in the low-risk fifth category, and the remainder would

agree that it is impracticable. Given the wide availability of acceptable alternatives to most U.S. technology, only a coordinated policy of economic warfare both within and outside CoCom (Coordinating Committee for Multilateral Export Controls) could impose such a blockage. There is clearly little hope for support of such a policy in Western Europe and Japan. Thus, the most difficult risk assessment lies with the dual-use technologies in categories two and four.

ASSESSING MILITARY RISK

Two fundamentally different techniques can be used to restrict trade involving technologies with potential military significance. In the first system, decisions are made case-by-case. Each proposed technology sale is subjected to careful analysis to determine the possible military uses to which it might be put, and to decide the significance and likelihood of these military applications occurring. Decisions are made on the merits after detailed consideration of the individual case, using some standard of acceptable risk. The second basic approach proceeds deductively. It first establishes lists of specific and generic capabilities that are deemed militarily significant and of the technologies that are instrumental to these capabilities. The sale of any item on this list becomes, by definition, detrimental to the security interests of the United States, and is therefore prohibited. The difference between these approaches is largely one of basic orientation; actual licensing systems combine elements of both. Nevertheless, the fundamental orientation of a licensing system towards either a case method or a list strategy shapes its possibilities and its weaknesses. The following pages consider these alternative approaches in more detail.

THE CASE METHOD APPROACH

Ideally, a case method or ad hoc system of export control includes a comprehensive sys-

tem of risk assessment in which a number of characteristics of the technology in question and of the circumstances of its sales are ascertained, the importance and implications of each piece of information are weighed, and a decision is made on the basis of a complete understanding of both the technology and its probable end use. At least seven different considerations may enter into this kind of assessment:

1. **The capabilities of the technology must be thoroughly understood; its various uses must be identified; and the ease or difficulty with which it might be modified and diverted must be assessed.** The task can be performed only by experts thoroughly conversant with the technology and its range of applications.

2. **For each application of the technology, the comparative capabilities of the United States and the recipient nation must be assessed.** This assessment involves determining technical leads and lags and estimating the rate of change in the differentials. In this connection, the possibility cannot be dismissed that a technology which is obsolete in the West may still have a significant impact on the military capacity of, for instance, the Soviet Union or the People's Republic of China (PRC).

3. **The mechanisms of transfer must be considered.** In 1976, the Defense Sci-

ence Board produced an analysis of the transfer of technology and U.S. national security.[2] The resulting document, commonly known as the Bucy report, assessed selected areas of high technology, their impact on U.S. strategic requirements, the full range of mechanisms through which they may be transferred (see chapter VI), and the effectiveness of current export control restrictions. One principal finding of the Bucy report (others are summarized below) is that the effectiveness of a technology sale varies according to the relationship between seller and buyer; the more active and continuing the relationship, the better the chance that the technology will be assimilated. The report ranks transfer mechanisms according to this criterion and concludes that the most effective—and therefore the most risky— transactions are the sale of turnkey plants, licenses with extensive teaching efforts, joint ventures, and training and exchanges that involve prolonged contact between buyer and seller and the provision of technical information. As a rule, these should be subject to closer scrutiny and tighter controls than less active mechanisms, like product sales, which do not usually transfer current design and manufacturing technology.

4. **Knowledge of the recipient environment is required.** There is a common misconception that technology transfer is "something like a pass from a thrower to a receiver. In fact, it has more of the characteristics of an organ transplant, with all the attendant requirements of compatibility with the environment, plus the surgical (i.e., managerial) skills necessary to establish all the intimate working relationships between the transplant and the connecting parts of

the system."[3] The Soviet system, for instance, is often characterized as inefficient and inflexible, and therefore unable to make optimum use of imported technologies. Chapter X discusses the fact that the ability of the U.S.S.R. to absorb, diffuse, and duplicate Western technology has been hampered by a system in which enterprise managers, who are responsible for introducing new technology, often have no incentive to do so. Such impediments make it less likely that the technology will be used in a manner that produces results similar to those achieved in the country of origin. Even more importantly, the U.S.S.R.'s ability to improve imported technologies through domestic R&D and innovation is constrained by the difficulties of translating new concepts into serial production. For instance, according to recent testimony by Rauer Meyer, former Director of the Office of Export Administration (OEA), the Western-built Volga automobile plant has not revolutionized the Soviet motor vehicle industry; instead, the Soviets are currently conducting negotiations with Western firms to modernize other car plants. Meanwhile equipment similar to that used at Volga is currently being supplied to a tractor plant at Cheboksary. Despite the restructuring of civilian R&D activities in the Soviet Union during the late-1960's, the link remains weak between the economic incentives and material rewards of research institutions and the economic contributions of the new technologies developed by them. It would be a mistake, however, to extrapolate too easily from the civilian to the military sector. In the Soviet Union, the military takes priority; resource allocations are made first to the military, regardless of short-

---

[2]*An Analysis of Export Control of U.S. Technology—a DOD Perspective* (Washington, D.C.: Defense Science Board Task Force on Export of U.S. Technology, Feb. 4, 1976).

[3]Herbert Fusfeld, Director of Research, Kennecott Copper, quoted in National Academy of Sciences, *Review of the U.S./U.S.S.R. Agreement on Cooperation in the Fields of Science and Technology,* National Research Council, May 1977.

ages elsewhere in the economy, and the military sector receives the best manpower and equipment. The consequences of the inefficiencies that permeate other parts of the society, therefore, are not necessarily as serious in that sector.

5. **The risk of diversion of a dual-use technology is substantially affected by the requirements, priorities, and intentions of the recipient.** If, for instance, the technology is essential to meeting a need that has a very high civilian priority, the chances diminish that it will be diverted for direct military use. Large computers that direct oilfield operations are low-risk items from this perspective, not simply because they would be cumbersome and difficult to move and reprogram without detection, but because oil and gas production is extremely important to the Soviets. Similarly, it is unlikely that equipment installed in plants to manufacture drill bits will be put to any other use; the bits are sorely needed in the Soviet oil industry. There is always a chance, however, that priorities may change. Risk of diversion will therefore fluctuate over time, and for reasons not always apparent to Western observers.

6. **The existence and effectiveness of techniques to prevent conversion of civilian technologies to military use must be considered.** In some cases the technique is one of deterrence; while diversion to military use remains possible, the likelihood that the United States would learn of such diversion and react by cutting off future technology transfers diminishes Soviet incentives to divert a given technology. The ability to conduct on-site inspections, to monitor plant output, or to incorporate in the technology devices designed to prevent reverse engineering or alteration all have this effect. It must be recognized, however, that no deterrent is infallible. The large computer installed in the Kama River

truck factory, for instance, is monitored by the American firm that supplied it as part of its contractual agreement with the U.S.S.R. Periodic reports showing the allocation of computer time are made to the Department of Commerce, but there is good reason to believe that, if it is analyzed at all, this data (which arrives in the form of voluminous computer printouts) is subject only to spot checks. Another approach is to render diversion physically difficult or impossible; e.g., to seal electronic components in a medium that will destroy the component if any attempt is made to disassemble it. Such attempts are expensive and according to technical experts are rarely, if ever, infallible.

7. **Finally, individual sales of technology cannot always be evaluated in isolation.** Sometimes the impact of a technology transfer can only be appreciated in the context of a number of related sales that may have preceded it; the importance of a single item may derive from its position as one of a series of items that, taken together, enhance an existing capacity or provide a new one. There is, for instance, a qualitative as well as a quantitative difference between providing 5 computers and 5,000; similarly, a relatively small piece of machinery may only assume its proper importance after it is perceived as a link in the chain of an entire process that has been acquired piecemeal.

If all the preceding factors could be weighed, knowledge of the chances of a technology's enhancing the military capabilities of an adversary would be substantial. But the effectiveness of this kind of risk assessment, and the case-by-case decisions that it entails, is vitiated by two problems. The first is the absence of clear policy guidelines. The Export Administration Act of 1969 (see chapter VII) declares that it is the policy of the United States "to restrict the export of goods and technology which would make a significant contribution to the military potential of any other nation or nations which would prove detrimental to the national security of the United States." The law does not define "significance;" presumably this is left to the officials who administer the licensing process. The Department of Defense (DOD) maintains a list of criteria to be applied to the potential sale of any dual-use technology. These criteria form the basis of judgments about the probability of diversion taking place and being detected, and the consequences of such a diversion:

1. Is the item appropriate in quantity, quality, demonstrable need, design, etc., to the stated civilian end use?
2. Is there any evidence that the stated end user is engaged in military or military support activities to which this item could be applied?
3. How difficult would it be to divert this item to military purposes?
4. Could such diversion be carried out without detection?
5. Is there evidence of a serious deficiency in the military sector which this item, if diverted, would fill?
6. Is technology of military significance, which is not already available, extractable from this item?[4]

The answers to these questions would provide much of the information needed to make an accurate judgment of the probability of military diversion and impact of this diversion on the adversary's military capacities. This information does not, however, help those making the decisions on export licensing to evaluate the "significance" of such impacts against the objectives of U.S. strategic policy. But a yardstick against which "significance" may be measured will probably never be forthcoming.

The significance of any given improvement in Soviet or Chinese military capability

and potential is almost impossible to define in the abstract. An improvement in the capacity to acquire a new military capability may not matter unless the country concerned makes the effort to translate potential into real capabilities. The significance of actual military hardware depends on whether a situation arises in which it can be put to use. How useful a given military capability may be—whether in battle or for political intimidation—may depend on the way in which the United States structures its own foreign policy or military objectives and the capabilities demanded of U.S. forces. Even if the United States were to articulate a detailed array of political/military objectives, and the force characteristics necessary to achieve these objectives, doubt would remain about the significance of incremental improvements in the military forces of potential opponents. In the absence of such a clear and explicit set of objectives, the officials who administer an export-control system must to some extent rely on commonsense and conventional wisdom.

A second problem lies in the actual administration of the case-by-case approach. Limitations on the resources available to administer export controls and the complexity of the procedures may make the system so inefficient as to be counterproductive. As chapters III and VII document, industry criticism has centered on the delays in the processing of export license applications by OEA. The volume of cases that must be handled, the volume of information that must be assembled on controversial cases, and the diversity of interests represented in the process have resulted not only in delays, but in decisions which have been subject to intense retrospective criticism on the grounds that military risk or foreign availability were improperly assessed or that foreign policy interests improperly outweighed serious military implications. Examples of such criticism may be found in the Kama River truck plant, the Bryant grinder case, and the controversy surrounding the Dresser drill-bit plant.

---

[4]See Jonathan B. Bingham and Victor C. Johnson, "A Rational Approach to Export Controls," *Foreign Affairs*, April 1979, p. 889.

THE CRITICAL TECHNOLOGY APPROACH

The recognized need both to clarify policy and to simplify and sharpen the licensing procedure has caused Congress to endorse the DOD investigation of the critical technology approach to export control. This exercise, which represents a systematic effort to confront the problem of military risk through a comprehensive reappraisal of the Commodity Control List (CCL), could potentially shift the weight of present U.S. export-licensing policy from the case method to the "list approach." The critical technology approach grows directly from the findings and recommendations of the Bucy report, which may be summarized as follows:

- Design and manufacturing know-how are the most important elements in strategic technology control. Therefore, the categories of export that should receive primary emphasis are arrays of design and manufacturing know-how; keystone manufacturing, inspection and test equipment; and products accompanied by sophisticated operation, application, or maintenance know-how.
- The more active the participation of the transmittor, the more effective the technology transfer mechanism. Therefore, more active mechanisms of transfer must be tightly controlled, but product sales may be largely decontrolled since these usually do not transfer current design and manufacturing technology. Control of product sales should stress their intrinsic utility.
- The United States should preserve its strategic leadtime by denying all exports of technology that represent revolutionary advances to the receiving country. Transfers may be approved if the technology represents only an evolutionary advance, unless both nations are on the same evolutionary track. In this case, the receiving country's immediate gain from the acquisition of the technology should be assessed.

- Current U.S. export control laws should employ simplified criteria in order to expedite the majority of license requests. Currently, the absence of established criteria for evaluating technology transfers requires a cumbersome case-by-case analysis of all export applications. DOD should, therefore, develop policy objectives and strategies for the control of key high-technology fields that specifically identify the key elements of technology, including critical processes and key manufacturing equipment.
- These key elements of technology should be released only to other CoCom nations. Any CoCom nation that allows this technology to pass to any Communist country should be prohibited from receiving any further strategic know-how.
- Techniques meant to discourage diversion of products to military applications are not a meaningful control mechanism when applied to key design and manufacturing know-how, and should not be relied on to prevent diversion to military use.

The critical technology approach is predicated on the assumption implicit in the Bucy report that "one can select the subset of technologies of significant military value on which our national military technology superiority can be presumed to be most dependent."[5] It goes on to assume that this set of critical technologies will be small in number and relatively stable over time; that they can be subjected to stringent export controls that deny them automatically to any Communist country; and that the development of a Military Critical Technology Product and Information List, which will replace the present Controlled Commodity and CoCom lists, will allow the decontrol of many products

---

[5]Testimony of Dr. Ruth M. Davis, former Deputy Under Secretary of Defense for Research and Advanced Technology, before the Subcommittee on International Economic Policy and Trade, Committee on Foreign Affairs, U.S. House of Representatives, Mar. 22, 1979.

and processes which currently appear on the latter, but which are not in fact "critical." The approach is intended both to enhance the protection of U.S. technological leadtime and to make the export control process simpler, quicker, and more predictable by eliminating most of the present need for case-by-case review; the goal is to protect the military leadtime of the United States with minimal interference with trade.

As originally conceived, the methodology employed by DOD in the critical technology exercise may be summarized as follows:

1. **Determinations of critical technology areas.** The first list of military critical technologies was completed in January 1979. It identifies 15 broad areas of applied science or engineering that will serve as indicators of the fields in which the specific critical technologies to be controlled will be found. The 15 areas are:
   — computer network technology;
   — large computer system technology;
   — software technology;
   — automated real-time control technology;
   — composite and defense materials processing and manufacturing technology;
   — directed energy technology;
   — LSI-VLSI design and manufacturing technology (LSI refers to large-scale integration and VLSI to very large-scale integration in microelectronics);
   — military instrumentation technology;
   — telecommunications technology;
   — guidance and control technology;
   — microwave componentry technology;
   — military vehicular engine technology;
   — advanced optics technology (including fiber optics);
   — sensor technology; and
   — underseas system technology.

2. **Determination of specific component technologies within each of these 15 areas of applied science and engineering.** Various degrees of progress have apparently been made in 9 of the 15

areas listed above. This work has been accomplished by Critical Technology Expert Groups (CTEG) composed of volunteers from industry working with DOD and other Government officials. CTEGs are examining such areas as computer networks, LSI manufacturing design technology, ray processors, acoustical rays, lasers, wide-body aircraft, etc., but not all of the groups have reported their findings. DOD has testified that the date of completion of this step will be determined by the budgetary allocation, and has made no prediction of when activities in all 15 areas might be completed.

3. **After completion of step 2 for each of the 15 broad areas, analysis of the military critical technologies to determine the elements of design, manufacture, utilization, testing, and maintenance functions that can be subjected to export controls.** This step recognizes the fact that it may be impossible to fully control all critical technologies because some mechanisms of technology transfer may be difficult or impossible to contain; e.g., information in the public domain.

4. **Recommendations as to which products, technical information, or other controllable features of each military critical technology should be placed on a list of embargoed items.** This step will utilize criteria that correspond to those employed by DOD and listed above. They include the determination of foreign availability; the technological capability in and military reliance on the critical technology by the potential recipient; and the comparison of these capabilities and dependencies with those in the United States, including the rate of change of this comparative differential.

5. **Formulation of a Military Critical Technology Product and Information List of items not to be exported, accompanied by a list of technology transfer mecha-**

nisms effective for each of the critical items which should be subject to Government control.

At the end of this process, DOD will presumably have arrived at a list of critical products and information that should be barred from export and a list of technology transfer mechanisms that should be subject to Government control. Some assume that the United States will propose CoCom's adoption of this list in place of its present one and that it will also replace the existing CCL with the list of critical technologies. These changes may or may not be accompanied by a transfer of the main responsibility for export control from the Secretary of Commerce to the Secretary of Defense.

It is by no means clear, however, that this will be the outcome. Since its inception in the summer of 1976, progress on the critical technology approach has been slow. High DOD officials have, in the past, attributed this to inadequate resources, asserting that there have been "no technological or institutional hurdles which would prevent the implementation of the Critical Technology Approach."[6] This assertion is somewhat controversial. Discussions within DOD have indicated a lack of consensus on the aims and probable results of the critical technologies exercise. This uncertainty, as well as the conceptual difficulties inherent in the enterprise, has almost certainly contributed to the delay.

Some in DOD regard the critical technologies approach primarily as an in-house exercise. They expect that the product will not be a new form of CCL, but rather an enhanced internal capability for assessing the military impact of dual-use technologies. A variety of offices within DOD perform technical assessments for license applications. In the past, these offices have not always applied uniform criteria to the cases under their consideration.

In August 1979, those offices in DOD responsible for export licensing and those engaged in the critical technology exercise were reorganized (see chapter VII) and their activities centralized. This should provide an excellent opportunity, not only for strengthening and rationalizing the Department's role in the export-licensing process, but for defining with more precision the Department's practical expectations for a critical technologies list. At the least, an important product of the critical technologies approach should be refined and internally consistent guidelines for assessing the strategic capabilities of technologies.

It would be premature at this stage in the development of the critical technology approach to speculate on the difficulties that may arise in attempts to implement it, or on the possible consequences of its implementation. Several observations are in order, however. First, whatever the procedural outcome of the current exercise, DOD will profit from the detailed information it has gathered and the insights it has gained on the military capabilities of many technologies. On the other hand, it would be both misleading and unwise to regard the development of a critical technology list as a panacea to the difficult problem of protecting U.S. military technology leads. Skepticism already exists, both in Government circles and within the business community, as to whether the revised lists will indeed be shorter than present ones; there is fear, in other words, that reluctance to decontrol items or a broad definition of criticality will result in similar or longer lists. This might further inhibit East-West trade and could also provoke objections among some members of CoCom. From the other side, there are fears that a critical technology list will be too short, i.e., that items of marginal, but potentially important, military utility will be decontrolled to the ultimate detriment of the United States.

It is highly likely that, whatever the outcome, the list will be criticized, either for the items it includes or excludes. The belief that a critical technology list can ever be entirely noncontroversial rests on the assumption that definitive, highly refined, empirical judgments can be made regarding the mili-

---

[6] Ibid.

tary utility of a myriad of products and processes. This is unlikely. In the final analysis, inclusion on such a list requires judgments on the part of policymakers; the issues are not purely matters of technical or scientific "fact." Moreover, it is dangerous to assume that the existence of a critical technologies list can in itself obviate the case-by-case review. Considerations of both foreign availability and end use can never be entirely eliminated; simply because an item appears on a U.S. list of embargoed technologies does not prevent its export from abroad; and simply because an item does not appear on this list does not mean that under certain circumstances it could not constitute a significant improvement in the strategic position of an adversary. Finally, because the cutting edge of technology moves so rapidly, any list must be subject to constant review and update.

Those involved in the critical technology effort recognize these problems. One important aim of their activities is to substantially decrease the volume of cases that are presently subjected to detailed case analysis so that resources may be concentrated on those cases involving difficult judgments. In order for this to occur, however, methods must be devised to screen export applications to the Communist world—not necessarily in the exhaustive manner that pertains now, but in some way that will "catch" potentially troublesome cases involving technologies that do not appear on the critical technologies list, assuming that this list is generally viewed as comprehensive without being overly inclusive. One might imagine a system that proceeded roughly in the following manner:

1. All requests for export licenses to the Communist world would be subjected to an initial screening process. The criteria applied here would reflect the concerns of all the executive departments involved in licensing and might ask such questions as:

   —Is the item on the critical technologies control list? If so, presumably no further inquiry is necessary. If not,
   —Is the stated end use plausible?
   —Are large amounts of training entailed in the sale?
   —Is there any obvious military relevance, even if this is not "critical"?
   —Have inordinately large quantities of this equipment been exported?

   The object is to raise a "red flag;" to catch-out potentially troublesome cases. In this way, the volume of cases that require further review should be greatly reduced and the serious "log-jam" which presently plagues the licensing procedure substantially eliminated.

2. Any case in which a red flag appears would then be subject to a moderate degree of examination specifically targeted to answer the particular objection raised in the first cursory screening.

3. Should this moderate examination not resolve the problem, an analysis similar to the intensive case-by-case review presently conducted by OEA should be conducted.

4. In addition, random checks should be made to ensure that the procedure is producing the desired results. These might take the form of periodically selecting isolated applications that otherwise would have been granted after step 1 and subjecting them to the deeper consideration of steps 2 or 3.

5. In cases where threat of reverse engineering or diversion appears to be a major problem, an analysis should be conducted of the procedural or technical mechanisms that could minimize the dangers.

SUMMARY

The process outlined above is intended as nothing more than a suggestion for a way of thinking of export control in a manner that combines the case study and list approaches discussed here. It should be apparent that unless both are utilized, no one formula can resolve the immensely complex issue of determining which technologies make "significant contributions" to the military capacities of our adversaries, and no simple procedure is likely to soon be instituted to protect such technologies. The protraction of the critical technology exercise itself indicates the extreme difficulty which confronts even the Nation's foremost technical experts in making recommendations in these areas, and should caution all observers of the folly of expecting magical automatic solutions to such complex problems. Western technology has undoubtedly contributed to Soviet military capabilities in the past and it will continue to do so in the future, regardless of any unilateral efforts the United States could undertake. There is no reason to believe that drastic changes in DOD's efforts in the area of export control will materially alter this situation. In the final analysis, the national security of the United States is most surely protected by its maintenance of technological leads in those areas that have been deemed militarily critical.

CHAPTER VI

Technology Transfer: Definition and Measurement

CONTENTS

Technology Transfer: Definition and Measurement

Discussions of the economic consequences of trade in technology for both the United States and the Communist world have been hampered by conceptual and practical difficulties in gathering and interpreting data. There is no universally accepted definition of "technology," and in many critical instances, useful data is simply unavailable. Any attempt to assess the economic importance of this trade must therefore include a discussion of the nature of technology and technology transfer and the ways in which they can be measured.

DEFINITIONS

Technology must be differentiated from science on one hand and from products on the other. Science is the pursuit of knowledge, whereas technology is the specific application of knowledge to the production of goods and services. Science flows freely across international boundaries, and even if it were possible to effectively control this flow, the prospect of doing so raises at the very least grave Constitutional questions. Some control of technology, however, is both desirable and necessary in the interests of national security because of the military or strategic capabilities it may provide.

The distinction between technology and products is more troublesome. If technology is broadly defined to mean the knowledge necessary to design, create, or implement a process; the process itself; or any services related to the process, the problem of how to treat the resulting product remains. Often this will be a "technology intensive" product, one that might be said to "embody" technology or from which the technology may be extracted through a process known as "reverse engineering"—the deduction of the techniques of manufacture from examination of the product itself. Often too tech-

nology-intensive products have military applications that cause them to pose as severe a problem to national security as the design and manufacturing know-how that went into them.

For commercial purposes, "technology" usually refers either to equipment and processes that transform raw materials into goods and services, to the training that accompany these, or to final products like computers that embody high technology. But there is little agreement, in the United States or abroad, as to exactly which products and process should be included in these categories. There are, furthermore, problems of measurement within each category. The cost of equipment or of the licenses for rights to processes, for instance, may not necessarily reflect the value to the buyer in terms of the quality, output, innovativeness, and profitability of the final product. The value of a purchase, which includes the skills of the workplace—the training required to operate machines, to achieve practical familiarity with the theoretical aspects of equipment, and to become able to adapt and extend the operation of the equipment—is difficult to quantify. Finally, there is disagreement over which products qualify as "high technology" items.

To these empirical problems must be added the difficulties engendered by the fact that a number of both commercial and non-commercial vehicles exist through which technology of potential economic value is exported to the East. Commercial vehicles of technology transfer include turnkey factories (i.e., a factory built in the recipient country by a foreign firm, which is turned over to the recipient only when it is ready to "turn the key" and start production); licensing (with and without training programs); joint ventures; technical exchanges; training in high-technology areas; sale of processing equipment; provision of engineering docu-mentation and technical data; consulting; proposals (documented and undocumented); and sale of products that embody technology. Noncommercial vehicles include visits in both directions of students, scientists, and businessmen or managers; the use of unclassified published technical data and patents; the reverse engineering of single machines or components; and clandestine activities. All of the latter modes of technology transfer cost negligible amounts of hard currency and, for the most part, have been beyond Government control. Communist states have made the most of these techniques, although they are by no means unique in this regard. These channels of technology transfer have historically been and will continue to be of great importance to market and nonmarket nations alike.

PROBLEMS OF MEASUREMENT

COMMERCIAL TRADE IN TECHNOLOGY

The most common forms of commercial technology transfer are the direct sale of products embodying high technology and various forms of industrial cooperation agreements.

High-Technology Products

The U.S. Department of Commerce recently attempted to isolate trade in high technology through the examination of exports in selected categories of the Standard International Trade Classification (SITC). This classification scheme summarizes trade information for approximately 10,000 different items by organizing it into commodity groupings. The Commerce study selected 25 categories of products which, it contends, contain all those goods that reflect best practice in critical technology sectors—machinery and transport equipment and professional, scientific, and controlling instruments (see table 14). This effort is by far the most precise and comprehensive attempt to use trade statistics to measure technology transfers.

There are problems with the Commerce list, however. Aside from quarrels over what constitutes a "high technology" good, no list based on trade data can be sufficiently detailed to precisely distinguish between levels of technology. This could be accomplished only through a case-by-case examination of individual exports in light of an accepted set of criteria defining "high technology." The Commerce Department classifications are therefore overly inclusive; they "catch" items which do not in fact embody "high" technology, if by that is meant state-of-the-art or items unobtainable in the East. This means that calculations of high-technology trade based on these categories are inflated. Second, techniques used to value and describe exports at point of origin in the United States cannot reflect the contribution of third nations. U.S. technology embodied in products originating from American subsidiaries in Europe or Japan appears in the trade statistics of these countries and

Photo credit: Bureau of East-West Trade, U.S. Department of Commerce

U.S.-U.S.S.R. technology transfer through the mechanism of trade fairs

Table 14.—High-Technology Items

SITC	Description
71142	Jet and gas turbines for aircraft
7117	Nuclear reactors
7142	Calculating machines (including electronic computers)
7143	Statistical machines (punch card or tape)
71492	Parts of office machinery (including computer parts)
7151	Machine tools for metal
71852	Glassworking machinery
7192	Pumps and centrifuges
71952	Machine tools for wood, plastic, etc.
71954	Parts and accessories for machine tools
71992	Cocks, valves, etc.
7249	Telecommunications equipment (except TC & radio receivers)
72911	Primary batteries and cells
7293	Tubes, transistors, photocells, etc.
72952	Electrical measuring and control instruments
7297	Electron and proton accelerators
7299	Electrical machinery, n.e.s. (including electromagnets, traffic control equipment, signaling apparatus, etc.)
7341	Aircraft, heavier than air
73492	Aircraft parts
7351	Warships
73592	Special purpose vessels (including submersible vessels)
8611	Optical elements
8613	Optical instruments
86161	Image projectors (might include holograph projectors)
8619	Measuring and control instruments, n.e.s.

SOURCE: *Quantification of Western Exports of High Technology Products to Communist Countries*, prepared by John Young, Industry and Trade Administration, Office of East-West Policy and Planning, U.S. Department of Commerce, Project No. D-41.

not in those of the United States. Finally, customs valuations are determined by the price of the sale. Price does not necessarily reflect the full market value of the commodity, however; some firms deliberately underprice an initial sale in order to break into Eastern markets.

With these reservations, and in the absence of alternative superior measures, the Commerce system has been used in chapter III to analyze U.S. and industrialized world exports of high-technology products to the Communist nations.

Industrial Cooperation Agreements

Industrial cooperation agreements have become increasingly common in East-West trade. In its most general sense, the term refers to a broad charter extending over a number of years to conduct commercial relations between a Western firm and a centrally planned economy. Industrial cooperation includes a wide variety of possible relationships, ranging from the sale of licenses and patents to coproduction agreements and turnkey plant sales. The comprehensive list incorporated into table 15 summarizes the basic mechanisms and techniques utilized in these ventures. These frequently involve relationships between trading partners which extend beyond simple sales of goods and services, to continuous and close contacts between trading partners, training, and technical assistance programs. It can be expected that these agreements lead to considerable communication of technical know-how congruent with sales of plant and capital equipment.

Activities in this area are extremely difficult to measure. Cooperation agreements are often complex and their values particularly difficult to establish because many East-West transactions involve countertrade rather than cash (see chapter III).

Countertrade is particularly attractive to Eastern nations with scarce hard-currency resources and a need to foster exports to the West. But while its importance in Communist countries is becoming increasingly apparent, little data on such agreements exist. The U.S. Department of Commerce estimates that in Poland, 40 to 50 percent of electrical products and machinery exports to the West in the 1980's will be part of countertrade agreements; and 38 percent of Soviet trade turnover between 1976 and 1980 will be generated through countertrade.[1] There are no comprehensive studies of the full range of countertrade transactions, although the Organization for Economic Cooperation and Development (OECD) has studied individual categories of contracts.[2]

[1] See U.S. Department of Commerce, *East-West Countertrade Practices: An Introductory Guide for Business*, Industry and Trade Administration, August 1978.
[2] Organization for Economic Cooperation and Development, *Countertrade Practices in East-West Economic Relations*, Paris, Mar. 23, 1978.

Table 15.—Types of Contractual Arrangements Included in Different Definitions of East-West Industrial Cooperation

1. Sale of equipment for complete production systems, or turnkey plant sales (usually including technical assistance).
2. Licensing of patents, copyrights, and production know-how.
3. Franchising of trademarks and marketing know-how.
4. Licensing or franchising with provision for market sharing and quality control.
5. Cooperative sourcing: long-term agreement for purchases and sales between partners, especially in the form of exchanges of industrial raw materials and intermediate products.
6. Subcontracting: contractual agreement for provision of production services, for a short term and on the basis of existing capabilities.
7. Sale of plant, equipment, and/or technology (1-3 above) with provision for complete or partial payment in resulting or related products.
8. Production contractings: contractual agreement for production on a continuing basis, to partner specifications, of intermediate or final goods to be incorporated into the partner's product or to be marketed by him. In contrast to subcontracting, production-contracting usually is on the basis of a partially transferred production capability, in the form of capital equipment and/or technology (on basis of a license or technical assistance contract).
9. Coproduction: mutual agreement to narrow specialization and exchange components so that each partner may produce and market the same end product in his respective market area. Usually on the basis of some shared technology.
10. Product specialization: mutual agreement to narrow the range of end products produced by each partner and then to exchange them so that each commands a full line in his respective market area. In contrast to cooperative sourcing, product specialization involves adjustment in existing product lines.
11. Comarketing: agreement to divide market areas for some product(s) and/or to assume responsibilities for marketing and servicing each other's product(s) in respective areas. Joint marketing in third markets may be included.
12. Project cooperation: joint tendering for development projects in third countries.
13. Joint research and development: joint planning, and the coordinated implementation of R&D programs, with provision for joint commercial rights to all product or process technology developed under the agreement.
14. Any of the above in the framework of a specially formed mixed company or joint venture between the partner firms (on the basis of joint equity participation, profit and risk-sharing, joint management).

SOURCE: Office of Technology Assessment.

Table 16 summarizes one of the most recent attempts to classify types of cooperation agreements by frequency. It shows that in 1976 coproduction based on the principle of specialization accounted for more than 38 percent of East-West agreements. This kind of transaction involves the transfer of an entire production activity to a new location, usually in Eastern Europe. After coproduction, the next most common agreements were turnkey plant sales and the sale of licenses.

Coproduction.—Under this kind of agreement, each partner specializes either in the production of certain parts of a finished product, which is then assembled by one or both partners; or in the manufacture of a limited number of articles in the production range, which are exchanged so that each partner can offer a full range of products. The technology is usually provided by one of the partners, but in some cases may be the culmination of joint R&D effort. Generally, coproduction and specialization agreements also include cooperative marketing arrangements. Usually the product bears the trademark of both partners, each of which has exclusivity for the market in its own area but shares the market in other countries. In cooperative agreements with the Soviet Union, the Western partner usually has priority for selling in the industrialized West, and the Soviet Union confines its sales to Warsaw Pact nations and possibly certain developing countries.

The attraction of such agreements for both the Western and Eastern partners is obvious. The Western firm may acquire raw materials and/or labor in the East. The Eastern country expands its repertoire of manufacture, its markets, and often its potential for earning hard currency.

Turnkey Plants.—Of all cooperation agreements, turnkey transactions are perhaps the most effective means of technology transfer. Although technology may in many cases be purchased or leased through straightforward transactions in the marketplace, turnkey projects afford the possibility of acquiring whole production systems—from feasibility studies, construction, and training through technical assistance during the initial run-in period. Further, most trans-

Table 16.—Classification of East-West Industrial Cooperation Agreements by Percent

	Total	Supply of license[a]	Delivery of plant	Specialization coproduction	Subcontracting	Joint venturing and other
Survey of June 1, 1976						
Bulgaria	100.0	17.1	25.7	31.4	11.4	14.4
Czechoslovakia	100.0	27.3	—	22.7	9.1	40.9
East Germany	100.0	—	23.5	14.2	7.1	33.8
Hungary	100.0	29.5	16.3	32.6	9.6	12.0
Poland	100.0	21.7	24.2	32.3	6.4	15.4
Romania.	100.0	19.4	25.5	14.2	7.1	33.8
U.S.S.R..	100.0	3.2	20.4	61.5	4.7	10.2
Total CMEA countries						
1972.	100.0	28.2	11.9	37.1	7.9	14.9
1975.	100.0	26.1	21.7	33.3	6.8	12.1
June 1, 1976	100.0	17.1	20.5	38.3	7.4	16.7

CMEA = Council for Mutual Economic Assistance or Comecom.
[a]Supply of license in exchange (in part at least) for products or components.
SOURCE: Economic Commission for Europe, United Nations.

actions guarantee an ongoing relationship with the supplier, opening the possibility of access to developing technology. The continuity of these relationships is universally regarded as the most important single element affecting the success of a technology transfer.

Turnkey projects in their pure form, involving purchase of an entire installation from one firm or one country, are relatively rare—at least in the case of the Soviet Union. Most often, a Communist nation contracts with many Western firms for particular components of a complex, including marketing and subsidiary services. The Soviet Kama River truck plant is a good example. Here, the U.S.S.R. dealt with Western firms in several countries, assembling its own sophisticated mixture of goods and services to fit its own specifications.[3]

Licenses and Patents.—The acquisition of technology through licenses accelerates indigenous technological progress and enhances potential export capabilities in the East. According to one estimate, the purchase of a license may cause technological progress in the affected field to leap by 7 to 8

years, compared to only 3 to 5 years with the purchase of know-how and 1 to 2 years for coproduction.[4] Often the acquisition of a license creates requirements for other improvements, more imports, further licenses, and the promotion of exports. Licenses may be paid for in either currency or in products through countertrade arrangements. In Eastern Europe, the latter predominate.[5]

Licensing arrangements are varied, ranging from a straightforward authorization to exploit an individual patent to complex agreements on industrial cooperation. These may provide for the grant of licenses for using patents linked with the importation of certain capital goods; of licenses to use know-how and technical assistance in building turnkey plants or other industrial installations; and of licenses to use trademarks.

It is apparent that the diversity of modes through which technology is transferred and the complex interdependence of activities, which are directly or indirectly involved in the process, make it extremely difficult to accurately measure the value of technology that flows to the East in commercial transac-

---

[3]See Harlan S. Finer, Howard Gobstein, and George D. Holliday, "KamAZ: U.S. Technology Transfer to the Soviet Union," in Henry R. Nau, ed., *Technology Transfer and U.S. Foreign Policy* (New York: Praeger Publishers, 1976).

[4]See Jozef Wilczynski, "License in the West-East-West Transfer of Technology," *Journal of World Trade Law*, March-April 1977.
[5]*The U.S. Perspective on East-West Industrial Cooperation*, International Development Centre of Indiana University (Bloomington, Ind., 1975).

tions. No extensive statistical analysis of the transfer function in this respect has been made, and available data can support only crude analyses of overall volumes and trends. Any comprehensive assessment of the economic importance of these transactions would require data of a sophistication presently unavailable.

NONCOMMERCIAL TECHNOLOGY TRANSFER

Open and regular contacts between the scientific and engineering communities of the United States and the Soviet Union have received official encouragement through a number of bilateral agreements. In July 1959, a formal agreement was concluded between the U.S. National Academy of Sciences (NAS) and the Academy of Sciences in the U.S.S.R.; in the same year the International Research and Exchanges Board (IREX) began a program that sent American graduate students and young instructors to the U.S.S.R. In 1972, the U.S./U.S.S.R. Agreement on Cooperation in the Fields of Science and Technology (S&T) was completed, instituting bilateral cooperative programs in a number of scientific fields. The S&T agreement is predicated on the idea of building and maintaining a world scientific community through open channels of communication. More recently, exchanges with the People's Republic of China (PRC) have begun.

The role that such contacts have in transferring American technology with potential commercial value is the subject of considerable disagreement.

Two recent studies of the S&T agreements and the exchanges program by NAS have attempted to assess the value to both sides of the information exchanged in these programs.[6] Both concluded that exchanges with

the Soviet Union were worthwhile, although their value to U.S. participants may be limited by American scientists' lack of familiarity with the Soviet Union's unique style of science and engineering and by the lack of Soviet candor regarding weaknesses in many areas of its research. Both programs were plagued by the rigidity of the Soviet bureaucracy (although problems with the U.S. bureaucracy seemed to rank a close second) and by erratic attendance on the Soviet side. In 1978, for example, NAS extended invitations to 44 Soviet scientists; only 4 participated.

A review of the two studies indicates that while the initial contacts provided some useful information about Soviet research (especially in the fields of medicine, weather forecasting, accelerated drug testing, nuclear fusion, magnetohydrodynamics, superconducting magnets, and earthquake prediction), the primary value of the U.S./U.S.S.R. exchanges to America has been one of educating the scientific and engineering community about the nature of the Soviet scientific system:

> Not only do U.S. scientists and engineers have the opportunity of acquiring at first hand new ideas and new perspectives from their Soviet colleagues, they also become more familiar with the relevant Soviet scientific literature and are alerted to particular Soviet scientists and engineers whose future publications likely merit special attention . . . . [The Soviets] have probably received more technical value in computer topics, in econometrics, and in management science than has the U.S., largely because the U.S. is more advanced in these areas. But the most significant value to the U.S. . . . lies in better U.S. understanding of the Soviet planning and management process, and of Soviet status and approaches in economics, management science and computer science.[7] It is nevertheless true that the United States has, on the whole, taught the Soviets more than it has learned from them. The NAS expects the future balance to shift toward greater equality.[8]

---

[6]National Academy of Sciences, *Review of the U.S./U.S.S.R. Agreement on Cooperation in the Fields of Science and Technology,* National Research Council, May 1977, and *Review of U.S./U.S.S.R. Interacademy Exchanges and Relations,* National Research Council, September 1977.

---

[7]Ibid., *Agreement on Cooperation,* pp. 7, 43.

[8]Ibid., *Interacademy Exchanges and Relations,* p. 3.

According to NAS, the risk of inadvertently communicating important technology through scientific exchange is minimal. The Commerce Department's Office of Export Administration regularly briefs U.S. scientists on topics they should not discuss in the exchange programs, and "except in certain narrow and well-delineated fields, problems of technology do not loom large . . . The Soviets have not managed to translate into practice the wealth of American technical data already available to them through the open literature [and as a result] their technology is unlikely to benefit greatly from any further technical data we might disclose except certain specific data which are proprietary or classified."[9]

A different cost/benefit balance may exist in the student exchanges between the United States and the U.S.S.R. These can result in the transfer of technology that is difficult to quantify or even identify. Since about 1972, Soviet "students," who are usually experienced engineers, scientists, and managers of R&D establishments, have concentrated on study programs in the United States in semiconductor technology,

[9]Ibid., *Interacademy Exchanges*, p. 4; *Agreement on Cooperation*, p. 43.

American magnetohydrodynamic (MHD) technology arrives in the Soviet Union as part of the U.S./U.S.S.R. Cooperation Program

computers, and other fields related to problems of applied research. Large numbers of Chinese "scholars" are similarly beginning to appear in the West. Data reflecting the number of such students and the institutions they attend tell little of the nature and amount of the technology they carry back with them. It has been alleged that this information carries potential military significance. As far as can be determined, however, no systematic attempt has ever been made to quantify its value in either military or commercial terms. Any complete assessment of such exchanges must weigh both strategic and potential commercial losses against their political and cultural value.

CHAPTER VII

The East-West Trade Policy of the United States

CONTENTS

The East-West Trade Policy
of the United States

The present state of U.S. trade policy toward Communist nations reflects the ambivalence and dissension which, for the past three decades, have characterized U.S. posture toward the Eastern bloc. Much of the basic structure of present programs was designed at the height of the cold war, when suspicions about the capabilities and intentions of the Communist world ran very deep. The early legislation was intended to impose a virtual trade embargo on these countries. Since that time, however, three major changes have occurred: 1) the United States has lost much of its leverage with its Western trading partners and is now unable to impose a unified trading posture within the Western bloc; 2) it is no longer possible to treat Communist nations as a monolithic bloc; and 3) there has been an overall improvement of relations with the Eastern world. Together, these developments have led to a series of alterations in the basic policy of the United States, beginning with the Export Administration Act of 1969. This policy, however, is implemented through an administrative structure which was fashioned some 30 years ago. The product of years of incremental modification, this system embraces a cumbersome and sometimes confusing set of procedures that reflect diverse and frequently conflicting interests.

U.S. policy that has a direct impact on trade with the East can be divided into three categories:

1. export-licensing controls that govern the export of products or technologies appearing on a list of controlled commodities, and nuclear equipment and nuclear fuels;
2. controls over export and import facilities that regulate the use of credits, loan guarantees, or other incentives for trade with certain categories of nations; and
3. control over tariffs that allows the United States to levy higher rates of duties on imports from countries to whom "most-favored-nation" (MFN) status has not been granted.

Apart from these measures, there is a range of possible legislation which if adopted could facilitate trade with the Communist world. This includes tax, patent, and antitrust law. There appears to have been little interest in Congress in reformulating U.S. tax, patent, and antitrust policy in ways that might expedite East-West transactions. These issues are touched on elsewhere in this report (see chapters II and III).

EXPORT-LICENSING CONTROLS

LAWS AND AMENDMENTS

The Export Control Act of 1949

The decision to exert strict peacetime controls over U.S. exports to certain countries in the name of national security marked the advent of the modern era of U.S. foreign and national security policy. Export controls as such were no innovation. The Trading With the Enemy Act of 1917, for instance, had granted the President power to impose such controls in time of war or, with the consent of Congress, national emergency. But until the end of World War II and the beginning of the cold war, the idea of continuing national security controls over trade on a regular peacetime basis was unprecedented. Such a notion was premised on the thesis that trade and other economic transactions constituted a "weapon" of considerable potential impact; and that technology was an increasingly important determinant of national power. The development of nuclear weapons provided a potent example of the military significance of technology.

The Export Control Act emphasized the danger to U.S. national security of the unrestricted export of materials without regard to their potential military significance and declared it to be the policy of the United States to "exercise the necessary vigilance" over exports to ensure this security. Its purpose was to deny militarily useful exports to the Soviet Union and its allies. The effect of the Export Control Act was to make exporting a privilege and not a right, and it signaled a policy in which national security considerations took precedence over the economic advantages of foreign trade.

The Act was broadly worded. It empowered the President to prohibit or curtail the export of "any articles, materials, or supplies, including technical data, except under such rules and regulations as he shall prescribe," (sec. 3(a)). Any materials or technology could come under the purview of the Act, so long as the President determined

that their export would contribute to the military potential of any country threatening the national security of the United States. Thus, the language of the Act was clearly consistent with the control of items with only indirect military utility. The President was authorized to delegate the power to determine which articles to control to the appropriate executive departments and Federal agencies concerned with the domestic and foreign policy aspects of trade.

Theoretically, the Export Control Act extended equally to all countries. As it was administered, however, licenses were usually easily obtainable for exports to Western nations and, from the start, Communist destinations were singled out for the proscription of exports. The early list of controlled commodities was long and comprehensive and the Act was rigidly enough interpreted that items of economic, as well as military significance, came under its purview.[1] Responsibility for the administration of export controls was lodged in the Office of Export Control of the Department of Commerce, which had already operated similar controls to ensure the availability of supplies during World War II. These controls came to extend to three categories of items: exports of commodities and technical data; reexports of U.S.-originated commodities and technical data from one foreign country to another; and U.S.-originated parts and components used in a foreign country to manufacture a foreign end product for export. Despite the basic changes in policy over the next 30 years (see below), the apparatus that grew to administer these controls has survived nearly intact to the present.

---

[1]U.S. House of Representatives, *Investigation and Study of the Administration, Operation and Enforcement of the Export Control Act of 1949, and Related Acts,* 87th Cong., 1st sess., October and December 1961. See also R. J. Carrick, *East-West Technology Transfer in Perspective,* Policy Papers in International Affairs (Berkeley, Calif.: University of California, 1978), p. 25.

The Mutual Defense Assistance Control Act of 1951

The Export Control Act represented the unilateral response of the United States to the Communist threat that confronted the free world in the aftermath of World War II. It was clear, however, that if a policy of virtual trade embargo was to succeed for long the cooperation of America's allies was vital. The need for the United States to solicit this cooperation had become apparent at least as early as 1947 when Congress discussed the use of U.S. foreign aid as a lever to ensure allied accord in limiting exports to the Soviet bloc. In both the Economic Cooperation Act and Foreign Assistance Act of 1948, aid was tied to trade. In the former, the prospect of U.S. aid was used to encourage cooperation; the latter Act made the Marshall plan hostage to restraint in exports to the East. These were not popular policies in Japan or in Western Europe, which not only had different perceptions of the nature of the Communist threat, but which also had relied much more extensively than the United States on trade with Eastern Europe.

The United States first attempted to enforce a united Allied approach to trade with the Communist bloc in October 1951, with the passage of the Mutual Defense Assistance Control Act or Battle Act (Public Law 87-195). The Battle Act had a dual thrust. First, it reaffirmed the objectives of the Export Control Act by clearly stating a policy in which trade was to be used as a weapon against the Soviet Union and its satellites. This is apparent in title II, which declared it to be U.S. policy to regulate the export of commodities other than arms, ammunition, implements of war, etc, "to oppose and offset by nonmilitary action acts which threaten the security of the United States and the peace of the World" (sec. 201).

Even more important, however, the Battle Act formally announced the intention of the United States to seek multilateral cooperation in the implementation of this policy. It created sanctions through which such cooperation might be enjoined and provided a leg-islative mandate for active U.S. participation in multilateral organizations designed to realize the embargo. Title I of the Act empowered the President of the United States to terminate all forms of military, economic, and financial assistance to any nation that knowingly permitted the sale of U.S. embargoed goods to a prohibited destination; i.e., to any country "threatening the security of the United States." These embargoed goods included arms, ammunition, implements of war, atomic energy materials, petroleum, transportation materials of strategic value, and items of strategic significance used in the production of arms, ammunition, and implements of war (sec. 101). Moreover, Congress stipulated that the United States negotiate with those countries receiving its aid "to undertake a program controlling exports of items (other than arms, etc.) . . . which should be controlled to any nation or combination of nations threatening the security of the United States, including the Union of Soviet Socialist Republics and all countries under its domination" (sec. 202).

Allied response to the Battle Act was never enthusiastic. As a 1969 House Banking Committee report noted,

> From the outset, few West European or Japanese statesmen or businessmen shared the underlying assumption, or for that matter, the ultimate objective of the embargo. Only under the most intense pressure and coercion did Europe and Japan accede to this restrictive policy . . . . A chain of legislation followed stipulating that nations receiving U.S. aid had to conform to rules laid down by the United States concerning exports to Communist countries. The threats of these laws to cut off American aid became the main bargaining weapon with which Western European governments were brought to cooperate in the embargo policy.

Allied differences with the United States rested both on policy and economic interests. Europeans simply could not accept the view that denying trade would put an end to communism or even curtail the Communist countries' development. In more pragmatic terms, trade with Eastern Europe was a matter of no small consequence to our West European partners . . . .[2]

At this time, however, U.S. economic and military aid far outweighed such trade in economic importance to Western Europe. Thus, although export controls there and in Japan would never be as severe as those in the United States, America's allies did evince some willingness to join in a coordinated trade policy. This can be seen in the founding in 1950 of the Coordinating Committee for Multilateral Export Controls, or CoCom. CoCom, which remains a functioning body today, is an informal multilateral organization made up of the United States and its principal allies. It attempts to coordinate the national export controls of its members into a unified policy that limits strategic trade with the Communist bloc. The history and operations of CoCom are discussed in chapter VIII.

The 1960's – The Beginnings of Moderation

By the early 1960's, pressure from Europe and from some parts of the U.S. business community led to a major reevaluation of U.S. export policy. Discussion in the United States over the shape and future of export controls began to emphasize a search for a proper balance between the economic benefits of expanded trade with the East and the threat to U.S. national security posed by this trade. The history of export controls since the 1960's has been a gradual movement from an exclusive emphasis on the security aspects of trade toward relaxation of controls. There was, however, no major change in policy during the 1960's.

President Kennedy, for example, in his January 30, 1961 State of the Union address, requested greater discretion for using "economic tools . . . to help reestablish historic ties of friendship" between the United States and the Eastern bloc whenever this

was "clearly in the national interest."[3] In order to facilitate any resulting trade, Kennedy established by executive order the Export Control Review Board, a cabinet-level body which considered the merit of applications for exports to the Communist world.

Congress, on the other hand, took no initiative to formulate a less restrictive policy. On the contrary, a 1962 amendment to the Export Control Act explicitly broadened the criteria for adding items to the list of controlled commodities by formally including exports of economic as well as military significance under its aegis. The language of the Act was thus altered to read that "unrestricted exports of materials without regard to their military *and economic* significance may adversely affect the national security of the United States." (Emphasis added.) Licenses for any export making a "significant contribution to the military or economic potential" of nations threatening this national security were to be denied. This amendment may not have substantially affected the number and kinds of commodities under control; criteria for inclusion on the list were already broadly interpreted. But the spirit of the amendment implies the declaration of outright economic warfare on the Communist world.

In an attempt perhaps to ameliorate the effect of this declaration, President Johnson in 1965 created a Special Committee on U.S. Trade Relations with East European countries and the Soviet Union. Its task was to explore "all aspects of expanding trade" in support of the President's policy of "building bridges" between the United States and the countries of Eastern Europe and the U.S.S.R.,[4] a policy which the President reaffirmed in a State of the Union address in which he announced that the Government was "now exploring ways to increase peaceful trade with the countries of Eastern Europe and the Soviet Union."[5] Immediate

---

[2]Hearings on H.R. 4293 to extend and amend the Export Control Act of 1949, Committee on Banking and Currency, 1969, p. 4.

[3]Department of State, *The Battle Act in New Times*, 15th Report to Congress, p. 5.
[4]Department of State, *The Battle Act Report*, 18th Report to Congress, p. 49.
[5]Ibid.

increases in such trade were not forthcoming, however. The members of the Special Committee—labor, business, and financial leaders—were generally "hardliners" on East-West relations and the committee's recommendations did little to encourage expansion. Although the committee recommended that the President be given discretionary authority to grant or withdraw MFN tariff treatment to and from individual Communist countries when he determined this to be in the national interest, it also felt that trade with Communist countries should neither be subsidized nor receive artificial encouragement.[6] No basic alterations were proposed to the system of export controls, but it was suggested that the role of the Department of Defense (DOD) be expanded and that it, rather than the Department of Commerce, become the primary agency responsible for identifying strategic goods.

The Export Administration Act of 1969

By the time of the first Nixon Administration, however, the policy of "economic warfare" had come under increasing attack. The economic leverage on which the Battle Act relied had been greatly diminished by the rapid reconstruction of the Japanese and West European economies and the consequent reduction of their need for U.S. aid. The Battle Act, in fact, had never been invoked to enforce sanctions. This by no means indicated that the export controls of all nations receiving U.S. aid were as stringent as those of the United States, or that America's allies were willing to pursue policies of economic embargo. Often, the executive used its waiver authority under the Battle Act to countenance European exports to the East that otherwise would have violated the law. During the mid-1950's Western European exports to the Warsaw Pact nations increased while the amount of American aid to Western Europe decreased (see chapter VIII). Furthermore, American manufacturers had begun to complain that over-

ly restrictive legislation placed them at a competitive disadvantage with Japan and the countries of Western Europe, whose trade with the Soviet Union, the People's Republic of China (PRC), and Eastern Europe was expanding.

Congress was responsive to these pressures. In the face of the national weariness with the cold war and changing perceptions of superpower relations, a burgeoning balance of payments deficit, and recognition of the growing commercial value of an East-West trade in which the United States was not participating, the initiative to liberalize export controls began to come from Congress. This resulted in the passage of the Export Administration Act of 1969 (Public Law 91-184). This Act symbolized the attempt to achieve a new emphasis for export controls— away from a restrictive and strategic embargo toward a careful expansion of exports.

This is not to say that the Export Administration Act had the wholehearted support of a unanimous Congress. On the contrary, the controversies that surrounded its passage are significant, for the disagreements that surfaced in 1969 over the future of export controls have yet to be resolved. They reflect differing perceptions of the nature of the threat to the United States posed by the Soviet Union and of the ways in which this threat should be faced.

Extension of the Export Control Act in the House came under the jurisdiction of the Committee on Banking and Currency, which originally reported out a bill extending the existing 1949 legislation with only minor changes in administration. This was consistent with the position of the Nixon Administration at the time. But a dissent to the majority report of the committee argued for a fundamental change in the law, asserting that the basic premises that underlay the Export Control Act were no longer valid: the Sino-Soviet bloc was no longer monolithic; goods withheld from the Soviet Union by the United States could be obtained elsewhere; and the attempts of the United States to impose unilateral export controls more severe

---

[6]Ibid., pp. 67-69.

than those of its allies were divisive. According to one Representative, the United States had

> ... moved into a period in which the Congress should maintain a close, in-depth review of our export control laws with a view to reshaping them in light of political, economic, and technological changes taking place in Western Europe, Japan, and the Communist countries of Eastern Europe .... At this stage of development, the United States has at least as much to gain as the Communist countries from mutual trade and the barring of this trade today is hurting us more than them .... Controls on commercial goods continue not only as an irritant to our allies but as a loss in business to U.S. firms .... The Export Control Act should be amended to include a finding that expanded trade in peaceful goods and technology with all countries with which we have diplomatic or trading relations can further the sound growth and stability of the U.S. economy as well as further our foreign policy objectives.[7]

Another proposed amendment would have introduced the criterion of foreign availability in the disposition of export license applications. The President, it was suggested, should "take into consideration the availability of an export from any nation with which we have a defense treaty commitment in determining whether or not an export license shall be denied or granted to one of our own exporters."[8]

These minority views in the House of Representatives were consonant with the prevailing opinion in the Senate. In its report on the Senate version of the export control legislation, the Senate Committee on Banking and Currency asserted that since 1949, "virtually every circumstance which made the Export Control Act both advisable and feasible has changed." It was no longer possible to "impede the development of Russia by refusing to sell goods to it," for the Soviets could obtain what they desired elsewhere.

The competitive disadvantage suffered by American businessmen and the U.S. balance of payments deficit were added reasons for overhauling the existing legislation. In sum,

> The attitude apparent in the language of the Export Control Act is one of open hostility, which is an accurate reflection of the prevailing attitude 20 years ago. The committee believes that it will be helpful in the attempt to reach greater understanding with Russia and nations of Eastern Europe if the legislation which deals with the regulation of exports accurately reflects current attitudes.[9]

The minority view in the Senate, on the other hand, resembled that of the majority in the House. At the heart of the disagreement both within and between the two Houses were the protagonists' assumptions about the nature and future of East-West relations. Some Senators were in favor of maintaining the existing export control legislation with minor administrative changes. They wrote of the proposed Senate revisions:

> The proposal which would replace the present Export Control Act is based on the assertion that factors which brought about the enactment of the Export Control Act no longer exist. We cannot agree with such an assertion. It is suggested that we are now living in an era in which the Soviet Union presents a reduced threat to the security of th: United States. We find no evidence that such a new era has been ushered in. In fact, we consider the Soviet Union as a much greater threat to the security of the United States than it was when the Export Control Act of 1949 was passed.[10]

The views of the majority of the Senate Banking Committee eventually prevailed in Congress and "export control" was replaced by "export administration." The new Act attempted to reconcile an encouragement of trade with the East with the maintenance of national security concerns by declaring it to be the policy of the United States both "to

---

[7]Representative Thomas Ashley, in U.S. Congress, House Committee on Banking and Currency, Export Control Act Extension, report no. 91-524, Sept. 29, 1969, pp. 9-11.

[8]Representative Gary Brown, ibid., pp. 18-19.

[9]U.S. Congress, Senate Committee on Banking and Currency, report no. 91-336, July 24, 1969, pp. 2-3.

[10]Senators Wallace F. Bennett and John G. Tower, ibid., p. 22.

encourage trade with all countries with which we have diplomatic or trading relations," and "to restrict the export of goods and technology which would make a significant contribution to the military potential of any other nation . . . detrimental to the national security of the United States" (sec. 3). The Act specifically noted the negative impact of unwarranted regulation of trade on the U.S. balance of payments and the impediment of this regulation to the efforts of businessmen to expand trade. The employment of America's technological resources abroad was no longer to be regarded merely as an instrument of foreign policy and national security; in the Act, trade also became an instrument to "further the sound growth and stability" of the U.S. economy, and Congress' intention to promote trade in peaceful goods was clearly expressed. All language implying that trade restrictions might be used to pursue policies of economic warfare was deleted. Under the Act, therefore, export controls became exceptions limited to three basic purposes: to protect the national security, to protect goods and commodities in short supply, and to further foreign policy aims as determined by the President. The presumption had now shifted in favor of more normal economic relations with the Communist world.

In order to implement this intention, the Secretary of Commerce was authorized to undertake the organizational and procedural changes necessary to revise control regulations and shorten lists of controlled commodities by removing items of purely economic or marginal military use; only goods and technologies that would make a significant military contribution, were in short supply, or would in the view of the President further a foreign policy aim of the United States were subject to control. Exporters were given the right to obtain information on the criteria for export licenses, to learn the reasons for denials or delays in granting licenses, and to present evidence to support their applications in regulatory proceedings. Finally, the administrative agencies responsible for export control were enjoined to consult among themselves and with affected industries for information and advice on the revision of the controlled commodity lists.

The 1972 Amendments: The Equal Export Opportunity Act

The Export Administration Act expired in 1972, at which time it was amended and extended until 1974 by the Equal Export Opportunity Act (Public Law 92-412). Its provisions reflect two problems with the implementation of the 1969 Act—reviews of the unilateral and multilateral commodity control lists, and the proposed consultations among agencies and with affected industries. Although 2½ years had passed since this legislation was enacted, the House Committee on Banking and Currency found that the required reviews and revisions had not been made and that consultations with industry still left much to be desired. According to the committee, this situation stemmed, first, from a shortage of Federal agency manpower and technical expertise regarding the "state of the art" of many products available both in the United States and in Europe. Second, sufficient procedures for consultation with domestic producers who knew the product, the competition, and the "state of the art" had not been developed.[11]

Although the House report voiced these concerns, the committee recommended extension of the 1969 Act without alteration, rejecting not only amendments addressed to these problems but also one which would have declared it to be "the policy of the United States to use export controls to oppose the denial by any country of the rights of its Jewish and other citizens to free emigration and the free exercise of religion." In this connection, the committee expressed its sympathy in the plight of Soviet Jews, but felt "that the amendment in question might not be the best approach to the resolution of the problem at this time."[12]

[11]U.S. Congress, House Committee on Banking and Currency, *International Economic Policy Act of 1972*, report no. 92-1260, July 27, 1972, p. 4.

[12]U.S. Congress, House Committee on Banking and Currency, *International Economic Policy Act of 1972*, report no. 92-1260, pt. II, Aug. 3, 1972, p. 2.

The amendments which would eventually be embodied in the Equal Export Opportunity Act ultimately evolved from the report of the Senate Committee on Banking, Housing, and Urban Affairs, which shared the House Committee's concerns over the necessity for review of unilateral and multilateral controls, and over consultation with industry. Of particular interest to the Senate was the possible handicap the Act might pose to American businessmen in competing with other CoCom countries for markets in the East:

> At this time the United States controls 495 classifications of goods and technology by multilateral (CoCom) agreement with our Allies. In addition, the United States chooses to retain unilateral controls on 461 classifications of goods and technology. The United States is the only CoCom country which controls the export of a significantly greater number of items than those which the CoCom agree to control multilaterally. . . . Items which are available in comparable quality and quantity from foreign sources shall be removed from unilateral controls unless the Secretary gives adequate evidence that such decontrol would threaten the national security.[13]

The committee also found that the establishment of Technical Advisory Committees (TACs) would "enable the Government to utilize more effectively the technical and commercial expertise which only representatives of industry affected by export controls can provide,"[14] and it recommended the creation of such committees.

These proposals encountered the same opposition as had the Export Administration Act 3 years before. Senators Tower and Bannett particularly felt that the Act as it stood was flexible enough for a policy of expanding trade without jeopardizing national security interests. They objected to the establishment of TACs, for instance, on the grounds that the judgments required in export-licens-

ing decisions were governmental responsibilities that industry experts were ill-equipped to make; that TACs would introduce new administrative burdens to the licensing system; that the informal consultation arrangements already adopted by the Department of Commerce were adequate; and finally, that the requirement of consultation with TACs would inhibit the Commerce Department from placing new items under control.

These objections notwithstanding, Congress passed the Equal Export Opportunity Act and gave legislative mandate to the views of the majority Senate report. First, the new Act emphasized the adverse effect on U.S. balance of payments of excessive export controls, particularly those which are more restrictive than those imposed by America's CoCom allies. It directed the Secretary of Commerce to remove, so far as the national security of the United States permitted, unilateral U.S. controls over commodities available ". . . without restriction from sources outside the United States in significant quantities and comparable in quality to those produced in the United States" (sec. 4(b)(2)(B)). This made foreign availability—the existence of significant quantities of comparable goods outside the United States—a formal reason for granting a U.S. export license.

A second provision ensured the use of private sector expertise by requiring the Government to consult with qualified private industry experts on all licensing decisions. To accomplish this, the Secretary of Commerce was directed "upon written request by representatives of a substantial segment of any industry" that produces commodities subject to export controls to appoint TACs consisting of representatives of U.S. industry and Government. The TACs were to be "consulted with respect to questions involving technical matters, worldwide availability, and actual utilization of production and technology and licensing procedures which may affect the level of [unilateral U.S. and CoCom] export controls" (sec. 5(c)(1), 5(c)(2)). Since 1972, the Secretary of Commerce has established eight TACs in the following

---

[13]U.S. Congress, Senate Committee on Banking, Housing and Urban Affairs, *Equal Export Opportunity Act* and the *International Economic Policy Act of 1972*, report no. 92-890, June 19, 1972, pp. 2-3.

[14]Ibid., p. 4.

fields: semiconductors; semiconductor manufacturing and test equipment; numerically controlled machine tools; telecommunications equipment; computer systems; computer peripherals, components, and related test equipment; and electronic instrumentation.

The Export Administration Act Amendments of 1974[15]

Consideration of the Export Administration Act in 1974 occurred in the aftermath of the OPEC oil embargo, economic recession, and serious domestic shortages in several commodities. Although the discussions in both Houses were understandably dominated by the issue of short supply controls, Congress also passed amendments that had an impact on the transfer of technology through national security and foreign policy controls.

In the House Banking and Currency Committee, the central concern arose from the bilateral exchange agreements in science and technology that had been signed after the 1972 U.S.-U.S.S.R. summit. As a result of these agreements and the Joint Commission established under them, many U.S. companies entered into technical cooperation agreements with the Soviet Union, some of which called for the exchange of pure, unembodied technology. Under the existing legislation the Department of Commerce and the other agencies of Government concerned with export control were not informed of the details of these technical cooperation agreements until they led to application for export licenses. This made it difficult for the Government to effectively discharge its export control responsibility. The Subcommittee on International Trade of the House Committee on Banking, therefore, considered an early notification system for technical cooperation agreements. It also investigated charges that technical secrets that would endanger

national security were being exported to the U.S.S.R. through the agreements. Testimony from expert public witnesses, as well as from representatives of the Departments of Defense, State, and Commerce, discounted these charges. And although Government officials testified that new reporting requirements would give them better control of exports to protect national security, no provision on prior notification of technical cooperation agreements appeared in the Senate version of the bill or in the final legislation.

As before, the amendments to the Export Administration Act adopted by the full committee originated in the report of the Senate Committee on Banking, Housing, and Urban Affairs. The Senate discussions of national security controls resurrected the twin concerns of prior Congresses that U.S. businesses not be unduly penalized in the administration of export controls and that the national security not be jeopardized by the transfer of sensitive technologies. On the subject of the administration of licensing provisions, the Senate proposed a new section to the Export Administration Act requiring that applications for export licenses be acted on within 90 days of their submission. If the deadline could not be met, the applicant was to be informed of the reasons for the delay and an estimate given of the time needed for decision. Other amendments required the Secretary of Commerce to report to Congress on the steps taken to expedite the licensing process; the Departments of Commerce, Defense, and State, as well as other appropriate agencies, to be represented on TACs; and disclosure to the House and Senate of information on the reason for export controls already in effect or contemplated.

A final amendment called for review by the Secretary of Defense of all exports to "controlled" countries (i.e., Communist countries). The Military Procurement Authorization Act of 1974 (Public Law 93-365) had already mandated such a review for technologies developed directly or indirectly as a result of R&D funded by DOD. This

---

[15]The material in this and the following section is drawn from Patricia Wertman, "A Brief Overview of the Amending of the Export Administration Act of 1969, With Special Emphasis on National Security and Foreign Policy Controls," Congressional Research Service, Feb. 8, 1979.

oversight was now extended to all license applications. The Secretary was empowered to recommend to the President disapproval of any export if it would significantly increase the military capability of a controlled country. A Presidential decision to override the Secretary of Defense was to be submitted to Congress, which had 30 days in which to overrule the President's decision by majority vote. These provisions were designed "to insure that DOD has an adequate opportunity to consider the military and national security implications of exports to Communist countries and that the Congress has a voice in the decision in the event of White House and DOD disagreement."[16]

The Export Administration Act Amendments of 1977

In 1977 there was no serious disagreement between the two Houses on the substance of the proposed amendments to the Export Administration Act. In both the mood was clearly in favor of facilitating the expansion of East-West trade so far as this was consistent with national security. One important area of discussion was the issue of foreign availability. Section 4(b)(2) of the existing legislation stipulated that "goods freely available elsewhere shall not be controlled for export from the United States unless it is demonstrated that the absence of controls would damage the national security." The existing legislation allowed Presidential discretion in imposing national security controls "regardless of the availability" (sec. 4(b)(1)). The House wanted to ensure that the necessity for control was justified, i.e., it wanted to make exemptions from control on the grounds of foreign availability the basic policy of the Act. The Senate agreed, providing in its bill that in the cases where "adequate evidence has been presented to the President demonstrating that the absence of such controls would prove detrimental to the national security of the United States . . . such evidence is to be included in the annual

report required by the act."[17] Moreover, the Senate version required the President to initiate negotiations with other countries to eliminate foreign availability in such instances. Both of the latter requirements were included in the enacted legislation.

In addition, the bills voted out of committee in both Houses sought to alter the ideological classification of countries to which exports should be controlled. Under the existing legislation the Secretary of Defense was directed to review applications for exports to "controlled" countries, i.e., Communist countries as defined in section 620(f) of the Foreign Assistance Act of 1961.[18] Both Houses substituted for "controlled country" the phrase "country to which exports are restricted for national security purposes." According to the Senate, the previous approach was both "straitjacketed" and "inconsistent," little serving the Nation's interest in maintaining flexibility in the scope and application of export control. It was crucial that export control policy reflect the changing complexion of international relations, yet existing legislation foreclosed or diminished new market opportunities in Eastern Europe. At the same time, it ignored the possibility, however remote, of potential threats to the Nation's security from entirely different parts of the world. As the Senate observed, one of the major purposes of the amended legislation was to "promote and encourage a continuing reexamination of export control policies and practices to insure that they reflect changing world conditions and the changing dimensions of national security . . . The bill is intended to diminish the tendency of rigid cold war perceptions of national security to dominate the export control process."[19]

---

[16]U.S. Congress, Senate Committee on Banking, Housing, and Urban Affairs, *Export Administration Act Amendments of 1974*, report no. 93-1024, July 22, 1974, p. 9.

[17]U.S. Congress, Senate Committee on Banking, Housing, and Urban Affairs, *Export Administration Amendments of 1977*, report no. 95-104, Apr. 26, 1977, p. 29.

[18]These countries include the Soviet Union, Albania, Bulgaria, Czechoslovakia, Estonia, East Germany, Hungary, Latvia, Lithuania, Poland, Romania, People's Republic of China, Yugoslavia, Tibet, Outer Mongolia, North Korea, North Vietnam, South Vietnam, and Cambodia.

[19]Senate report no. 95-104, op. cit., p. 9.

The House similarly desired to reduce emphasis on Communist countries as the focus of export controls, recognizing that Communist and non-Communist countries alike might vary in the extent to which they constituted a threat to the national security of the United States. But implicit in this reduction of emphasis on specific countries as the basis for export controls was the need to put greater emphasis on the nature of commodities to be exported.[20] What both Houses sought, therefore, was a more flexible approach to export controls that would shift emphasis from the country of destination to the exported commodity. Both bills expressed this in identical language: "In administering export controls for national security purposes, United States policy toward individual countries shall not be determined exclusively on the basis of a country's Communist or non-Communist status but shall take into account such factors as the country's present and potential relationship to the United States, its ability and willingness to control retransfers of United States exports in accordance with United States policy, and such other factors as the President may deem appropriate."[21] The President was to periodically review policy toward individual countries.

The amended Act also limited the grounds on which the Secretary of Defense could recommend against export for national security reasons. Instead of restricting exports of products that "significantly increase the military capability" of a country, it became necessary to show that the exports would "make a significant contribution to the military potential of such country." In another section the original language, "significantly increase the military capability of such country" became "make a significant contribution, which would prove detrimental to the national security of the United States, to the military potential of such country" (sec. 4(h)(1)). The import of these changes was that it was no longer sufficient simply to

[20]House report no. 95-190, op. cit., pp. 3-4.
[21]Senate report no. 95-104, op. cit., p. 57; House report no. 95-190, op. cit., p. 32.

show that an export in some way contributed to foreign military capabilities (presumably a very wide range of products and technologies contribute in some way to military uses). The Secretary must now stipulate that the military impact is detrimental to the security of the United States.

The 1977 amendments also embodied several procedural changes in the administration of export controls. Both the House and the Senate were displeased with the persistently slow processing of licenses for export, and both proposed bills reiterated the provision that all export licenses be approved within 90 days unless the applicant was notified in writing that additional time was needed. In addition, the applicant was enabled to respond fully in writing to the questions and considerations raised by the application. In the event of interagency review of a proposed export, the applicant could review the documentation to determine that it accurately described the proposed export. If the export license was denied, the applicant was to be informed of the specific statutory authority for the denial, i.e., national security, foreign policy, or short supply. (The Commerce Department had been denying export licenses on the nonstatutory grounds of "national interest.") In addition, the Secretary of Commerce was required to review the export regulations and lists in order to simplify and clarify them. Within 1 year the Secretary of Commerce was to report to Congress any actions to simplify the export rules and regulations.

The 1977 amendments also extended terms of service on TACs from 2 to 4 years, instructed TACs to review multilateral as well as unilateral controls, and the Secretary of Commerce to report to Congress semiannually on consultations with TACs. But while the Senate bill provided that TACs be informed of the reasons for the failure of the Government to accept their advice, the House bill stated that:

The committee notes that it considered and rejected recommendations by industry that the Government be required to justify directly to the TACs any refusal to accept their advice. The committee views such a requirement as an unwarranted intrusion of the private sector into governmental decisionmaking. The committee bill preserves the requirements that the Government be accountable for its actions, without creating a presumption that the Government is constrained to accept the advice of any single interest group.[22]

The final legislation incorporated the House version.

The Export Administration Act of 1979

The 1977 amendments extended the Export Administration Act until September 30, 1979. But by September 1978, attempts were already underway in the House of Representatives to produce legislation that would impose conditions restrictive to the growth of East-West trade. The Technology Transfer Ban Act (H.R. 14085), introduced by Representative Dornan on September 14, 1978, asserted that no coherent national policy controlled the transfer of technology to the Communist world; that actions taken by the Soviet Union, including human rights violations and enterprises in Africa, demonstrated that Soviet and American views of detente basically differed and belied the expectation that increased economic interdependence with the West would moderate Soviet military and political objectives; that trade with the West was being utilized by the Communist world to acquire strategic technology; and that current U.S. procedures did not adequately prevent the transfer of critical technology to the Soviet Union, Eastern Europe, and China nor adequately encourage America's allies to do the same. The bill therefore proposed to restrict the export of goods and technology that "could make any contribution to the military or economic potential" of any nation, which would prove detrimental to the national security of the United States.

This language could obviously be interpreted broadly enough to virtually embargo all trade with the Communist world. After the bill died in Committee, its supporters modified it, and in March 1979, the "Export Administration Reform Act of 1979" (H.R. 3216), introduced by Representative Lester Wolff and cosponsored by Representatives Miller, Ichord, Dornan, and 21 others was referred jointly to the House Foreign Affairs and Armed Services Committees. This bill sought to assign a larger role to DOD in export control proceedings by giving the Secretary of Defense the primary responsibility for identifying the types of technologies and goods to be controlled for national security purposes. It furthermore mandated the formulation of a list of critical technologies (see chapter V) and prohibited the export of "any critical technology or critical good to any controlled nation" as well as such exports to any other nation, except under validated license. "National security impact statements" were to be required by Congress in all cases of the President's deciding to overrule or modify classifications of technologies by the Secretary of Defense, or in any cases of licensing decision made on grounds of foreign availability. Congress was given the power to overrule such Presidential determinations by a resolution in either House.

The major alternative to this bill in the House was H.R. 2539, introduced by Representative Jonathan Bingham in March 1979. This bill emphasized the importance of exports to the U.S. national interest, and noted the detrimental effects of the present "uncertain" administration of export controls to the U.S. economy. It declared it to be the policy of the United States to use export controls to further the national security of the United States, but also to encourage trade with all nations with which the United States had diplomatic or trading relations, and to restrict exports only in exceptional circumstances after full consideration of the economic impact of such restrictions. The legislation sought to make the process of export licensing more accountable to the public

---

[22]House report no. 95-190, op. cit., p. 16.

and to Congress, and to encourage multilateral cooperation in the use of export controls.

The legislation that was ultimately reported out of Committee, H.R. 4034, most closely resembled the Bingham bill. Under H.R. 4034, the Department of Commerce retained the lead role in the administration of the export control system with the Departments of State and Defense providing principal supportive roles, but the working relationship between the three was formally defined. DOD, whose responsibility lay in the national security aspects of the export-licensing system, would continue to conduct technical evaluations of the military implications and potential military diversion of proposed exports. Its concurrence was formally required in any changes in the commodity control list. The Secretary of Defense was also given authority to appeal directly to the President on any licensing decision inconsistent with national security. The critical technologies approach was encouraged and the Secretary required to report annually on its progress.

The State Department continued to have responsibility for recommending the use of export controls for foreign policy purposes and for representing the United States in CoCom, although authorization for participation in CoCom was transferred from the Battle Act to the Export Administration Act. In addition, the Secretary of State was given the authority to appeal directly to the President on any licensing decision inconsistent with U.S. foreign policy interests.

In the hearings on H.R. 4034, a considerable amount of testimony focused on the inefficiency of the present system in processing export license applications. The bill had provided for a series of "suspense points" that would automatically elevate undecided cases to higher policy levels, although the President was given authority to waive these time limits in important cases for purposes of renegotiating with the seller or foreign customer. The bill also provided for congressional veto over the intended (or expanded) use of export controls for foreign

policy purposes. In addition, a new form of export license, a "qualified general license" (see chapter II) was instituted to expedite the licensing process by allowing exports of certain categories of previously licensed exports, obviating repeated validated license application.

H.R. 4034 explicitly distinguished the criteria and procedures in the use of national security and foreign policy export controls. National security controls were designed to

> . . . prevent the acquisition or delay [the acquisition] by hostile or potentially hostile countries of goods and technology which would significantly enhance their military capabilities to the detriment of U.S. national security.[23]

This statement had clear implications for the issues of both foreign availability and of the role of Congress within the licensing process. By their very nature, licensing decisions in the national security arena are highly technical; this tends to preclude major congressional involvement; foreign availability makes U.S. controls ineffective in any case.

The bill was less precise on the use of controls for foreign policy reasons, the purposes of which can range from the human rights policy of another country; to inhibiting another country's capacity to threaten countries friendly to the United States; to associating the U.S. diplomatically with a particular group of countries; to disassociating the United States from the policies of a repressive regime. Because foreign policy controls involve political—as opposed to technical—decisions, congressional involvement was deemed more appropriate. Thus, H.R. 4034 contained a provision for congressional veto on the Presidential use of foreign policy controls.

On the issue of foreign availability, the House report noted that U.S. ability to unilaterally deny goods and technology to the

---

[23]House Committee on Foreign Affairs, *Report on the Export Administration Act Amendments of 1979,* report no. 96-200, May 15, 1979, p. 7.

Eastern bloc has been eroded by an array of factors, including the increased competition in the high-technology marketplace. In acknowledgment of this fact, the bill required the Secretary of Commerce to establish a capability within the Office of Export Administration (OEA) to continuously monitor the issue of foreign availability, utilizing in part the Government-industry TACs and seeking a more unified CoCom response to the foreign availability problem.

The primary legislation considered in the Senate was S. 737, introduced on March 22, 1979, by Senator Adlai Stevenson and referred to the Committee on Banking, Housing, and Urban Affairs.[24] The bill was reported out on May 7, 1979, with committee amendments. Under S. 737, a new export control statute—the Export Administration Act of 1979—was to be established, superseding the Export Administration Act of 1969, as amended.

As with the House bill, findings and policy declarations stressed the importance of exports to the U.S. economy. Particular attention was given to the minimization of "uncertainties in export control policy" as a means of encouraging trade with all countries with which the U.S. has diplomatic or trading relations, except in cases where such trade would be against the national interest.

As introduced, S. 737 had referred to the "right to export." The bill was amended to substitute the word "ability," the intention being not to denote a constitutional or otherwise legally enforceable right to export free from Government restriction, but rather to reinforce the strong presumption that citizens should be free to engage in international commerce except in instances where regulation is clearly needed to advance important public interests. Thus, the control of exports should be the exception and not the rule.

The bill reaffirmed the notion that export controls administered for national security purposes should give special emphasis to controlling exports of technology (and goods that contribute significantly to transfer of such technology) that could make a significant contribution to the military potential of any country, detrimental to U.S. national security, and also declared that the United States should cooperate with its allies for this purpose. The President was required to annually review unilateral national security controls and to review multilateral export controls maintained for national security purposes every 3 years. High administrative priority was given to the prevention of exports of critical goods and technology, and the Secretaries of Commerce and Defense were enjoined to revise controls to ensure they are focused on and limited to militarily critical goods and technology.

The criteria for using export controls for foreign policy purposes as set forth in S. 737 included the following:

> (1) alternative means to further the foreign policy purposes in question; (2) the likelihood that foreign competitors will join the United States in effectively controlling such exports; (3) the probability that such controls will achieve the intended foreign policy purpose; (4) the effect of such controls on United States exports, employment, and production, and on the international reputation of the United States as a supplier of goods and technology; (5) the reaction of other countries to the imposition or enlargement of such export controls by the United States; and (6) the foreign policy consequences of not imposing controls.[25]

The President was required to reconsider annually export controls maintained for foreign policy reasons and to report the results to Congress. Thus, foreign availability was to be assessed in both the foreign policy and national security cases. The Department of Commerce's OEA would be responsible for assessing foreign availability. Review and revision of export control lists were also required.

---

[24]On Apr. 23, 1970, S. 977 was introduced by Senator Proxmire at the request of the Administration. Its provisions offered no major reforms in the administration of export controls.

[25]See Senate Committee on Banking, Housing, and Urban Affairs, *Report on the Export Administration Act of 1979*, report no. 96-169, May 15, 1979, pp. 5-6.

Like its counterpart in the House, S. 737 instituted a qualified general license,to be used to permit multiple shipments to a particular consignee or for a specified end use. Similarly, it established a timetable for export license review, placing a 90-day limit on review of a license if referral to other agencies or CoCom was not required, and a 180-day limit where referral was necessary.

Finally, S. 737 contained a subsection that superseded the Battle Act and required the President to initiate negotiations with CoCom members for the purpose of reaching agreement on: 1) publishing the CoCom control list, 2) modification of controls to obtain full acceptance and enforcement by CoCom members, and 3) adoption of more effective enforcement procedures.

Debates in both the House and the Senate on their respective bills underlined the two major themes which, since 1969, have surrounded the passage of and amendments to the Export Administration Act: the threat to U.S. national security posed by the sale of dual-use technologies to the Communist world; and the importance to the U.S. national interest of a positive trade balance and therefore of a healthy export sector. The Act that ultimately emerged, the Export Administration Act of 1979 (Public Law 96-72), closely follows both H.R. 4034 and S. 737 and therefore leans to the latter preoccupation. This Act, which expires on September 30, 1983, is reproduced in its entirety in the appendix to this volume. A selection of its major provisions may, however, be summarized as follows:

- The Act finds that the ability of U.S. citizens to engage in international commerce is a fundamental concern; that exports contribute significantly to the national security and well-being of the United States; and that over-restriction or uncertainty in the exercise of export controls can be detrimental to the interests of the United States. On the other hand, export of goods or technology without regard to whether they make a significant contribution to the military potential of recipient countries may adversely affect the national security of the United States.

- The Act declares it to be the policy of the United States to minimize uncertainty in export controls and to encourage trade. Export controls are to be utilized only after full consideration of their economic impacts and only to the extent necessary to protect U.S. national security, to further significant foreign policy goals, or to protect the domestic economy in cases of short supply.

- A qualified general license, as proposed in the Senate bill, is established and a detailed procedure for processing export-licensing applications, including deadlines, provisions for multiagency consultation, and applicant notification and consultation is outlined. Qualified general licenses, in lieu of validated licenses, are to be encouraged to the maximum feasible extent.

- The Battle Act is superseded and authorization provided for U.S. participation in CoCom. The President is enjoined to enter into negotiations with other CoCom governments with a view toward reducing the scope of export controls, publishing the CoCom lists and other pertinent documents, and holding periodic high-level meetings on CoCom policy.

- U.S. firms or enterprises (excepting educational institutions) entering into commercial agreements with controlled countries must now report these agreements to the Secretary of Commerce if they cite an intergovernmental technical cooperation agreement and will result in the export of unpublished technical data.

- In cases where reliable evidence shows diversion of dual-use items to military use, the Secretary of Commerce is authorized to deny all further exports to the end user responsible for the diversion until such time as it ceases.

- Foreign availability shall be continuously reviewed by the Secretary of Commerce in consultation with other agencies and with the TACs, and an office established to gather information and engage in ongoing monitoring activities.

- Validated licenses may not be required in cases where foreign availability has been demonstrated, except in cases where this provision is waived by the President. In these cases, the Secretary of Commerce must publish the details of the basis and estimated economic impact of the decision.
- The Commodity Control List (CCL) may be indexed, i.e., annual increases in performance levels of items subject to controls identified and items automatically deleted on the basis of these stipulations.
- The President is enjoined to consider alternative actions and the following criteria before curtailing exports for foreign policy purposes:
  —the probability that such controls will achieve the intended foreign policy purpose in light of other factors, such as foreign availability;
  —the reaction of other countries;
  —the likely effect of the controls on the U.S. economy; and
  —the ability of the United States to effectively enforce the controls.

Summary

The present Export Administration Act is the embodiment of a policy of encouraging trade with the Communist world in a manner that nevertheless protects U.S. national security and allows the President flexibility in the use of export controls to further foreign policy aims. The fact that these aims may not always be entirely consistent is reflected in the content of the congressional debates that have surrounded the Act since its passage in 1969, and in the nature of the amendments to it. These amendments have sometimes pulled in different directions— some attempting to facilitate the expansion of trade, others expressing concern at the strategic implications of that trade. The general drift of the legislation has, however,

been toward liberalization of export controls of goods and technologies to the East.

As the nature and quantity of the goods and technologies permitted for export to Communist nations have expanded and the climate of detente has encouraged American businessmen to seek new markets in the East, the system for the administration of export controls has had to contend with growing numbers of cases involving increasing technological variety and complexity. In addition, it has had to balance the sometimes conflicting demands of facilitating trade and protecting national security. The following sections describe the operation of this system and discuss the problems that it has encountered.

THE ADMINISTRATION OF U.S. EXPORT CONTROLS: THE LICENSING SYSTEM

In U.S. law, the freedom to export is a privilege and not a right. All U.S. exporters require permission from the Government to ship their goods. In accordance with the requirements of the Export Administration Act, the licensing system through which this permission is granted is administered by the Department of Commerce, which has jurisdiction over most commodities and unclassified technical data. The only exceptions, which fall under the jurisdiction of other Federal agencies, are munitions exports, which are controlled by the Department of State; nuclear materials by the Nuclear Regulatory Commission; and gold and foreign currency by the Treasury Department. Commerce's authority extends to the reexport of commodities and data to third countries; to the utilization of American technical data overseas; to the use of U.S.-origin parts and components in commodities manufactured abroad and destined to a third country; and to exports of commodities and data by any person subject to the jurisdiction of the United States.

As the orientation and objectives of U.S. trade policy with Communist countries have shifted since the end of World War II, so too have variations occurred in trade levels and in the number and kinds of items on the list of controlled commodities. But the procedures and institutions of the export control administration system established over 30 years ago persist. This phenomenon was summed up in a recent Government study which noted:

> In the aftermath of World War II, in response to problems of the Cold War, *security defined in military terms became the overriding purpose abroad—both in concept and in organizational form. Today, the concept has somewhat changed, but the organizational form mostly remains.* (author's emphasis)[26]

The Commodity Control List

There are two types of export license—general and validated. A general license permits the export of certain commodities and technical data without the need to submit a formal application or obtain a license document for each transaction. These apply to most commercial transactions, and approximately 90 to 95 percent of all U.S. exports are shipped under their authority. The remaining 5 to 10 percent, sent under validated license, are subjected to a rigorous application process.

Technologies requiring a validated license are specified by the U.S. Department of Commerce in the CCL (see figure 3). The present CCL is the descendant of the lists of controlled commodities that the Secretary of Commerce, with the advice of the Secretaries of State and Defense, was first enjoined to compile under the Export Control Act. Under the terms of this Act and Department of Commerce regulations, the CCL contains those technologies, products, or commodities that fall into the following general categories:

- products and technical data that the U.S. Government determines capable of contributing significantly to the design, manufacture, and utilization of military hardware, or that fall under the CoCom strategic control system; petroleum and other products or commodities in short supply; and
- some devices related to nuclear weapons and explosive devices; certain nuclear power facilities; and crime control and detection equipment, that is controlled for foreign policy reasons.[27]

Most items on the list are also on the CoCom list of controlled commodities; but at present 38 items are unilaterally controlled by the United States,[28] according to the interpretation of Commerce, Defense, and State officials of the general criteria provided in the law. U.S. industry, anxious to expand exports to Communist-bloc countries, is particularly critical of this fact, and as will become apparent below, much of the criticism of the export-licensing system as a whole is directed at the composition of the list itself.

At the present time, the CCL contains some 200 entries, many of which embody high technology. These are grouped in the following 10 categories:

Group	Types of commodities
0	Metalworking machinery
1	Chemical and petroleum equipment
2	Electrical and power-generating equipment
3	General industrial equipment
4	Transportation equipment
5	Electronics and precision instruments
6	Metals, minerals, and their manufacture
7	Chemicals, metalloids, and petroleum products
8	Rubber and rubber products
9	Miscellaneous

---

[26]Graham T. Allison, "Overview of Findings and Recommendations from Defense and Arms Control Cases," in appendixes: *Commission on the Organization of the Government for the Conduct of Foreign Policy*, June 1975, vol. IV (Washington, D.C.: U.S. Government Printing Office, 1975), p. 21.

[27]See the testimony of Rauer Meyer, Director of the Office of Export Administration, U.S. Department of Commerce, before the Subcommittee on Domestic and International Scientific Planning, Analysis, and Cooperation, Committee on Science and Technology, U.S. House of Representatives, Oct. 4, 1978.

[28]See the testimony of Stanley Marcuss, Senior Deputy Assistant Secretary for Industry and Trade, U.S. Department of Commerce, before the Subcommittee on International Economic Policy and Trade, Committee on Foreign Affairs, U.S. House of Representatives, Mar. 7, 1979.

Figure 3.—Sample Page of the U.S. Commodity Control List

Commodity Control List—399.1   CCL-1

Export Control Commodity Number and Commodity Description	Unit	Process-ing Code	Validated License Required	GLV $ Value Limits		
				T	V	Q

GROUP O—METAL-WORKING MACHINERY [1]

Forming Machines:

1072A   Presses and specialized controls, accessories, ||------²|| MG || QSTVWYZ || 1,000 || 1,000 || 0
and parts therefor, as follows:

(a) Presses (stabilized equipment using rams) for applying high impact energy work forces through use of explosives or compressed gases including air;

(b) Presses specially designed or re-designed for the working or forming of metals, alloys, or other materials with a melting point exceeding 3,452°F (1,900°C);

(c) Hydraulic presses, as follows:
(1) vertical presses having a total rated force of over 10,000 tons; or
(2) horizontal presses having a total rated force of over 5,000 tons;

(d) Isostatic presses, as follows (isostatic presses are those capable of pressurizing a closed cavity through various media (gas, liquid, solid particles, etc.) to create equal force in all directions within the cavity upon a workpiece or material) :
(1) capable of achieving a maximum working pressure of 20,000 of psi (1,406 kg/cm²) or greater and possessing a chamber cavity with an inside diameter in excess of 16 inches (40.6 cm) ; or
(2) capable of achieving a maximum working pressure of 5,000 psi (351 kg/cm²) or greater and having a controlled thermal environment within the closed cavity, *except those possessing a chamber cavity with an inside diameter of less than 5 inches (127 mm) and which are also capable of achieving and maintaining a controlled thermal environment only between +176°F (+80°C) and −31°F (−35°C); and*

(e) Control equipment, accessories, and parts which are specially designed for the above presses.

1075A   Spin-forming and flow-forming machines, ||------²|| MG || QSTVWYZ || 1,000 || 1,000 || 0
double support or three roller types, as follows:

(a) Horizontal spindle type designed to have and having a drive motor of 80 hp (59kW) or more; or

(b) Vertical spindle type designed to have and having a drive motor of 50 hp (37kW) or more; and

(c) Specially designed parts and accessories therefor.

Other Metal-Working Machinery:

1080A   Machines and equipment, including special- ||------²|| MG || QSTVWYZ || 1,000 || 1,000 || 0
ized tooling and fixtures, and specially designed
parts and accessories therefor, specially designed for making or measuring gas turbine blades, *including but not limited to* the following:

(a) Blade belt grinding machines;

(b) Blade edge radiusing machines;

(c) Blade aerofoil milling and/or grinding machines;

(d) Blade blank pre-forming machines;

(e) Blade rolling machines;

(f) Blade aerofoil shaping machines, *except metal removing type;*

(g) Blade root grinding machines;

(h) Blade aerofoil scribing equipment;

(i) Blade aerofoil and/or root automatic measuring equipment;

(j) Precision vacuum investment casting equipment;

(k) Small home drilling equipment for producing holes less than 0.030 inch (0.76mm) in diameter; and

(l) Directional solidification casting equipment.

---

[1] See § 370.10 for commodities which require export authorization from other U.S. Government Departments and Agencies.
² Report machines in "number."

Each entry on the CCL contains a general description of the technical commodity, the countries for which validated licenses are required, and in some cases, value limtations on exports which set restrictions on the number or dollar value of items that may be exported.

For export control purposes, all foreign countries except Canada, which is subject to minimal restrictions, are divided into seven separate country groups, designated by alphabetic symbols (see table 17). Most Communist-bloc countries are included in a single group, but Poland and Romaina both have MFN status and are treated separately. Communist countries to which most trade is embargoed (North Korea, Vietnam, Cambodia, and Cuba) also form a distinct group.

The need for an exporter to apply for a validated license therefore turns both on the commodity or data to be exported and the country of destination. Exporters may have to consult the CCL for guidance on each separate transaction. Validated licenses are required for most high-technology exports to Communist destinations and for strategic materials and products to all destinations except Canada. The Commerce Department issues and updates a series of regulations that lay out these requirements.[29]

---

[29]U.S. Department of Commerce, *Export Administration Regulations,* June 1, 1978. This is a looseleaf publication that includes Export Administration regulations and supplementary Export Administration bulletins.

Table 17.—Export Administration System: Country Grouping

Country Group O

Romania

Country Group S

Southern Rhodesia

Country Group T

North America:
Greenland
Miquelon and St. Pierre Islands

Southern Area:
Mexico (including Cozumel and Revilla Gigedo Islands)

Central America:
Belize
Costa Rica
El Salvador
Guatemala
Honduras (including Bahia and Swan Islands)
Nicaragua
Panama

Bermuda and Caribbean Area:
Bahamas
Barbados
Bermuda
Dominican Republic

French West Indies
Haiti (including Gonave and Tortuga Islands)
Jamaica
Leeward and Windward Islands
Netherlands Antilles
Trinidad and Tobago

South America:
Northern Area:
  Colombia
  French Guiana (including Inini)
  Guyana
  Surinam
  Venezuela

Western Area:
  Bolivia
  Chile
  Ecuador (including the Galapagos Islands
  Peru

Eastern Area:
  Argentina
  Brazil
  Falkland Islands (Islas Malvinas)
  Paraguay
  Uruguay

Country Group V

All countries not included in any other country group (except Canada)

Country Group W

Poland

Country Group Y

Albania
Bulgaria
Czechoslovakia
Estonia
German Democratic Republic (including East Berlin)
Hungary
Laos
Latvia
Lithuania
Outer Mongolia
People's Republic of China (excluding Republic of China)
Union of Soviet Socialist Republics

Country Group Z

North Korea
Vietnam
Cambodia
Cuba

NOTE: Canada is not included in any country group.
SOURCE: U.S. Department of Commerce, *Export Administration Regulations,* supplement no. 1, pt. 370, June 1, 1978.

The Executive Agencies in the Licensing Procedure

The Department of Commerce.—Exporters seeking validated licenses must enter a system primarily administered by the Department of Commerce, which works in cooperation with the Departments of State, Energy, and Defense. Its export control responsibilities give Commerce a somewhat contradictory mandate. On one hand, it is charged with the general promotion and encouragement of U.S. exports; on the other, it is required to administer an elaborate system designed to limit certain of these transactions.

The Department has attempted to reconcile these activities by keeping export restrictions to the minimum necessary to fulfill the objectives of the Export Administration Act, thus causing the least negative impact on U.S. trade. Jurisdictional ambiguities and conflict cannot be entirely avoided. Reorganizations in the past few years have been designed to minimize their effects, and recent discussions within the Administration have once again raised the prospect of further reorganization. Offices responsible for the promotion of East-West trade and those charged with the control of such trade were once housed together in the Department's Bureau of East-West Trade, which was established in 1972 to encourage

and facilitate the trade resulting from the U.S./U.S.S.R. trade agreement. The two functions have since been separated. In 1977, a new Industry and Trade Administration was established. It contains two parallel bureaus with functions relevant to East-West trade and its administration: the Bureau of Trade Regulation, which houses OEA and is responsible for administering export controls; and the Bureau of East-West Trade, which retains trade promotion operations. Figure 4 illustrates this organization.

The Department of State.—The Department of State is primarily concerned with the foreign policy, as opposed to the economic and commercial, implications of export control. Its involvement in export control is threefold. First, State advises the Commerce Department on any foreign policy considerations arising from U.S. export license applications. These foreign policy issues may cover matters as diverse as U.S. national security, virtual embargoes on trade with certain Communist countries, U.N. sanctions on trade with Rhodesia, selected restraints on trade with South Africa, the former embargo on trade with Uganda, and controls for human rights, antiterrorism, and regional stability purposes. State is also involved with nuclear nonproliferation cases and the export of hazardous substances. Sec-

Figure 4.—Industry and Trade Administration, U.S. Department of Commerce

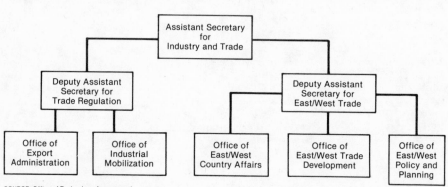

SOURCE: Office of Technology Assessment.

ond, the State Department assumes the leading role in U.S. efforts to implement multilateral export controls. In this connection, it represents the United States in all CoCom sessions, including CoCom list reviews and exception cases (see chapter VIII). Third, under the Battle Act the State Department has primary responsibility for the development of a list of items completely embargoed to the Communist world. This list, the Mutual Defense Assistance Control List, includes arms, ammunition, implements of war, atomic energy materials, and certain dual-use items.

All of these activities are handled in the Office of East-West Trade in the Bureau of Economic and Business Affairs. The Office's seven officers, two of whom reside in Paris where CoCom is headquartered, all have full-time responsibility for export controls.

The Departments of Energy and Defense. —The Departments of Energy and Defense play important roles in the system by providing technical expertise and advice on national security matters. The Department of Energy advises on energy-related exports, such as oil-extractive equipment, and reviews all cases involving nuclear materials.

As would be expected, DOD has been heavily involved with the export control system from the outset. In 1962, the Assistant Secretary of Defense for International Security Affairs was charged with the responsibility for DOD's role in the implementation of trade control policies, and subsequent amendments to the Export Administration Act have further delineated the role of the Secretary of Defense within the control system (see above). DOD's task is to evaluate the military and strategic potential of items under review. This entails a complex consulting system within the Department that may involve the technical and intelligence arms of the military services as well as a number of other offices and agencies.

Until recently, responsibility for export licensing within DOD was diffused. Although the processing of applications came

under the jurisdiction of the International Security Affairs Branch, policy functions, including obtaining technical evaluations, were carried out in the Office of the Deputy Under Secretary of Defense for Research and Advanced Technology. Both functions have now been centralized in the Office of the Deputy Under Secretary for International Programs and Technology. The reorganization is intended to streamline the Department's role in processing export license applications.

Other Agencies.—In addition to the Departments discussed above, any other agency with pertinent expertise may be asked to contribute technical advice on individual applications. Bodies sporadically involved in the licensing process include the Treasury Department, the Central Intelligence Agency (CIA), the National Aeronautics and Space Administration, and the National Bureau of Standards.

The Mechanics of the Validated Licensing Procedures

Entry to the export control licensing system is by way of OEA, which receives all export license applications. Figure 5 delineates its present administrative structure. A license application requires detailed technical information on the product or process to be exported, the quantity to be exported, unit selling price, and total sales receipts. Details concerning the foreign buyer, including intermediate and final consignees, and on the end use of the product must also be provided (see figure 6). In theory, the prospective exporter may obtain an advisory opinion on the likely disposition of the application, and thus avoid costly pre-sale promotion. In some instances, however, this process has been as time-consuming as the formal application itself. Furthermore, the advisory opinion is delivered orally and is not binding.

An application is initially received in the Operations Division where it is logged in, entered into a computerized information system, and briefly reviewed against a list of known or suspected violators of export con-

Figure 5.—Office of Export Administration, U.S. Department of Commerce

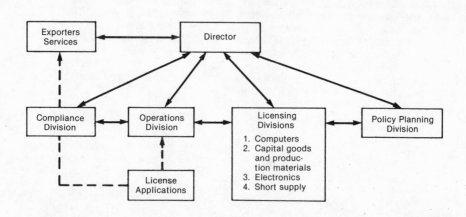

SOURCE: Comptroller General, *Export Controls: Need to Clarify Policy and Simplify Administration,* Report to the Congress, March 1, 1979, p. 34.

trol laws. It then moves to one of the three Licensing Divisions—Computers, Electronics, or Capital Goods and Production Materials. At this point the application is given a careful technical review by OEA staff. This review focuses on the following items: the function and use of the equipment; its level of sophistication; normal military/civilian use in the United States and the country of destination; foreign availability of comparable equipment in terms of both quantity and quality; suitability of the equipment for the proposed end use; the known activities of the end user; likelihood of diversion; and the economic and commercial implications of the proposed export. In investigating these points, the staff may draw on outside resources, including TACs, the Nation's intelligence agencies, and the Export Information Service of the Department of Commerce. Specifically, they attempt to answer the following questions:

• Is the item designed or intended for military purposes? Does it have significant military use?

• If the item has both military and civilian uses, will the transaction involve only the latter?

• Does the item contain advanced or unique technology of significance in terms of the export control program's objectives?

• Is there a shortage of the item in the area of destination that affects military potential?

• For strategically significant nonmilitary items, can non-U.S. sources supply a comparable item or an adequate substitute? What is the normal use in the country of destination?

After this information has been gathered and evaluated, the request moves to the Policy Planning Division where it is reviewed in terms of general OEA policy. At this stage the Division must determine whether it has sufficient data on which to base a decision and sufficient authority to unilaterally make that decision, or whether consultation with other Federal agencies is required. If the latter, it must also determine what kind of consultation is called for.

Figure 6.—Export License Application

FORM DIB-622P (REV. 3-75)
(FORMERLY FC-419)
Form Approved: OMB No. 41-R.0735

CONFIDENTIAL — Information furnished herewith is deemed confidential and will not be published or disclosed except in accordance with provision of Section 7 (c) of the Export Administration Act of 1969, as amended.

U.S. DEPARTMENT OF COMMERCE
DOMESTIC AND INTERNATIONAL
BUSINESS ADMINISTRATION
BUREAU OF EAST-WEST TRADE
OFFICE OF EXPORT ADMINISTRATION
WASHINGTON, D.C. 20230

APPLICATION FOR

EXPORT LICENSE

DATE RECEIVED *(Leave Blank)*

CASE NO. *(Leave Blank)*

DATE OF APPLICATION

APPLICANT'S TELEPHONE NO.

1. APPLICANT'S NAME

STREET ADDRESS

CITY, STATE, ZIP CODE

2. PURCHASER IN FOREIGN COUNTRY
(If same as ultimate consignee, state "SAME AS ITEM 3"; if same as intermediate consignee, state "SAME AS ITEM 4.")

NAME

STREET ADDRESS

CITY AND COUNTRY

3. ULTIMATE CONSIGNEE IN FOREIGN COUNTRY

NAME

STREET ADDRESS

CITY AND COUNTRY

4. INTERMEDIATE CONSIGNEE IN FOREIGN COUNTRY.
(If none, state "NONE", if unknown, state "UNKNOWN.")

NAME

STREET ADDRESS

CITY AND COUNTRY

5. COUNTRY OF ULTIMATE DESTINATION

6. APPLICANT'S REFERENCE NUMBER

7. (a) QUANTITY TO BE SHIPPED	(b) COMMODITY DESCRIPTION AS GIVEN IN COMMODITY CONTROL LIST *(Include characteristics such as basic ingredients, composition, type, size, gauge, grade, horsepower, etc.)*	(c) EXPORT CONTROL COMMODITY NUMBER AND PROCESSING NUMBER	(d) TOTAL SELLING PRICE AND POINT OF DELIVERY *(Indicate F.O.B., F.A.S., C.I.F., etc.)*	
			UNIT PRICE	TOTAL PRICE
			TOTAL	

8. FILL IN IF PERSON OTHER THAN APPLICANT IS AUTHORIZED TO RECEIVE LICENSE

NAME

STREET ADDRESS

CITY, STATE, ZIP CODE

9. IF APPLICANT IS NOT THE PRODUCER OF COMMODITY TO BE EXPORTED, GIVE NAME AND ADDRESS OF SUPPLIER.
(If unknown, state "UNKNOWN.")

10. END USE OF COMMODITIES COVERED BY THIS APPLICATION. DESCRIBE FULLY.

11. IF APPLICANT IS NOT EXPORTING FOR HIS OWN ACCOUNT, GIVE NAME AND ADDRESS OF FOREIGN PRINCIPAL AND EXPLAIN FULLY

12. ADDITIONAL INFORMATION *(Attach separate sheet if more space is needed.)*

13. APPLICANT'S CERTIFICATION. — The undersigned applicant hereby makes application for a license to export and certifies as follows: That all statements herein, and in any documents or attachments submitted in support hereof, are true and correct to the best of his knowledge and belief; and that (a) he has read the instructions on the fifth copy of this application and is familiar with the U.S. Department of Commerce Export Administration Regulations; (b) this application conforms to such instructions and regulations; (c) unless Item 14 is completed, he negotiated with and secured the export order directly from the purchaser or ultimate consignee or through his or their agents abroad; (d) all parties to the export transaction, the exact commodities and quantities, or the exact technical data, and all other terms of the order and other facts of the export transaction are fully and accurately reflected herein; (e) documents and records evidencing the order and other facts of the export transaction to which this application relates will be retained by him for 2 years from whichever is later: the time of (i) the export from the United States, or (ii) any known reexport, transshipment, or diversion, or (iii) any other termination of the transaction, whether formally in writing or by any other means, and made available to the Department of Commerce upon demand; (f) any material or substantive changes in the terms of the order or other facts of the export transaction as reflected in this application or any certification made in connection therewith, whether the application is still under consideration or after a license has been granted, will be reported promptly by him to the Department of Commerce; and (g) if the license is granted, he will be strictly accountable for its use in accordance with the Department of Commerce Export Administration Regulations and all terms and conditions specified on the face of the license.

Type or Print _____ (Applicant *(Same as Item 1)*)

SIGN HERE _____ (Signature of person authorized to execute this application.)

Type or Print _____ (Name and title of person whose signature appears on the line to the left)

14. ORDER PARTY'S CERTIFICATION (See § 372.6 (c) of the *Export Administration Regulations.*) — The undersigned order party certifies to the truth and correctness of Item 13 (d) above, and that he has no information concerning the export transaction that is inconsistent with, or undisclosed by the application and agrees to comply with Items 13 (e) and 13 (f) above.

Type or Print _____ (Order Party)

SIGN IN INK _____ (Signature of person authorized to sign for the Order Party)

Type or Print _____ (Name and title of person whose signature appears on the line to the left)

This license application and any license issued pursuant thereto are expressly subject to all rules and regulations of the Department of Commerce. Making any false statement or concealing any material fact in connection with this application or altering in any way the validated license issued, is punishable by imprisonment or fine, or both, and by denial of export privileges under the Export Administration Act of 1969, as amended, and any other Federal statutes.

FOR OFFICIAL USE ONLY

ACTION TAKEN	VALIDITY PERIOD	AUTHORITY	RATING		DV	TECH. DATA			
☐ APPROVED									
☐ REJECTED	MONTHS		END USE CHECK	RE-EXPORT	SUPPORT DOCUMENT	TYPE OF LICENSE	(Licensing officer)	(No.)	(Date)
DOCUMENTATION							(Review officer)		(Date)

NOTE: Submit the first four copies of this application, Form DIB-622P (with top stub attached), to the Office of Export Administration, Room 1617M, Domestic and International Business Administration, U.S. Department of Commerce, Washington, D.C. 20230, retaining the quintuplicate copy of the form for your files. Remove the long carbon sheet from in front of the quintuplicate copy. Do *not* remove any other carbon sheets. See Special Instructions on back of quintuplicate. Reproduction of this form is permissible, providing that content, format, size, and color of paper and ink are the same.

ORIGINAL
O E A FILE COPY

In theory, all applications for validated licenses should be subjected to a formal interagency review process; in practice, the Commerce Department often takes unilateral licensing decisions, with the consent of the other agencies involved in the system, i.e., the Departments of State, Defense, and Energy. This is a practical necessity: unless the vast majority of cases were decided without prolonged multiagency review, the system would be overwhelmed by the number of applications to be processed. The requirement for multiagency consideration is met, therefore, in any one of three ways, each involving progressively more active and formal interagency consultation. These are first, a unilateral decision by OEA based on prior delegations of authority from the other agencies; second, informal consultation which is usually bilateral; or third, full-scale formal multiagency review.

OEA decides which of these routes to take on the basis of internal guidelines called "policy determinations." A policy determination establishes procedures governing the disposition of particular categories or types of export applications. Policy determinations specify limitations or conditions respecting commodity, destination, and use or end uses that determine how an application decision must be resolved. In instances, therefore, where a new license application falls within the technical specifications, end use, and destination criteria previously established as acceptable for export, OEA may make a unilateral decision without the necessity of actual interagency consultation. The performance characteristics of the product or its destination will therefore indicate when OEA should discuss a case with another agency, even if it itself feels confident that the export can proceed. The majority of cases are decided by OEA alone. These usually involve exports to free world destinations. Most exports to Communist countries require some explicit consultation.

There are several kinds of informal consultation, but no rigid or explicit criteria govern the choice of one over another. In some instances, a single phone call to another agency (or, if the problem is the application itself, to the prospective exporter) may be sufficient. Alternatively, the Policy Planning Division may send a memorandum and supporting documents, which summarize the case on the basis of the technical evaluation conducted in the Licensing Division, to another agency, requesting its opinion or its concurrence in Commerce's recommendation. OEA may also send "waiver memoranda." These outline the case and request the acquiescence of each consulting agency to Commerce's decision on an application. Usually, this concurrence is forthcoming. For example, of the approximately 12,000 memoranda sent through the first 6 months of 1976, DOD recommended denial of only 9 applications that Commerce had favored. Defense concurred in 76 Commerce recommendations for denial and in 6 cases for partial approval.[30]

In some cases, however, the mechanisms of day-to-day contacts among staff-level personnel, bilateral agreements, and waiver memoranda cannot resolve a case, and more extensive interagency consultations are required. The applications subjected to this procedure nearly always involve dual-use items—nominally civilian products or processes that nevertheless have military applications which could enhance the strategic capabilities of a potential adversary. The product lines most frequently involved here include numerically controlled machine tools, semiconductor processing equipment, high-strength materials, high-temperature polymers, nuclear-related materials, computers, electronic testing and measuring equipment, magnetic recorders, and integrated circuits.

The formal interagency review process begins with the Operating Committee (OC). This is a senior staff-level group which meets weekly and is chaired by an executive director, a Commerce employee in OEA. Every effort is made to resolve interagency differences informally before a case is brought be-

---

[30]Report of the President's Task Force to Improve Export Administration Licensing Procedures (draft), Sept. 22, 1976, p. 23.

fore OC. Referral there requires careful and often protracted consideration of the strategic significance of the proposed transaction, the foreign availability of comparable commodities or data, and the licensing history of past applications for like or similar commodities or data.[31] All interagency decisions must be unanimous.

Obviously, the most complex and controversial cases reach this stage. It is these cases that cause delays in the system. At the weekly meetings, an average of 5 cases are discussed, although the agenda may contain up to 30 cases. One of the principal barriers to the consideration of more cases is the frequent inability of member departments of OC to arrive with prepared agency positions on applications. This is due in many instances to the complexity of the procedures for receiving technical advisory guidance within these departments. Each consulting agency in the system may decide whether to refer an application to any of a number of its own offices for technical evaluation. Thus, licensing responsibilities are not only diffused among executive branch agencies and departments; they are also diffused within departments. This point is further discussed below in the context of criticisms of the present system.

It is difficult to obtain an accurate count of the number of cases handled by OC. Estimates from senior personnel[32] have ranged from 250 to 300 cases for 1977 and 1978, but these figures are at variance with information supplied for the public record by other Department of Commerce officials. Congress has heard testimony, for example, that during 1977 "608 transactions were submitted to the Operating Committee for formal review."[33] For Communist-bloc countries only, other public testimony indicated that "374 [cases] required full-blown multiagency review within the Operating Committee," in 1978.[34] One reason for these discrepancies may be the fact that the lower numbers refer to categories on the OC agenda, where groups of similar cases may be handled together. Be that as it may, it is clear that only a small percentage of validated license applications actually reach the OC level. Furthermore, very few cases proceed beyond this level.

Cases that cannot be resolved at the OC level theoretically move through a series of committees involving progressively higher level decisionmakers (see figure 7). This formal structure is rarely utilized, however. Until recently, the sub-Advisory Committee on Export Policy (sub-ACEP), which is made up of Department Deputy Assistant Secretaries, was inactive; and even when it was operative, it met infrequently. For example, between 1975 and 1976, sub-ACEP met five times, discussing a total of six disputed cases. It has been suggested that the group meet monthly, in an effort to provide more policy guidance for the overall system on a continuing basis and to review unresolved cases, but meetings remain irregular.

ACEP itself, which is composed of Assistant Secretaries, has not met since at least 1975. Instead, issues still remaining in dispute above the sub-ACEP level have been referred to the Cabinet-level Export Administration Review Board (EARB). EARB exists to assure the highest level consideration of trade control policies and actions. In 1975 it met twice to discuss these policy matters and to deal with four disputed cases.

---

[31]See Arthur T. Downey, in *Export Licensing of Advanced Technology: A Review,* hearing before the Subcommittee on International Trade and Commerce of the Committee on International Relations, House of Representatives, 94th Cong., 2d sess. (Washington, D.C.: U.S. Government Printing Office, 1976), p. 72.

[32]Including Lawrence Brady, former Deputy Director, Office of Export Administration, and Thomas A. Hoya, Chairman of the Operating Committee.

[33]Testimony of Rauer H. Meyer, Director, Office of Export Administration, before the Subcommittee on Domestic and International Scientific Planning, Analysis, and Cooperation, Committee on Science and Technology, U.S. House of Representatives, Oct. 4, 1978, p. 15 (typed manuscript).

[34]Testimony of Stanley J. Marcuss, Senior Deputy Assistant Secretary for Industry and Trade, Subcommittee on International Economic Policy and Trade, Committee on Foreign Affairs, U.S. House of Representatives, Mar. 7, 1979, p. 8 (typed manuscript).

Figure 7.—Multiagency Advisory Committee on Export Policy Structure

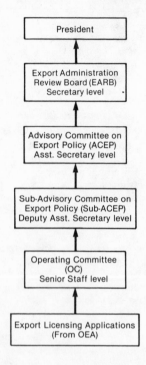

SOURCE: Comptroller General, *Export Controls: Needs to Clarify Policy and Simplify Administration*, Report to the Congress, Mar. 1, 1979, p. 35.

Through September of 1976, the Board met once to review policy issues and three cases. It has not met since, although it has handled a few matters via exchange of memoranda. In the event of EARB's inability to resolve a case, the President is the final arbiter. The recent proposed sales of a Dresser drill-bit factory and a Sperry-Univac computer to TASS, the Soviet news agency, are examples of cases decided at the Presidential level.

An applicant whose export request is rejected at any point in the review process receives a "negative consideration" letter. The exporter may respond formally to any of the reasons given for denial, and unofficial discussions may also continue between the applicant and the licensing officer. In some instances, however, the Government's sensitivities about strategic or foreign policy interests and the exporter's proprietary interests may circumscribe forthright and frank exchanges, and the reason given for denial may be as vague as "national security considerations." This is one source of the dissatisfaction of some parts of the business community with the licensing process.

A similar and parallel structure (see figure 8) exists to carry out U.S. multilateral responsibilities under the Battle Act. These include the periodic CoCom list reviews and the processing of the individual exception requests.

The Department of State handles interagency involvement in CoCom cases in a manner similar to Commerce's administration of U.S. unilateral export controls. That is, it relies on interagency advisory mechanisms. In this system, the counterpart of OC is Working Group I, and the formal higher policy level entity dealing with CoCom matters is the Economic Defense Advisory Committee (EDAC). It is composed of Assistant Secretaries from the Departments of State (the chair), Defense, Commerce, Energy, and Treasury. CIA acts in an advisory capacity.

The bulk of the workload of EDAC falls on Working Group I, a senior staff-level interagency group. An executive committee, chaired by the Director of the Office of East-West Trade, provides its operational guidance. Cases that remain unresolved here are passed along to successively higher policy levels and, ultimately, if necessary, to the President. Technical Task Groups (TTGs), composed of interagency technical experts and the private industry technical experts on TACs, both provide input and advise on decisions affecting CoCom list reviews and exceptions. CoCom itself is the subject of chapter VIII.

Compliance

Efforts to enforce domestic compliance with export controls are centered in OEA's

Figure 8.—Economic Defense Advisory Committee Structure

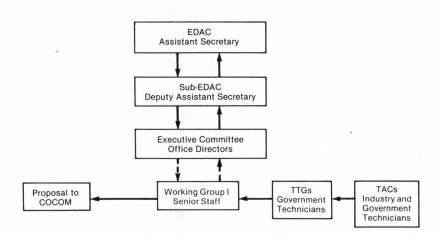

SOURCE: Comptroller General, *Export Controls: Need to Clarify Policy and Simplify Administration*, Report to the Congress, March 1, 1979, p. 19.

Compliance Division; overseas enforcement activities are handled through the Department of State. OEA's Compliance Division has three branches—facilities, intelligence, and investigations—with personnel in Washington and New York. OEA's computerized list of approved applications and discrepancies, such as overshipments, are forwarded to the Compliance Division. Although it also enforces controls made for short supply and foreign policy purposes, its major efforts lie in the area of strategic goods and technologies. These are concentrated in five kinds of activities: prelicensing checks, physical inspections of cargo shipments, postshipment document reviews, general license review, and validated license comparisons. Violations of export controls are punishable under the Export Administration Act, which provides for civil and criminal penalties, including fines and/or imprisonment.

The intelligence branch provides information to licensing officers on those potential exporters singled out during a preliminary screening process. This screening is accomplished through checking all applications against the Department of Commerce's Economic Defense List, a comprehensive index of firms and individuals previously denied export privileges or listed as suspect on the basis of allegations received by Commerce. The intelligence branch also provides information at this stage on individuals or firms that have supplied insufficient documentation for their applications. In both cases, information is gathered through cooperation with the Department of State, the intelligence agencies, and trade and industry sources.

Spot checks are made on cargo to physically verify that the contents of shipments correspond to licenses. There are also reviews of postshipment documents: the Bureau of Customs verifies that every shipment over $250 is listed on an outgoing carrier's manifest, a service for which it is reimbursed by OEA. In addition, OEA reviews declarations of cargo shipped under general license and searches for discrepancies between declarations and the data appearing on approved validated licenses. General license reviews ascertain, through the descriptive data on the customs declarations, whether previous shipments should in fact have been made under validated license. Finally, for items sent under validated license, the Bureau of Customs collects information on shippers' export declarations.

The State Department administers two programs in foreign countries to inspect commodities that have been licensed by the Department of Commerce. One, the safeguard program, applies to computers. It comes under the jurisdiction of CoCom, and is implemented by industry representatives in Communist countries. The other is a U.S. program performed by U.S. overseas posts. The purpose of both programs is to ensure that dual-use items are being used for the purpose designated on the license. Violators who fall under U.S. jurisdiction are subject to administrative and criminal penalties specified in the Export Administration Act; others may be denied future U.S. shipments. It is obvious, however, that monitoring third-party transfers of technology of U.S. origin is virtually impossible without the active cooperation of foreign states.

The safeguard program has been in effect since 1976 and applies to those licenses granted under a CoCom stipulation that:

> Responsible Western representatives of the supplier will have the right to access to the computer facility and all equipment wherever located during normal working hours or at any time when the computer is operating and will be furnished information demonstrating continued authorized application of the equipment.[35]

Individual licensing agreements designate the frequency of required inspections. In addition to these inspections, the Commerce Department may request that U.S. officials overseas examine strategic commodities

that have previously been licensed for export by the United States. This usually occurs in cases of suspected diversion from the original end use.

Criticism of the System

The export-licensing system has often been the object of criticism from U.S. industry, America's CoCom trading partners, Congress, and members of the academic community. The criticisms may be summarized as follows:

• The system suffers from needless delays and an excessive amount of uncertainty. This is exacerbated by the cumbersome interagency review process and the time-consuming case-by-case approach to applications.
• The diffusion of responsibility among and within agencies results in a lack of administrative responsiveness and forthrightness in the system. This further discourages efforts by American business to expand exports.
• The system suffers from a lack of adequate policy guidance. This results in a situation in which too much discretion is allowed to midlevel administrators and too little influence is exerted by those who are both technically qualified and attuned to both the changing environment of U.S./U.S.S.R. relations and the vital role of exports in the U.S. economy.
• The U.S. CCL is too restrictive. By attempting to control items unilaterally, it incurs the antagonism of foreign exporters who need reexport licenses for technology of U.S. origin. Moreover, the unilateral control puts U.S. businessmen at a competitive disadvantage. Not only are more products controlled than in Western Europe and Japan, but neither the list itself nor the system that administers it takes adequate consideration of foreign availability.

Delays.—The first of these criticisms is the most frequently heard and most universally recognized. U.S. exporters testifying

---

[35]Quoted in the Comptroller General, *Export Controls: Need to Clarify Policy and Simplify Administration,* op. cit., p. 55.

before Congress have been virtually unanimous in their condemnation of a system that subjects their license applications to long and often seemingly arbitrary delays. Cases that are subjected to holdups—sometimes of many months or even several years—create serious problems for the potential exporter and may damage the credibility of American suppliers abroad. The law has set a 90-day limit on the entire licensing process, from application to approval or denial; but in 1978 almost 2,000 applications exceeded this limit, about twice the 1977 figure (see table 18). As might be expected, licenses involving the export of high-technology dual-use items, like sophisticated electronic equipment and computers, to Communist destinations accounted for the largest portion of these delinquent cases. In 1975, of the 1,105 delayed cases, 923 were destined for the Communist world. In all, 957 cases involved computers and electronic equipment, 827 of which were for export to Communist countries. Similar patterns have prevailed since.

However, delayed cases represent only about 3 percent of all license applications. It is not the number, therefore, but the visibility of these cases that is disturbing. The fact that they usually represent large orders of high-technology items, and that they involve highly sensitive issues of national security and foreign policy, make them the subject of the bulk of the publicity concerning export licenses. It is worth noting here that, in cases held for national security considerations, the applications are requesting exceptions for items that have serious potential military uses. The very fact that the cases are considered at all indicates a willingness on the part of the Government to assist exporters and permit sales wherever possible. Obviously, the quickest response would be a simple "no." This is not necessarily the case when applications are delayed for foreign policy reasons.

A second important point to be made about the delayed cases is that, although their numbers may be relatively small, they are growing. This points to a disturbing trend. Between the enactment of the Export Administration Act in 1969 and 1974, the number of applications received by OEA declined steadily—from over 145,000 to less than 66,000, a drop of 55 percent. This decrease is a reflection of the liberalization of U.S. policy toward East-West trade, and can be attributed to the fact that items of purely economic significance were dropped from the CCL, thus reducing the number of required validated license applications. After 1975, however, the downward trend in application numbers ceased. The workload has since been growing at an increasing rate. In 1976, OEA processed 54,359 cases; in 1977, 58,967; and in 1978, 63,476. The total for the 1979 calendar year is expected to exceed 77,000.

A comparison of authorized funding and staff levels of OEA with workload estimates over the past 11 years indicates the problem facing OEA staff (see table 19). In the past 5 years, the number of authorized positions has risen about 18 percent;[36] if one assumes that 68,000 cases will be processed in FY 1979, the number of license applications will have risen more than 27 percent during the

Table 18.—Export License Applications Pending for More Than 90 Days

Ending period	Number of applications
1975[a]	1,105
Apr. 30, 1976[a]	294
Sept. 2, 1976[b]	298
Sept. 2, 1977[b]	454
June 1978[c]	603
Sept. 1, 1978[b]	645
Feb. 2, 1979[c]	585
1976[d]	689
1977[d]	1,032
1978[d]	1,988

SOURCES:
[a]Report of the President's Task Force to Improve Export Administration Licensing Procedures, Sept. 22, 1976 (draft).
[b]Office of Export Administration data.
[c]Testimony of Stanley Marcuss, Senior Deputy Assistant Secretary for Industry and Trade, Office of Export Administration, before the Subcommittee on International Economic Policy and Trade, Committee on Foreign Affairs, U.S. House of Representatives, Mar. 7, 1979.
[d]Export Administration Act Amendments of 1979, *Report,* Committee on Foreign Affairs, U.S. House of Representatives, 96th Cong. 1st sess., May 15, 1979, p. 4.

---

[36]168 positions are authorized for FY 1979. OEA presently employs 145 professionals and clerical personnel and is, according to recent testimony, actively seeking additional staff.

Table 19.—Funding, Authorized Positions, and License Applications Volume

| Fiscal year | Funding ($000) | Positions | Export license applications | |
			Received	Processed
1979	$5,670	168	—	68,000*
1978	5,550	162	—	63,476
1977	5,256	167	—	58,967
1976	4,626	142	—	54,359
1975	4,400	142	—	52,600
1974	4,634	114	65,883	—
1973	4,103	154	64,070	—
1972	6,111	256	78,561	—
1971	5,900	256	107,615	—
1970	5,358	256	132,498	—
1969	5,358	272	145,369	—

*Estimated.
NOTE: After 1974 the Department of Commerce changed its method of counting license applications. Instead of tabulating the number of applications received, it now reports the number actually processed. The two may not be identical, as some applications are withdrawn before a decision is made.

SOURCE: Office of Export Administration, U.S. Department of Commerce.

same period. Moreover, the system's administrators contend that the complexity of the issues surrounding export license applications has increased tremendously since the passage of the Export Administration Act, so that a simple comparison of workload measured by number of applications processed is misleading. There has also been an increase in the number of CoCom exception cases, which tend to be the most complex. To this must be added the fact that the overall workload is clearly growing at an accelerated pace.

Similar or heavier workloads may be found in the other Departments involved in the license process. The number of cases referred to the State Department's Office of East-West Trade because of their foreign policy implications more than doubled between 1977 and 1978, rising from 1,200 to 2,500, but the Washington staff level of five employees has remained unchanged. Before the recent DOD reorganization, the Office of Strategic Technology and Munitions Control of DOD had a staff of three to four professionals and two secretaries who were responsible both for coordinating and developing all DOD positions on U.S. license applications, and for the Department's contribution to CoCom list reviews and exceptions. The workload of each professional employee in the Office was enormous; in 1975 alone, it handled 2,200 cases. This excessive workload is one of the primary reasons for license decision delays.

The previous Deputy Assistant Secretary responsible for the licensing process in Defense did institute a "Guillotine Closure" system designed to speed-up the system. Its goal is to provide the Secretary of Commerce with an advisory opinion regarding the national security aspects of an export license application within 30 days. The procedure operates on the assumption that a license can be issued unless the technician recommends otherwise within a specified period of time. Before the guillotine was put into place, only 72 percent of the cases handled by Defense between January and September 1978 were closed within 30 days. With the system in-place, from October to December 1978, 98 percent of the cases were closed within the stipulated period. In addition, the average time per case was reduced from 29 days to 12 and the age of the most delinquent case from 165 days to 35 (see figures 9 and 10). It would appear from this that the present system is sufficiently flexible to respond to calls for increased efficiency.

In addition to the sheer volume of paperwork which comes before DOD, holdups are often occasioned by the need to tap additional technical resources in the Army, Air Force, and Navy; in Department-wide research laboratories and facilities; and in other Pentagon offices. For instance, in 1975 virtually all computer cases were referred to the Institute for Defense Analysis. In addition, most computer, electronics, and technical data cases went to the Office of the Under Secretary of Defense for Research and Engineering, while the Office of the Assistant Secretary of Defense for Installations and Logistics screened most machine tool and technical data cases. As the system has been administered, the processing of export control requests gets low priority among technical experts whose main tasks lie elsewhere. Whether the recent reorganiza-

Figure 9.—Age of DOD Export Control Cases
(January-September 1978)

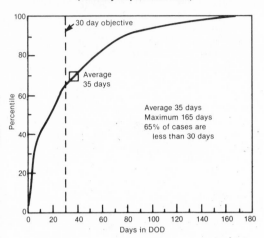

SOURCE: Testimony of The Honorable William J. Perry, March 5, 1979,
Senate Committee on Banking, Housing, and Urban Affairs.

Figure 10.—Age of DOD Export Control Cases
(October-December 1978)

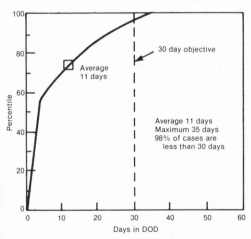

SOURCE: Testimony of The Honorable William J. Perry, March 5, 1979, Senate
Committee on Banking, Housing, and Urban Affairs.

tion, described above, will alleviate this situation remains to be seen, but one immediate change will be staffing levels. The new export-licensing program will have a staff of 14 and an initial budget of $500,000.

Policy Guidance.—It is possible, however, that the quality of decisions may suffer from attempts to speed- up the system as it is presently constituted and that fundamental changes are needed to preserve both the responsiveness demanded by industry and the policy guidance and technical evaluation necessary to protect of U.S. national security. It is here that the charges of inadequate policy guidance arise.

A recent Presidential Review Memorandum on East-West Technology Transfer (PRM 31) investigated this and other problems in the present licensing system and concluded that the present organization should be augmented. It suggested that the National Security Council (NSC), the Office of Science and Technology Policy, and the Arms Control and Disarmament Agency all be given export control responsibilities, presumably to lend policy guidance. Implementation of this recommendation has begun. Representatives from the three agencies now may participate in OC discussions, and an ad hoc technology transfer group has been created within NSC.

After its own investigation of the system, however, the General Accounting Office (GAO) has declared that it does "not believe the problems associated with diffused management authority can be solved by adding more Government agencies to the licensing process.• This is not a regulatory activity which suffers from a lack of bureaucratic attention, and better attention, not more attention, is needed at the licensing level."[37]

Restrictiveness of Export Controls.—Many of the companies engaged in East-West trade are multinationals. Their ex-

[37]Comptroller General, *Export Controls: Need to Clarify Policy,* op. cit., p. 66.

perience in developing technology, moving personnel, and monitoring research activities around the world has convinced them that technology comparable to their own is almost always otherwise available. They contend that the administrators of U.S. export controls underestimate the degree of technical sophistication in the East and ignore technical developments in Western countries both within and outside CoCom.

Many examples illustrate the availability of technology from multiple sources:

- Hungary exhibited advanced computer-controlled machines at the 1978 machine tool show in Chicago. Many of the machines that it brought to the United States would not have been exportable by U.S. firms under Export Administration guidelines.
- The sale of a U.S. Univac computer to TASS was stopped during the summer of 1978 for foreign policy reasons. Although the decision was later reversed, TASS has now signed with a French computer manufacturer.
- A U.S. firm was unable to obtain a license for the sale of an aluminum production plant. Here too, a contract was signed with the French within a very short period of time.
- A U.S. firm was approached by the PRC regarding the sale of some electronic equipment. After preliminary discussions with OEA, the American firm withdrew because it feared it could not get the necessary export licenses. The Chinese were confident, however, that the equipment could be obtained elsewhere: the PRC had a close relationship with a Japanese firm that could supply the equipment and indeed had done so under similar circumstances in the past.

Many firms also contend that U.S. export controls extract high costs of compliance and discourage entry into Eastern markets. Accounting methods do not usually lend themselves to ready identification of the direct costs of validated license applications, but large corporations heavily engaged in trade with the East have estimated that they incur annual licensing-related costs of as much as $1 million, a figure that includes the salaries of several employees solely assigned to deal with OEA and the intermittent services of numerous executives.

But, the largest costs may be indirect and unmeasurable. U.S. businessmen contend that the current export control procedures have resulted in suspicion toward the United States and U.S. firms—suspicions that have caused Eastern nations to go elsewhere. Some firms report that over the past few years approaches to U.S. business officials in Eastern nations have markedly decreased. This is attributed to the fact that the delay and uncertainty associated with dealing with the United States has caused potential customers to turn to other suppliers. The perceptions of the potential customers are important. An illustrative example can be found in the case of a U.S. company that had been negotiating for over a year with the PRC. If consummated, the deal in question would have provided the PRC with access to the use of new communications technology, but not the technology itself. The day before the contract was to have been signed, a high-level U.S. Government official made a comment that was interpreted by the PRC delegation as an indication that the U.S. Government might not fully support the sale. Negotiations broke down and the PRC soon signed with another country—an agreement that gave them access to the same capabilities and to the underlying technology.

Moreover, as chapter IX demonstrates, firms in other countries can sometimes get license pre-approvals that eliminate a major area of doubt from negotiations with Communist countries. Differences in the administration of the export controls between the United States and its CoCom partners lead some American corporations to perceive the U.S. Government as the greatest obstacle to their international business activities.

The unilateral controls embodied in the U.S. CCL are believed to contribute to this problem. By reflecting unrealistic technical decisions for controlled items (i.e., setting technical parameters unnecessarily low), they allow firms in nations with different standards to capture business that would otherwise come to the United States. Finally many companies seem to feel that the administration of export controls does not sufficiently take into account the policing capabilities of the private sector. Corporations see it in their own interest to control the unauthorized transfer of goods by requiring that critical components be acquired from the original supplier.

Thus, industry argues that because of U.S. export controls, a considerable volume of trade with the Communist world is being administered by foreign subsidiaries and its benefits are being lost to the U.S. economy. Many of the administrative activities associated with East-West trade are carried out through divisions, subsidiaries or affiliates of U.S. firms located abroad. Reasons for this lie in the affinity between European businessmen and their Soviet-Eastern European counterparts; the reduced operating costs that result from geographic proximity; and the fact that relationships between the U.S. companies and centrally planned economies are closely related to their marketing efforts in Western Europe.

But U.S. executives sometimes also contend that companies operate in this manner because it avoids much of the "aggravation" associated with the administration of export controls in the United States. No body of data appears to be available on the extent to which U.S. export controls motivate multinationals to remove their operations from the United States. Given the relatively low volume of U.S. trade with the East it is unlikely that this is often a primary consideration.

Recommendations for Improvement.— These criticisms have sometimes generated tension between business and Government, and have resulted in a spate of proposals for reform of export-licensing procedures. Finding that the complex system for administering export controls is plagued by "vague criteria, insufficient funding, and low priorities," GAO has recommended, in part, that:

- The Export Administration Act should be amended to state that the President shall consider foreign availability when imposing export controls for foreign policy purposes.
- The President's semiannual report to Congress should discuss in more detail the uses and reasons for foreign policy controls.
- The foreign availability clause should be administered as a separate effort under a "foreign availability evaluator."
- A new procedure to process routine license requests should be established in OEA.
- A multiagency Export Policy Advisory Committee should be established at an appropriate administrative level. This would allow the abolition of the entire EDAC and ACEP referral structures.
- Export license application management responsibility should be centralized in the Department of Commerce's OEA so that responsibility for license applications is no longer diffused.
- Funding of technical evaluations should be centralized in OEA.[38]

The Export Administration Act of 1979 (see above and appendix) has gone some way toward meeting these suggestions, but it has not instituted fundamental alterations in the licensing system, and is unlikely to entirely still criticism from that part of the business world engaged in East-West trade.

CONTROL OF NUCLEAR EXPORTS

Since the Atomic Energy Act was passed in 1954, special controls have been imposed on the export of nuclear equipment and technology. These are presently governed by the

---

[38]Ibid., p. 18, vi, 42-49.

Nuclear Non-Proliferation Act (Public Law 95-242), enacted in March 1978, which established new procedures for controlling the export of nuclear facilities, equipment, materials, and technology; and dealt with the criteria, organization, and procedures for control of U.S. nuclear exports, both domestically and abroad. Under this Act, nuclear technology was added to the Department of Commerce's Control List, and primary responsibility for the controls divided between the Department of Energy, the Nuclear Regulatory Commission, the Department of Commerce, and the Department of State, depending on the specific materials and technology being licensed. In addition, special controls were placed on the export of components specially designed or prepared for use in nuclear facilities. Components for uranium enrichment facilities, fuel-reprocessing facilities, or heavy water production plants may now be exported only when specifically licensed for a cooperative agreement. Finally, the Act imposed additional conditions for Government approval of nuclear exports and expanded Government control over the export of component parts for nuclear facilities. It directs the Nuclear Regulatory Commission (NRC), in consultation with the Secretaries of State, Energy, and Commerce, and the Director of the Arms Control and Disarmament Agency, to determine which items should be subject to export controls because of their importance in nuclear explosions. Export licenses for these must be granted by NRC and only then if the following criteria are met:

1. International Atomic Energy Agency safeguards exist for the items;
2. No such item shall be diverted for any nuclear device; and
3. No such item shall be retransferred without prior U.S. consent.

In addition, NRC must certify in writing that the issuance of each export license will not result in adverse consequences for U.S. national security and defense.

The Department of Energy has primary jurisdiction over the control of transfer of certain types of nuclear technology. This control emanates from a provision of the Nuclear Non-Proliferation Act, which forbids any person to directly or indirectly engage in the production of any special nuclear material outside the United States, except as authorized by a determination by the Secretary of Energy that such activity would not be detrimental to the national security of the United States. The Department of Energy coordinates with other agencies, including the Arms Control and Disarmament Agency, in this process. Under the legislation, if NRC is unable to issue a license or takes too long to do so, the President has the authority either to block or to authorize exports.

CREDITS AND TARIFFS

THE EXPORT-IMPORT BANK OF THE UNITED STATES[39]

When the U.S. Government extended diplomatic recognition to the Socialist regime in Russia in 1933, the Soviet Union represented a vast, new, and badly needed market for American exports. To facilitate the U.S.-Soviet trade that was expected in the wake

of recognition, in February 1934, President Roosevelt issued Executive Order No. 6581 directing the Secretaries of Commerce and State to organize the Export-Import Bank (Eximbank) of Washington as a banking corporation under the laws of the District of Columbia. Although financing trade with the U.S.S.R. was the acknowledged primary purpose of Eximbank, the executive order also stated a general aim: "To aid in financing and to facilitate exports and imports and the exchange of commodities between the United States and other nations or the agen-

---

[39]For a complete history of the Bank, see George Holliday, "History of the Export-Import Bank," in Paul Marer, ed., *U.S. Financing of East-West Trade* (Bloomington, Ind.: International Development Research Center, 1975).

cies or nationals thereof." Ironically, it was to be 39 years before Eximbank extended any credit to the Soviet Union.

At first a relatively obscure institution with an initial total capitalization of $11 million, Eximbank has grown to participate in a variety of programs in over 140 countries. Presently it has an obligation ceiling of over $40 billion and over the past few years, has been responsible for the financing of approximately 10 percent of all U.S. exports. This support has been primarily concentrated in transportation and construction equipment and powerplants (especially nuclear), but manufactured goods and capital goods are also heavily supported. Approximately 18 percent of all U.S. manufactured exports and 21 percent of capital goods exports have been financed by Eximbank.

At first, Eximbank operated without official Government support, but in 1945 it received legislative mandate in the Export-Import Bank Act (Public Law 79-173). Under this Act, the Export-Import Bank of the United States became an independent agency of the Government, its management vested in a bipartisan Board of Directors. Four of these are appointed by the President with the advice and consent of the Senate; in addition, the Secretary of State sits on the Board ex officio. An Advisory Committee consisting of the Bank's chairman; the Secretaries of State, Treasury, and Commerce; and the Chairman of the Board of Governors of the Federal Reserve System advises on major policy questions. The U.S. Treasury provided Eximbank's original funds by purchasing $1 billion of the capital stock of the Bank and, in addition, by continuing at intervals to purchase part of its obligations. Eximbank receives no direct appropriation from Congress, however, although Congress does retain oversight control of the Bank's operations through the exposure ceilings it establishes on new and outstanding credit authorizations. In addition, Eximbank may borrow up to $6 billion directly from the U.S. Treasury to meet its short-term needs. It must satisfy its medium- and long-term requirements through borrowing from the Federal Financing Bank.

Eximbank carries out its mandate to promote U.S. exports through four programs:

- First, it can make direct loans in the form of dollar credits to foreign borrowers purchasing U.S. goods and services. These loans must be used to pay U.S. exporters and they must be repaid in dollars. This program is designed to supplement, not replace, private financing; it provides credit at favorable terms in cases where private institutions are unwilling to assume risks and it extends credit for longer terms than will private lenders. Eximbank usually demands a downpayment of at least 10 percent from the borrower. It then finances part of the loan through its own funds and requires private financing at commercial terms for the balance.
- Second, Eximbank can provide guarantees to private financial institutions that their loans financing U.S. exports will be repaid. These guarantees are backed by the full faith and credit of the United States and are designed to encourage private lenders to extend export credits, and to lower their interest costs. They are available for medium-term transactions (181 days to 5 years).
- Third, Eximbank, in cooperation with the Foreign Credit Insurance Association, a group of approximately 50 U.S. insurance companies, can insure U.S. exporters against the exceptional risks inherent in foreign transactions. In these cases, private insurance covers normal risks, and Eximbank extends coverage for extraordinary events such as war or expropriation.
- Finally, Eximbank can provide incentives for private banks to finance U.S. exports by administering a discount loan program. Discount loans are advance commitments to discount export debt obligations acquired by commercial banks. The commitment assures the private lender that additional funds will

be available should they be needed during the full maturity of the obligation, which is generally short term. Loans in which Eximbank is participating as a direct lender are ineligible for this program.

Eximbank activities in the Communist world have had a troubled and complex history. In 1934, the establishment of such an institution seemed desirable for several reasons. First, the Soviets were short of the hard currency necessary to pay for imports. Second, many U.S. private financial institutions were unwilling to risk providing credit to the U.S.S.R. Finally, Great Britain, Germany, Italy, and France had all successfully established similar organizations to provide foreign trade financing. The atmosphere of the time was one of intense competition for dwindling world markets and the Soviet Union seemed a potentially rich prize.

The barrier to Eximbank credit at first was the refusal of the Soviet Government to settle Tsarist debts to U.S. citizens. The Johnson Debt Default Act of 1934 (Public Law 80-772) had prohibited the extension of credits or financial assistance in any form to any foreign government in default on its obligations to the United States. In 1945, the Export-Import Bank Act expressly removed prohibitions on loans from Eximbank to governments in default as of April 1934. By this time, however, U.S.-U.S.S.R. relations had begun their rapid postwar deterioration. With the exception of Yugoslavia, the Bank made no loans to the Communist world through the cold war period of the 1950's. This self-imposed limitation was formalized in 1964, when the Foreign Assistance and Related Agencies Appropriation Act (Public Law 88-634) prohibited Eximbank from lending to or in any way participating in the extension of credits to any Communist country, except when the President determined that the extension of such credit was in the national interest. In 1964 and 1966, President Johnson did make such determinations as part of an effort to improve East-West relations, but with the exception of small loan guarantees to Romania

and Hungary in 1964, little Eximbank activity took place. In 1968, the prohibition of Eximbank involvement in trade with the Communist world was made absolute through an amendment to the Export-Import Bank Act that barred, without provision for Presidential waiver, the extension of Government credit to any country furnishing by direct Government action "goods, supplies, military assistance or advisors" to any nation engaged in armed conflict with the United States (Public Law 90-267). Thus, the Vietnam War had the effect of denying Eximbank credit to all Communist nations except Yugoslavia.

The absolute prohibition was lifted in 1971 for those countries not themselves in armed conflict with the United States (Export Expansion Finance Act, 85 Stat. 345), and once again the President was empowered to determine if a credit transaction with a specific Communist country would be in the national interest. President Nixon made such determinations for Romania in 1971 and Poland in 1972. Also in 1972, the formal inception of detente and the signing of the U.S.-Soviet trade agreement paved the way for a Presidential declaration in favor of allowing Eximbank credit to the U.S.S.R. The Soviet Union finally received its first Eximbank loan in February 1973, and in the next 15 months Eximbank exposure on the Soviet debt grew to over $460 million, an amount that supported over $1 billion in U.S. exports. Table 20 shows Eximbank exposure in the Communist world as of September 30, 1978. Eximbank was one of the

Table 20.—Export-Import Bank Exposure in Selected Communist Countries (Sept. 30, 1978)

Country	Millions of dollars
U.S.S.R.	$456.4
Poland	228.4
Romania	90.2
Yugoslavia	858.9
China	26.4*

*Granted before 1948.
NOTE: Exposure consists of the combined total of direct Eximbank credit, guarantees, and insurance programs. The bulk of this exposure is in direct loans.

SOURCE: Export-Import Bank of the United States, 1978 Annual Report.

Administration's major tools in detente, and the Soviets had been quick to take advantage of the credits newly available to them. All indications were that the amount of these credits and guarantees would continue to grow rapidly. Projects already partially funded by Eximbank included a $400 million chemical complex, the $342 million trade center, and a $36 million iron ore pellet plant. By 1974, applications were pending for help in financing $110 million in oil and gas exploration equipment and a $50 million tractor factory. $1.2 billion to $1.4 billion in new credits were projected through calendar year 1977. In 1974, the issue of Eximbank credits to the Soviets became enmeshed in a larger debate over the future of detente and the role of Congress in foreign policy. One focus of these concerns was the reaction of Congress against the use of subsidized credits. Testimony before the Senate International Finance Subcommittee, for instance, made it increasingly clear that Government-supported credits could no longer be regarded as a costless way of promoting exports.[40] Eximbank loans to the U.S.S.R. had, in accordance with Bank policy, been provided at the highly favorable fixed rate of 6 percent, but rapidly rising commercial rates were steadily increasing the export subsidy. Even after July 1974, when the Bank responded to severe congressional criticism by instituting a flexible interest rate, Eximbank's maximum was still below rates obtainable in the private sector.

The point of view that eventually held sway in Congress was summarized by Senator Stevenson:[41]

Both the level and rate of Exim assistance [to the U.S.S.R.] as well as the kind of projects involved raise serious questions about the policies being pursued. It is clear that detente is one of the goals. It is a goal which we all seek . . .

But it is far from certain that the United States can buy detente with credits. A genuine and lasting easing of tensions requires resolution of the difficult issues which divide the United States and the Soviet Union . . . difficult and long-standing problems which will not be resolved overnight and most certainly will not vanish at the first sign of American cash.

Unless the underlying factors which gave rise to these problems are solved, credits are unlikely to be of much avail. What is worse, they may have the effect of boosting Soviet military capability and in turn lead to a worsening of relations. It is significant, for example, that none of the Exim-assisted Soviet projects to date, and none of those which are planned, involve the export of U.S. consumer goods. Instead, all relate to capital construction or the development of productive capability by freeing Soviet resources for other purposes; the United States may be indirectly contributing to Russian military potential.

Thus, there was concern over actual or proposed projects—chemical complexes, the Kama River truck plant, oil and gas development projects, the construction of wide-bodied aircraft production facilities—which might have direct military possibilities. The completion of such projects would be tantamount to financing the military production capability of a long-standing adversary before the achievement of a permanent improvement in relations.

Congress was also concerned that rapid increases in Eximbank's financial exposure in the Soviet Union might unwittingly increase Soviet leverage over the United States. The possibility of Soviet threats to withhold payment on almost a half a billion dollars seemed an unacceptable risk. These doubts were enhanced by the apparent lack of need for such massive credits. West Germany had only recently declined to provide financing for a $1 billion iron and steel complex in Kursk, but the Soviet Union found the necessary cash itself. Similarly, it had agreed to pay $48 million in cash to a British firm for a new plastics factory, and still found itself with sufficient reserves to extend $600 million in credits to Argentina for an electric

[40]See Stanley J. Marcuss, "New Light on the Export-Import Bank," in Marer, op. cit., pp. 257-290.
[41]In the Report of the Committee on Banking, Housing, and Urban Affairs, U.S. Senate on Export-Import Bank Amendments of 1974, 93d Cong., 2d sess. (Washington, D.C.; U.S. Government Printing Office, Aug. 15, 1974).

power project. With such apparent financial capability, why should the Soviets require massive Eximbank assistance? An additional worry was the potentially adverse impact of Eximbank's worldwide activities on the competitive position of U.S. industry. The creation of production facilities abroad could result in the long-run export of U.S. jobs and decline of U.S. markets.

In response to these fears, Congress passed legislation that made the use of credit—and the withholding of credit—a political weapon. First, the Stevenson amendment to the 1974 Export-Import Bank Act put a $300 million limitation on new loans and guarantees to the Soviet Union. Within this ceiling, specific restrictions were placed on energy-related projects, reflecting Congress' particular concern with the potential military and practical relevance of Soviet oil and gas development. A maximum of $40 million was set for energy exploration and research, and no Eximbank loans were to be used for Soviet energy production and transmission. Congress was to periodically review these ceilings, which could be lifted by the President if he found it in the national interest to do so, and if Congress approved the increase by concurrent resolution.

The effect of this amendment was made moot, however, by the passage of the Jackson-Vanik amendment to the Trade Act of 1974 (Public Law 93-618). This resulted in the scrapping of the U.S.-Soviet trade agreement and effectively terminated all credits to the U.S.S.R. by linking trade benefits to nonmarket economies to the free emigration of their citizens. The Jackson-Vanik amendment is discussed in more detail below.

THE COMMODITY CREDIT CORPORATION

In addition to the financial assistance provided by Eximbank, credit is available to the Communist world through the Commodity Credit Corporation (CCC), created in 1933 by executive order. Since 1956 this organization has administered an export credit sales pro-

gram designed to help U.S. agricultural exporters expand their sales in foreign markets. During the early years of detente in the 1970's, CCC provided the Soviet Union with over $550 million in credits for the purchase of U.S. grains. Like Eximbank credits, CCC support was effectively suspended by the passage of the Jackson-Vanik amendment to the Trade Act of 1974. The provisions of that amendment have, however, allowed selected Communist-bloc countries to be deemed eligible for CCC programs (see below and table 21). CCC makes credits available to importers at somewhat better terms than could be obtained elsewhere. Although interest rates vary, they are usually slightly below equivalent market rates, and maturities are often longer than agricultural credits offered by private banks.

A related program was created by the Agricultural Trade Act of 1978 (Public Law 95-501). Under the provisions of this law, credit with commercially competitive repayment terms of up to 3 years is available to many foreign buyers of American farm products. Communist countries currently eligible for the 3-year credits—Yugoslavia, Romania, Poland, and Hungary—are also eligible for "intermediate" credit, with repayments of up to 10 years, for the following purposes:

> To finance purchases of grain for reserve stockpiling under international commodity agreements or other plans acceptable to the United States; to finance purchases of breed-

Table 21.—Communist Countries: Eligibility for U.S. Programs (as of July 15, 1979)

Country	MFN	Eximbank	CCC
U.S.S.R.	No	No	No
East Germany	No	No	No
Yugoslavia	Yes	Yes	Yes
Poland	Yes	Yes	Yes
Romania	Yes	Yes	Yes
Hungary	Yes	Yes	Yes
Bulgaria	No	No	No
Czechoslovakia	No	No	No
PRC	No	No	Yes

MFN = Most-favored-nation.
CCC = Commodity Credit Corporation.

SOURCE: Office of East-West Trade. Department of State.

ing livestock, including freight costs; to finance, where feasible, establishment of facilities for improved handling of imported farm products; and to meet credit competition from other countries, but not to initiate credit wars.

The Agricultural Trade Act further authorizes the Agriculture Department to offer CCC financing for up to 3 years on sales of commodities to the PRC. In a related action, the bill authorizes CCC credits up to 3 years to private U.S. exporters who make deferred payment sales to currently eligible countries, including China.

TARIFFS: MOST-FAVORED-NATION STATUS

Most-favored-nation status guarantees a nondiscriminatory U.S. tariff rate to the foreign exporter, i.e., a rate as low as that negotiated for any other American trading partner for any given commodity. Beginning in 1934 and continuing into the postwar era under the auspices of the General Agreement on Trade and Tariffs, a series of trade negotiations has resulted in the reduction or elimination of nearly all substantive tariffs levied on U.S. imports. Denial of MFN denies the potential foreign exporter any benefits flowing from these progressive relaxations of the tariff structure. Those States to which the United States denies MFN must attempt to market their products under the 1930 Hawley-Smoot Tariff, a system of relatively high tariff barriers which was constructed after the crash of 1929 to insulate the U.S. market.

The Trade Agreements Extension Act of 1951 withdrew MFN status from all Communist countries except Yugoslavia. The major obstacle at present to the granting of MFN to the U.S.S.R. and other remaining centrally planned economies is the Jackson-Vanik amendment to the 1974 Trade Act.

The Jackson-Vanik amendment proscribes the extension of any Government credits and/or investment guarantees, the signing of commercial treaties, or the granting of MFN to any nonmarket nation that:

1. denies its citizens the right or opportunity to emigrate;
2. imposes more than a nominal tax on emigration or on the visas or other documents required for emigration, for any purpose or cause whatever; or
3. imposes more than a nominal tax, levy, fine, fee, or other charge on any citizen as a consequence of the desire of such citizen to emigrate to the country of his or her choice.

Communist nations that already enjoyed MFN—Poland and Yugoslavia—were exempt from these provisions, but in order for any other Communist country to qualify for MFN, credits, or commercial treaties, the President must submit to Congress a report indicating that the country is not in violation of any of the above conditions, including information on the nature and implementation of emigration laws and policies and restrictions applied to those wishing to emigrate. The amendment further gives the President authority to waive its restrictions on reporting to Congress that:

- he has determined that such waiver will substantially promote the objectives of free emigration; and
- he has received assurances that the emigration practices of that country will eventually lead to free emigration.

A majority vote (within 90 days of receipt of the President's report) in either house of Congress can veto the extension of MFN status or U.S. Government credits, and the President retains the authority to suspend or withdraw the extension of MFN treatment to any country at any time.

Thus far, the President has used this waiver authority on two occasions—for Romania and Hungary, which both now enjoy MFN and Eximbank eligibility. In the case of Romania, confidential diplomatic discussions, which took place between senior U.S. and Romanian officials in 1975, resolved the major emigration issues to the satisfaction of President Ford. No written assurances were provided. In the case of Hungary, oral discussions in 1978 were supplemented by an exchange of letters containing assurances on emigration between the U.S. Ambassador to Hungary and the Hungarian Foreign Minister.[42]

The economic importance of MFN to recipient countries is discussed elsewhere (see chapters II and III), but it is important to note that the granting or withholding of

MFN has important symbolic value in the Communist world and may therefore affect the political climate as well as overall trade levels. This is true for both the PRC and the U.S.S.R. It is in this context that media speculation surrounding the granting of MFN to the Soviet Union must be understood. One key factor in any administration strategy to resume U.S. trade and tariff benefits to the U.S.S.R. must be the nature of the recent trade agreement with the PRC, which, among other things, proposes the extension of MFN. This agreement must still be ratified by Congress. Table 21 summarizes the present status of selected Communist countries with respect to their eligibility for U.S. credit and MFN.

---

[42]Executive Order 11854, Apr. 24, 1975, 40 F.R. 18391, applied to Romania; and Executive Order 12051, Apr. 7, 1978, 43 F.R. 15131, applied to Hungary.

CHAPTER VIII

Multilateral Export Control Policy: The Coordinating Committee (CoCom)

CONTENTS

Multilateral Export Control Policy: The Coordinating Committee (CoCom)

THE HISTORY OF COCOM

The Coordinating Committee for Multilateral Export Controls (CoCom) is the informal multilateral organization through which the United States and its allies attempt to coordinate the national controls they apply over the export of strategic materials and technology to the Communist world. It was originally conceived in postwar discussions between the United States, Britain, and France. By 1948, the U.S. Government had begun to enlist the cooperation of its West European allies for a coordinated embargo policy against the Communist bloc. Early negotiations on this matter were private and informal, but they were lent impetus by the events of 1948-49: the proclamation of the People's Republic of China (PRC), the Berlin crisis, the Tito-Stalin split, and the explosion of the Soviet atomic bomb. As East-West tensions grew, the coordination of export controls took on increasing importance.

Nonetheless, there was far from universal or enthusiastic agreement on the extent of the economic blocade that should be undertaken. After consensus had been reached by the United States, Britain, and France on the general direction of export controls, careful, delicate—and secret—discussions began with other European nations, several of which had doubts about the legality of the proposed embargo measures and many of which were pursuing neutral policies which seemed threatened by participation.

The formulation of the framework of the organization, completed in November 1949, is thus shrouded in secrecy. It is, in fact, doubtful whether any written understanding has ever existed; most likely, a "gentlemen's agreement" was undertaken, members agreeing to follow the licensing rules laid down by unanimous decisions among the group. CoCom began operations on January 1, 1950, with a membership consisting of the United States, England, France, Italy, the Netherlands, Belgium, and Luxembourg. In early 1950, Norway, Denmark, Canada, and West Germany joined, followed by Portugal in 1952 and Japan, Greece, and Turkey in 1953. Nonmembers, Western countries that chose not to compromise their neutrality, include Sweden, Switzerland, Iceland, Austria, and Finland.[1] Of these, Switzerland and Sweden are recognized as major alternative sources of some products and technologies on the CoCom lists. Their relations with CoCom involve an informal, albeit a somewhat unpredictable, cooperation. Neither seems desirous of allowing a sale that would push its relations with CoCom members to a serious confrontation, but neither is particu-

---

[1]See Gunnar Adler-Karlsson, *Western Economic Warfare, 1947-67, A Case Study in Foreign Economic Policy* (Stockholm: Almqvist and Wiksell, 1968).

larly interested in formalizing its cooperation. Generally, these non-CoCom countries favor a liberalization of multilateral export restraints, a view they share with important CoCom members (see chapter IX).

Initially, the organization had two operating entities, the Consultative Group (CG) and CoCom. The function of CG was to set the broad outlines of policy and settle issues of principle. It consisted of ministerial level officials or their personal representatives from all member countries. It was expected that CG would meet only rarely. In fact, it soon ceased to operate at all, and the entire multilateral organization has now come to be known simply as the Coordinating Committee or CoCom. Originally intended to be the operational arm of the system, CoCom was to implement the broad policy decisions made by CG. CoCom now meets in continuous session in Paris. Its representatives are midlevel diplomatic and technical specialists, who deal with the day-to-day problems of the export control system.

During the Korean war a special China Committee (ChinCom) was established to administer restrictions on trade with China and North Korea. For a time, these restrictions were more severe than those that applied to the Soviet Union and its European allies, but this "China differential" was eliminated in 1957 and ChinCom ceased to operate as a separate entity. Now the same CoCom controls apply to both the Soviet Union and PRC.

Because it is an informal and voluntary organization, CoCom has no power of enforcement. It is based neither on treaty nor executive agreement. Its members have no legal obligation to participate in its deliberations or to be bound by its recommendations and decisions. Furthermore, its operations have from the outset been highly confidential and its activities, at least in Europe, attract little or no publicity. It has been suggested that if this were not the case, some non-U.S. members might be forced to withdraw from CoCom, either because of internal domestic pressures or the incompatibility of individual country domestic laws with its controls.[2]

The force that initially brought CoCom into being and held .it together through its formative years was the enormous economic leverage the United States could exert on its Western allies in the immediate aftermath of World War II and the early years of the cold war. But as table 22 demonstrates, the amount of U.S. economic and military aid to the West began to decline seriously after 1955. As it decreased, Western European trade with the East was beginning to rise. In 1949, the combination of U.S. military and economic aid to Western Europe amounted to almost $6.3 billion while the total volume of trade of these Western European countries with Eastern Europe was only about $1.8 billion. It was not until 1955 that the balance shifted and this trade turnover exceeded the amount of U.S. aid. The Battle Act (see chapter VII) had attempted to use U.S. aid as a lever to compel allied compliance on export controls by providing for the discontinuance of U.S. financial assistance to countries that exported restricted commodities to Communist countries. Its sanctions have never been invoked, but whether this was due to its success in limiting East-West trade or to high-level policy decisions to avoid sensitive confrontations is unclear. In any case, the increasing interdependence of all CoCom members with the East at economic, political, and diplomatic levels has by now eliminated whatever leverage actually existed. It is significant, therefore, that in Europe and Japan CoCom is still perceived as a useful institution. This point is discussed further below.

---

[2]Richard T. Cupitt and John R. McIntyre, *CoCom: East-West Trade Relations, The List Review Process,* a paper presented to the International Studies Association Convention, Toronto, March 1979.

Table 22.—U.S. Aid to Western Europe Compared to East-West Trade 1949-55
(in millions of dollars)

	1949	1950	1951	1952	1953	1954	1955
Economic aid..................	$6,276.0	$3,819.2	$2,267.8	$1,349.1	$1,264.9	$ 636.6	$ 466.4
Military aid.....................	—	37.1	604.6	1,013.9	2,866.8	2,225.9	1,541.2
Total.......................	6,276.0	3,856.3	2,872.4	2,363.0	4,131.7	2,862.5	2,007.6
Western European exports to Eastern Europe	832.4	653.3	745.9	742.5	790.9	973.8	1,100.1
Western European imports from Eastern Europe	1,011.7	812.9	1,009.8	995.4	908.7	1,039.4	1,357.9

SOURCE: From Gunnar Adler-Karlsson, *Western Economic Warfare 1947-1967*, p. 46.

THE OPERATIONS OF COCOM

Obtaining a clear picture of CoCom's daily operations, to say nothing of assessing their effectiveness, is complicated by the secrecy which generally pervades the organization and its workings. Basically, the representatives of CoCom engage in three kinds of activities: the development of lists of technologies and products that will be embargoed, controlled, or monitored; weekly consultations on exceptions to these lists; and consultation on enforcement.

THE COCOM LISTS

There are three CoCom lists, organized according to the technical specifications and applications of the items contained on them:

1. a munitions list that includes all military items,
2. an atomic energy list that includes sources of fissionable materials, nuclear reactors, and their components, and
3. an industrial/commercial list.

Most of the activities of CoCom emanate from the last of these. By their very nature, munitions and nuclear materials have clear military purposes and strategic importance, and there is generally little debate over the wisdom of restricting their sale. The industrial list, on the other hand, contains those dual-use items (e.g., jet engines, air traffic control equipment, computers) that, although nominally civilian, have military potential. The technological content of these items is usually high.

The industrial list is subdivided into three categories: International List I (embargoed items); International List II (quantitatively controlled items); and International List III (exchange of information and surveillance items). List I contains those items that member nations agree not to sell to the Communist bloc unless permission is specifically granted after a request for an exception. List II contains items that may be exported, but only in specified quantities. Licenses to export more than the quantity specified for a given item—which may be expressed either in value or in number of units—require special exceptions. List III contains items that may be sold, but over which the exporting nation must maintain surveillance of end use. This information as well as the fact of the sale must be reported to CoCom.

Most of the dual-use items that pose the greatest problems for export controls are contained in List I, which is divided into 10 individual groupings. These conform closely to those on the Battle Act List and the U.S. Commodity Control List (CCL):

1. metalworking machinery;
2. chemical and petroleum equipment;

3. electrical and power-generating equipment;
4. general industrial equipment;
5. transportation equipment;
6. electronic and precision instruments;
7. metals, minerals, and their manufacture;
8. chemicals and metalloids;
9. petroleum products; and
10. rubber and rubber products.

The CoCom list itself is not public information, but it is virtually identical to the national lists of controlled items published by some CoCom members. Furthermore, the American CCL distinguishes between multilaterally and unilaterally controlled items, and the content of the CoCom industrial list can be inferred simply by subtracting the former from the latter.

At the outset, CoCom controls, at least as measured by the number of items on the lists, were quite stringent, although never as restrictive as U.S. unilateral controls. Debate among and within member countries on the relative weight that should be given to security concerns and trade advantages has been continuous. In order to accommodate this debate and to keep lists current, periodic list reviews for purposes of deletion, addition, and amendment were undertaken in 1954, 1958, 1961, 1964, 1967, 1971, and 1974-75. Another review is presently (1979) underway. No details of the decisions made in these reviews or the debates surrounding them are ever published. However, a comparison of U.S. and other national lists indicates that the overall trend in CoCom has been toward liberalization of controls.

Table 23 summarizes one attempt to estimate the changes in the magnitude of the list from its inception through the last completed review. This information was compiled from a variety of sources, including published national lists and private interviews.

As is evident from this table, the length of the list has fluctuated over the years, but the greatest single alteration was the elimina-

Table 23.—Number of Items on the CoCom Embargo List

List as of:	Number of items
November 1949	86
November 1951	270
January 1952	285
March 1954	265
August 1954	170
March 1958	181
July 1958	118
April 1961	NA
July 1962	NA
June 1964	150?
August 1965	161
March 1967	NA
September 1969	156
September 1972	151
March 1976	149

SOURCE: Cupitt and McIntyre, *CoCom: East-West Trade Relations, The List Review Process*, p. 23.

tion of large numbers of items after the cessation of hostilities in Korea and some easing of U.S./U.S.S.R. tensions. The changes in numbers may be attributed both to shifts in East-West relations and to increased pressures from European members to decrease the definition of strategic goods to a minimum in order to foster East-West trade. The stabilization of the numbers that has characterized the list since this time may be interpreted as a sign that drastic reductions are unlikely in the future. According to a 1978 administration statement:

> The process of considering which items meet the strategic criteria has been repeated many times over the years. As a result, changes during list reviews are now seldom dramatic. A few items are deleted and a few new ones added. But most of the changes consist of modernizing the technical descriptions to reflect technological progress.[3]

The procedures by which the 1979 review is being accomplished include the following steps:

• Original proposals for items to be added or deleted were due from member na-

---

[3]Special Report on Multilateral Export Controls, submitted by the President pursuant to sec. 117 of the Export Administration Amendments of 1977, printed in *Export Administration Act: Agenda for Reform*, hearings before Subcommittee on International Economic Policy and Trade, Oct. 4, 1978, p. 53.

tions not later than 4 months before the beginning of the review.

- Counterproposals could be submitted by any member on any item for which an original proposal was submitted, preferably 45 to 60 days before the beginning of the review.
- The review is then completed in two rounds. A draft document is distributed incorporating the proposals and counterproposals agreed to. If this is acceptable to the member nations, the revised list becomes effective after 60 days. Revisions to this draft may be submitted if two countries concur on an item.
- Additional proposals for changes that help achieve consistency among the items may also be submitted at this stage.

The main responsibility for the U.S. contribution to this procedure lies with the Department of State, but each principal department in the export control system is involved in developing U.S. proposals. Interagency Technical Task Groups (TTGs) provide technical analysis and evaluation of all items to be considered. This evaluation is based on the following:

- a laymen's description of items to be considered;
- a comparison of U.S., CoCom, neutral, and Communist country manufacturing capability and availability for each item;
- the potential civilian and military uses of the item and their significance, e.g., their strategic role;
- the technical feasibility of controlling the item including the possibility and ease of substitution;
- changes in use parameters resulting from technological progress and the rate of these changes;
- the present controls on design and manufacture of the equipment;
- the feasibility of reverse engineering, i.e., of extracting the technology from the product and the principal military and civilian uses of the result; and

- identification of critical technologies and keystone equipment.

TTGs draw on both public and private sources for their information, including the Government-industry Technical Advisory Committees (TACs) established to provide private sector guidance on the implementation of the Export Administration Act. One problem with this procedure, however, is that the apparatus of the review system tends to disappear between reviews, sometimes resulting in a substantial delay in generating U.S. positions for the list review.

EXCEPTION CASES

Once a week, representatives of each member nation meet in CoCom headquarters in Paris to consider exception requests. These are, in effect, petitions from firms in member countries to exempt from CoCom control, on a one-time basis, an item that appears on the CoCom embargo list and would otherwise be prohibited from sale. Each member government reviews all such requests and recommends full or partial approval or denial to CoCom, which in turn advises the petitioning nation. Decisions must be unanimous.

In the United States, exception requests from other CoCom countries, as well as applications for exceptions from U.S. exporters, are first sent to the Office of East-West Trade in the State Department. The administration of these requests is handled through the Economic Defense Advisory Committee (EDAC) structure described in chapter VII. This involves a multiagency review that begins in Working Group I with consultation among office director-level representatives of the Departments of State (which chairs the Group), Defense, Commerce, Energy, and the Treasury, with the Central Intelligence Agency (CIA) advising. If consensus on the application cannot be reached in the Working Group, the request is appealed to the sub-EDAC or Deputy Assistant Secretary level and ultimately to EDAC itself on which the Assistant Secretaries of each

Department sit. The criteria for granting these exceptions and the system for evaluating the requests are the same as those applied in the Advisory Committee on Export Policy (ACEP) structure for granting U.S. validated licenses, i.e., the case rests on the technical specifications of the proposed export, the proposed end use and end user, availability outside CoCom, etc.

As tables 24 and 25 demonstrate, both the value and number of CoCom-approved exceptions have risen substantially over the past several years.

The exceptions procedure has played a variety of roles in U.S. policy toward CoCom. In the early years of the organization, U.S. acquiescence in exceptions requests was a means of fulfilling the conditions of the Battle Act without actually having to impose sanctions on Western allies. The Battle Act allowed U.S. aid to be continued to those countries shipping embargoed items to the Communist bloc if a Presidential determina-

tion stated that "unusual circumstances indicate that the cessation of aid would be clearly detrimental to the security of the United States." Exceptions granted by CoCom thus became grounds for determinations which in turn obviated unpopular and possibly divisive U.S. actions against its allies. The fact that no U.S. aid was ever terminated under the Battle Act, therefore, reflects policy decisions on the part of the U.S. Government and does not necessarily indicate that no U.S. ally ever violated the terms of the Act by selling goods and technology unilaterally embargoed by the United States. At the same time, allied sales of U.S. unilaterally embargoed items were possibly deterred by the knowledge that details of the countries of destination, products, and values would be published in the State Department's annual *Battle Act Report*, which provides details of all Presidential determinations to Congress.

The character of the exception procedure has now changed; it is the channel through which export controls can be waived in favor of the requests of U.S. exporters to conduct business with the Communist world. As table 25 also demonstrates, the United States itself generates approximately half of all exception requests; at the same time, it has the reputation of being the CoCom member most concerned with maintaining the strictest possible embargo. Although other nations may have national control lists (see chapter IX) slightly more rigorous than CoCom's, the United States has the longest and strictest unilateral list. It is ironic, therefore, that since the early 1970's the United States has emerged as the major source of CoCom exceptions requests. One reason for this may be in the substantial worldwide technological lead the United States has in computers: computers and computer-related technology are among the most frequent entries on the CoCom embargoed list, and the United States is a preferred or sole supplier of many of these items. U.S. firms have therefore more often had the occasion to seek exceptions for these items than have firms in other countries.

Table 24.—CoCom Approved Exceptions (from all sources, in millions of current dollars)

Year	Value of exceptions
1967	$ 11
1968	8
1969	19
1970	62
1971	56
1972	124
1973	106
1974	119
1975	185
1976	162
1977	214

SOURCE: *Special Report on Multilateral Export Controls*, submitted by the President pursuant to sec. 117 of the Export Administration Amendments of 1977.

Table 25.—Volume of CoCom-Approved Exceptions

Year	Total	United States
1971-75	4,423	2,178
1976	884	432
1977	886	358
1978	1,035	500

SOURCE: Testimony of William A. Root before the Subcommittee on Research and Development, Committee on Armed Services, U.S. House of Representatives, May 24, 1979.

Table 26 shows the dominance of computers in the exception process.

The vast majority of exceptions requests submitted to CoCom are approved. For example, of the 1,380 cases considered in 1974, only 12 were disapproved; in 1977 of 1,087 requests, 31 were rejected. The United States objected in 30 of the 31 disapproved cases. This information for the years 1974-77 is summarized in table 27.

One of CoCom's major problems in dealing with exception cases has been the delays involved in reaching decisions—delays that originate most often in the United States. CoCom procedures call for a decision on an exception request 18 days after it has been submitted, with an automatic 2-week rescheduling in the absence of a decision, and additional weekly extensions at the discretion of the requesting member. The United States has been the major violator of these

procedural rules and the elaborate EDAC review process may very well preclude U.S. compliance with the time constraints.

U.S. delays in dealing with CoCom have involved both exception requests initiated by U.S. exporters and those initiated by other CoCom members. The latter category of cases includes not only exceptions but licenses for the reexport of U.S. origin products. This is the result of the fact that in U.S. export administration law, the Department of Commerce retains jurisdiction over the resale of goods and technologies even after they have been sold abroad. If, for instance, a French firm wishes to reexport an item that it obtained through a U.S. validated license, it must apply to the Office of Export Administration (OEA) for a new license, just as though it were a U.S. firm.

The draft report of the President's Task Force to Improve Export Administration Licensing Procedures has noted that the necessity to obtain CoCom approval added an average of 40 days to the processing time of most U.S. export cases.[4] Delays in deciding exceptions requests submitted by other nations may last even longer. The General Accounting Office (GAO) recently examined 76 test cases of CoCom exception requests submitted to the United States in 1977 and found not only that the United States is the perpetrator of inordinate delays, reserving

Table 26.—CoCom-Approved Computer Requests (in millions of dollars)

Year	Value	Percent of total
1971	$ 21	23
1972	66	39
1973	80	50[a]
1974	120	66[a]
1975	147	64[a]
1976	123	52
1977	168	63

[a]Omitting two exceptionally high-value cases in 1973 and one each in 1974 and 1975, which would distort the figures. None of the four proposed exports actually took place.

SOURCE: *Special Report on Multilateral Export Controls*, submitted by the President pursuant to sec. 117 of the Export Administration Amendments of 1977.

_____
[4]Report of the President's Task Force to Improve Export Administration Licensing Procedures (draft), Sept. 22, 1976, p. 112.

Table 27.—Disposition of CoCom Exception Cases[a]

Year	Approved	Disapproved	Withdrawn	Pending	Total
By number					
1974	1,243(-)	12(0)	16(-)	109(-)	1,380 (567)
1975	1,646(-)	22(0)	36(-)	94(-)	1,798 (798)
1976	884(432)	36(0)	36(7)	101(18)	1,057 (457)
1977	836(358)	31(0)	44(5)	176(55)	1,087 (419)
By value (dollars in millions)					
1976	$162(71)	$24(0)	$187(1)	$73(7)	$446 (79)
1977	214(55.2)	19(0.03)	9(0.03)	74(20.5)	317 (76)

[a]U.S. cases are in parentheses.

SOURCE: Cupitt and McIntyre, *CoCom: East-West Trade Relations*, p. 26.

opinion on more requests than other members and for longer periods, but that the U.S. CoCom delegation is often unable to explain why the system takes so long. It concluded that "the elaborate U.S. review process isn't designed to provide a response within the required time frame; indeed, some U.S. export control officials are not (even) aware of these deadlines."[5]

Delays not only fuel the criticisms of U.S. exporters, but subject the United States to antipathy from abroad. Foreign governments and businessmen have attributed delays by the United States in deciding their exception requests to a number of motives. These range from allegations that the U.S. Government is attempting to provide a commercial advantage for U.S. business by holding up the competition, to recognition of the fact that the cumbersomeness of the U.S. export control system itself is responsible for the tardiness with which the United States processes the requests. In neither case are friendly relations promoted within CoCom. This subject is discussed in greater detail in the concluding sections of this chapter.

ENFORCEMENT

Because no CoCom decision is legally binding on a member nation, all its decisions must be unanimous. Given the informal nature of the organization and the sensitivity of the issues with which it deals, however, little is known and less is publicized about the extent of overt or covert violations of CoCom

---

[5]Comptroller General, *Export Controls: Need to Clarify Policy and Simplify Administration*, Mar. 1, 1979, p. 11.

rules by members. Violations of individuals and firms can only be punished by national legislation; in the United States violations by other member countries are dealt with, if at all, out of the public eye in high diplomatic circles.

CoCom provides for the possibility that a country may determine that it must proceed with a particular export even in the face of a negative CoCom decision by allowing "national interest exceptions." These have been used very sparingly. According to a well-informed State Department official, they have been invoked about 12 times over the past 30 years, with each case having its own unique history and characteristics. It is far more common for problematic cases never to appear before CoCom at all, either because the Government determines that there is no need to submit them or because it prefers to deal with other member governments directly at the highest political levels. This occurred, for instance, when Great Britain first proposed selling Harrier jets to the PRC.

Language differences and the complex technical nature of same exports complicate the issue of compliance but, undoubtedly, covert exports do leak through the CoCom net. Although it is impossible to quantify the extent of this problem, there is a widespread perception among U.S. businessmen that CoCom regulations are not applied equally within all member countries. An often cited example is the sale of French semiconductor technology to Poland, a technology transfer that has probably already benefited Soviet efforts to manufacture semiconductors and that could not have taken place under U.S. export controls.

U.S. AND ALLIED VIEWS OF COCOM

Like any multilateral organization, CoCom is subject to the strains occasioned by the differing perceptions and interests of its sovereign members. But CoCom's unofficial status makes it particularly vulnerable to internal tensions. U.S. businessmen frequently remark on the "laxity" of export controls in other member nations, and charges of deliberate evasion of CoCom restrictions by firms in other countries are not uncommon. But America too is the subject of often heated criticism from its partners.

As part of an investigation of the differing export control policies of major U.S. allies (see chapter IX), OTA has also assembled information on the attitudes of businessmen and Government officials in West Germany, France, and Japan toward the utility of CoCom and America's role in it. The following discussion is necessarily impressionistic; no attempt was made to question a statistically significant sample of officials or industrial representatives in either the United States or abroad. The uniformity of response encountered, however, indicates that the generalizations that follow represent, if not every point of view, at least widely shared opinions.

UNITED STATES

Chapter VII has discussed the fact that many U.S. firms feel that they are under a competitive disadvantage in their dealings with the East. Not only does the export-licensing procedure sometimes subject them to excessive delays in the fulfillment of their contracts, but there is a widespread perception that the domestic control policies of America's major CoCom allies are significantly more liberal and more flexible than those of the United States. This is borne out in OTA's own studies of the export control policies of West Germany, France, Britain, and Japan that appear in chapter IX.

But the operations of CoCom are also discriminatory in the eyes of some U.S. executives who believe first, that the policies of other members give their industries advantages over the United States; and second, that CoCom regulations are not equally or consistently applied among all the member nations. Although American businessmen perceive the need for controls of militarily significant items, they contend that European and Japanese interpretation of this "significance" is much more liberal than that of the United States. This allows foreign firms to export to the Communist world items that U.S. exporters may not sell. In the view of U.S. business, the result is that the Communist nation gets the technology,

and the United States loses the sale. In addition, France, Japan, and West Germany are mentioned frequently as "unfair competitors" because it is thought that firms there often manage to evade or avoid CoCom regulations. Evidence for these allegations is usually circumstantial or anecdotal, and the charges may well be unfounded. Nevertheless, a significant number of highly placed U.S. executives appear to believe that firms in other Western countries consistently flout CoCom regulations, and that this often occurs with the active or passive connivance of the Governments involved.

Sources in several U.S. Government agencies have privately expressed similar views, but documentation of CoCom violations by member governments is universally regarded as highly sensitive material. The common position expressed by U.S. Government officials is that, given its informal nature and the constraints under which it operates, CoCom works surprisingly well. Problems between governments have traditionally been worked out quietly at very high policy levels, and this practice should continue. Attempts to formalize or significantly strengthen the organization are generally regarded as unwise, for they would almost certainly be resisted. As the following sections indicate, this apprehension was amply confirmed in discussions with individuals in other member countries. The impression gained is that pressure for any major changes in the organization might precipitate its demise.

WEST GERMANY

West German officials agree that CoCom is a necessary and useful organization, inasmuch as it prevents the export of strategic technology to Communist nations. No one in Bonn questions the need to embargo exports of military technology. Moreover, some officials feel that, were it not for CoCom, the Japanese would be less vigilant about preventing strategic technology exports to the Eastern European nations—Tokyo does not feel threatened by such distant countries.

The Germans also support the principle of coordinating Western export policies in a multilateral organization. Despite the approval of CoCom's aims, however, there is widespread feeling that CoCom must be reformed and made less cumbersome, that there must be more genuine equal competition on the industrial list, and that this list must be modified to take into account considerable technological progress within Eastern Europe.

The most frequent criticisms expressed were not of CoCom itself, but of the American role in CoCom, in particular the delays involved in reexport licenses, and inconsistencies in U.S. policy. The reexport-licensing process is often unpredictable because not only are the American national export control lists more restrictive than the CoCom lists, but licenses can be denied for *ad hoc* political reasons, which may be incomprehensible abroad. Officials cited with some bewilderment, for instance, President Carter's decision in July 1978 to deny Sperry-Univac an export license, and his subsequent reversal of this policy in April 1979, suggesting that these developments do not help the functioning of CoCom.

Moreover, there is a general feeling in Bonn that the various agencies of the American Government that deal with licensing questions are uncoordinated, further exacerbating the problem of delays and unpredictability and giving the impression of lack of policy direction, and that the American criteria for granting licenses are unclear. Regardless of the accuracy of these perceptions, the predominant view in West Germany seems to be that the American lists should be brought more in line with the CoCom lists; the license-granting procedure in Washington rationalized; and the criteria for making decisions on reexport licenses not be determined by the current state of human rights in the U.S.S.R.

Another German concern is the future treatment of the PRC within CoCom. There is a fear that if there is a new "China differential," this time to China's advantage, CoCom may well disintegrate. This is based on the feeling that, in order for CoCom to be a viable organization, all Communist countries must be treated equally. By the same token, therefore, there should be no differential for Eastern European countries (for instance, Romania) within CoCom, because any technology exported to Eastern Europe may find its way to the U.S.S.R. German spokesmen emphasize that they favor more discussion in CoCom of these issues. Finally, there is a general belief that German corporations, out of purely commercial motives, often act as unofficial watchdogs to ensure that controlled technology is not exported by a company from another country that has evaded CoCom rules. Thus, there are mechanisms other than CoCom for ensuring that the latest technology does not leak to the East, not the least of which is the Western companies' concern not to create potential competitors for their domestic markets in the U.S.S.R. This is an argument that has also been made by U.S. firms. As chapter III demonstrates, however, it is not indisputably clear that such "enlightened self-interest" always does protect domestic industry from economic, let alone strategic, repercussions.

German officials agree that if CoCom were to be made a formal organization it would probably collapse, because other members (notably the French) would not agree to acknowledge it publicly. If CoCom had to be ratified in national parliaments, the embargo policy would have to be openly discussed and justified, which would be counterproductive to its effectiveness. Thus, to the Germans at least, CoCom must remain informal.

FRANCE

Despite often vociferous criticisms made by French officials, the attitude towards CoCom in Paris is not uniformly negative. One CoCom official characterized it as a "useful organization because it acts as a brake on technology transfer." Most officials agree that it works as a "gentlemen's club," and that, in view of its unofficial status, it functions rather well. No Government official questions the need to control weapons exports to the Communist nations for reasons of national security; the bulk of the criticism of CoCom arises from its attempts to limit dual-use technology exports, particularly in light of the current industrial list reviews.

Apart from the more general problem of complying with American-inspired controls, French spokesmen make a variety of other criticisms of CoCom. It is, according to some, a cold war vestige, which is "against the spirit of detente." Moreover, they claim, it is politically counterproductive to have a multilateral strategic embargo directed against the U.S.S.R. More specifically the French regard the whole CoCom procedure as overly burdensome with far too many detailed technical discussions. The general French preference would be to change the system and make CoCom a loosely structured framework organization with general guidelines, avoiding case-by-case discussions of specific license requests. This would give the individual members of CoCom more autonomy within the organization and make it more acceptable. As one spokesman said, "We want to be able to judge for ourselves about how to protect France from the Russians."

Like the Germans, French officials are concerned about how CoCom will deal with China in the future, but in general the U.S.S.R. is a more important market for France than is China. There was some criticism of Britain's eagerness to sell "defensive" Harrier jets to China, since the difference between offensive and defensive planes was considered to be academic. There was a general feeling that it would be counterproductive to give China special treatment in CoCom, because this would make a mockery of the security interests that the organization is supposed to protect.

Most French officials feel that there is a large element of commercial rivalry that influences CoCom decisions. As one spokesman put it, "When the real interests of a country are concerned, CoCom doesn't matter." While other CoCom members cite France as the most frequent violator of CoCom regulations, French officials deny this charge and claim that other countries find ways of avoiding CoCom regulations—especially Great Britain. In addition, according to some French spokesmen, American multinational corporations use their subsidiaries in Europe to produce goods that are exported to Communist nations with European credit support and are not affected by the more stringent American national export controls.

Since much French technology is of American origin, French corporations are often dependent on American reexport licenses. A frequent complaint is that U.S. corporations have a double advantage—they can avoid American credit restrictions by exporting through European subsidiaries, while the U.S. Government can delay French exports through the reexport license system, thereby giving American corporations a competitive edge. Moreover, a Soviet predisposition towards American technology is thought to enhance the lead of U.S. firms. In light of these perceptions, it is perhaps surprising that there was so much French criticism of President Carter's initial decision to deny Sperry-Univac an export license for the Tass computer. The French, after all, profited from this decision when the contract eventually went to a French firm.

Some officials also claim that CoCom can be used by those whose products are less highly developed to prevent other members from gaining commercial leads. For instance, if Japan has a monopoly on a state-of-the-art piece of technology and wants to export it, another CoCom member may object to granting the Japanese CoCom permission to export the product. When that country itself has developed the technology, however, it may reverse its position. Thus the French feel that there is much hypocrisy in CoCom: "If I do not have what you have, I shall prevent you from selling it until I acquire the product too."

Finally, the French agree that commercial rivalry can be one of the best guarantees against exporting the most sophisticated technology to the U.S.S.R., and a more effective deterrent than CoCom. The threat of potential competition is a built-in economic incentive for corporations to guard against selling the U.S.S.R. the latest technology, and will cause them to carefully watch companies in other countries. There is therefore a self-regulating international mechanism based on commercial interest, which can detect violations of CoCom regulations. However, the motives for this vigilance are purely commercial; French corporations are not generally concerned about the security aspects of technology exports.

According to all those interviewed, the opinions expressed here represent the current consensus of interested parties, and a new government, whether leftist or rightist, would not implement fundamentally different policies with regard to the transfer of technology to Communist countries. The desire to secure French jobs through exports and to maintain good relations with Communist nations is shared by all elements of the political spectrum.

JAPAN

Japan has historically evinced a willingness to adhere to CoCom and to incorporate the CoCom list of embargoed goods into its own list of licensed exports. This policy can be traced in part to the forces that propelled Japan into CoCom at the outset. The role of U.S. aid in Japanese postwar economic recovery, the importance of the American political and military "umbrella" for Japanese security, and the psychological dependence bred by the trauma of wartime defeat left their impression on successive Japanese Governments until well into the 1960's. Each followed the American lead in basic matters of foreign policy, sometimes under other pressure, sometimes not. Such factors were reinforced by legitimate security concerns on the part of the Japanese, given the absence of a peace treaty with the Soviet

Union; ongoing political disputes with the U.S.S.R. over the Northern Territories and fishing grounds; and relations with the PRC, which were erratic at best.

But it is by no means certain that Japan will continue to accept CoCom constraints on its trade with the U.S.S.R. and China. The expansion of trade with both the Soviet Union and the PRC since 1968, the Nixon "shocks" of 1972, and the diplomatic rapprochement with China in 1978, have probably weakened many of the forces that led Japan to initially adhere to the CoCom system.

Another important consideration is the strong Japanese "policy sentiment" against exporting arms, nuclear weapons, or anything of obvious military significance. Such sentiments, rooted as they are in the postwar psychology of leaders and masses alike, are reinforced continually by eminently pragmatic considerations. The Asian nations, which in 1977 accounted for fully one-third of Japan's export market and over one-half of its sources of imported commodities, would react economically as well as politically to any signs of revived Japanese militarism or other elements of a "Greater East Asia Co-Prosperity Sphere." The point here is simply that CoCom or no, Japan probably cannot afford to export obviously militarily relevant products. Further, the "unilateral" controls imposed on any export mix by the characteristics of Japan's own history may mitigate whatever opposition to CoCom restrictions of exports otherwise exists.

A case can also be made that CoCom is useful to Japan in a variety of ways that have little to do with U.S. perceptions of security. It fits well with Japan's efforts to develop a policy of "even-handedness" in dealing with its two powerful Communist neighbors. In the event of pressure from either (more likely from the Chinese) for Japan to sell ships, aircraft, or machinery with clear strategic significance, the CoCom list provides a handy rationale for a refusal.

Second, the Japanese still have genuine policy interests that lead them to tread warily so far as the Soviet Union is concerned. If anything, to judge from Prime Minister Ohira's statements on the eve of his visit to the United States in May 1979, political concern over the U.S.S.R.'s military posture in the Far East is becoming increasingly pronounced among the Japanese leadership. The Northern Territories have taken on increasing strategic as well as economic importance in the eyes of both parties to the dispute. Thus, despite their principled refusal to use trade for political leverage, successive Japanese Governments have resisted the kind of long-term trade commitments that the Soviets have been urging on them. Further, according to Soviet sources, the Japanese have been dragging their heels in implementing an intergovernmental agreement on scientific and technical exchanges. The agreement was signed in 1974, but no timely steps were taken to implement it.

In the last resort, CoCom arrangements may be maintained, primarily because there is no effective domestic political coalition pushing for their elimination. The historically pro-Soviet stance of the Japanese Communist Party is now balanced by the anti-Soviet posture of the Japanese Socialists, and Japan's enormous quasi-public corporations present a consensual front regarding trade policy—at least as far as can be detected by the outsider. There is no Japanese equivalent of the public debate on East-West technology transfer that has occurred in the United States.

But other factors may be at work undermining the CoCom consensus from the Japanese perspective. Japan's economic recovery has generated new areas of economic competition between itself and other Western nations, especially in areas where U.S. goods and technology have been predominant (e.g., large computers). It is possible that the Japanese will be increasingly disposed to interpret American behavior in CoCom as motivated more by economic considerations (i.e., preserving U.S. markets in the East) than by intelligible security-strategic calculations. Meanwhile, growing Western protectionism against Japanese goods and import restric-

tions now arising in Western Europe may render the Communist bloc increasingly important as a market for Japanese exports.

In terms of domestic economic forces, the depressed state of Japan's shipbuilding industry may have already inclined certain Japanese firms to push for a relatively more liberal approach on the part of the Japanese Government towards the export of vessels with possible military application. Japanese firms as a whole have made a major contribution to the aggregate of the Soviet merchant fleet, as well as contributing to increases in the mean size of its vessels (one indicator of a successful modernization program). As American naval strategists have frequently noted, a major ingredient in Soviet strategic thinking in the area of naval warfare is provision for rapid conversion of its merchant marine to military purposes. Under such circumstances, controversies over naval exports may become more frequent.

Such factors, when combined with a diminished American presence in Asia (the return of Okinawa, the withdrawal from South Vietnam, the reduction of the military presence in South Korea) and Japanese rapprochement with the Chinese may sooner or later yield a situation in which traditional American arguments regarding the need for bolstering security arrangements against Communist States in all areas, including trade, will ring with decreasing conviction on Japanese ears. At the same time, growing Japanese self-reliance in defense matters may engender an increased R&D effort to develop a homegrown weapons technology and an adequate production base. Western experience in this area has shown that under such circumstances the impulse to recoup investments through the sale of related products, abroad as well as at home, subsequently becomes quite strong. How compatible this impulse might be with CoCom restrictions remains to be seen.

All the same, it is far too early to speculate that the balance will shift decisively in the direction just described. A number of countervailing factors still operate and their combined impetus promises to maintain the CoCom arrangements, including Japanese participation, intact. Whether or not such forces are adequate to sustain any U.S. attempts to formalize the consensus is a separate question. The first such factor is the comparatively small proportion of Japanese trade with the Communist countries. Although the absolute amount has been growing steadily, the percentages have remained generally stable at a low level (see chapter IX).

In sum, the future of the Japanese position on the subject of export controls, while laced with a number of uncertainties and contradictions, still shows no firm evidence of a collision course with American policy as it stands today. By the same token, there appears to be little or no sympathy for the idea of tighter controls over either end products or technology.

THE FUTURE OF COCOM

Any assessment of CoCom must acknowledge its surprising longevity. Despite its informal status and the increasingly divergent interests of its members, it has functioned—and functioned with reasonable success—for nearly 30 years. If many nations regard the continuance of CoCom with little enthusiasm, there at least appears to be similarly little enthusiasm for its dismemberment. In the face of the military risks that might accompany its disappearance, CoCom is tolerated.

But it is undeniably the predominant U.S. interest in CoCom that holds it together, and it has become an increasingly controversial organization in the last decade, its purpose and methods questioned. The preceding discussion has explored some of the important issues and strains within CoCom. It is useful to summarize the current state of debate within the organization in order to consider the directions it may take in the future.

Japan, Germany and, with more ambivalence, France continue to accept CoCom as a necessary and useful organization for the control of military exports to Communist nations.[6] Even if there were no CoCom, they

would restrict armament exports for national security reasons.

On the question of nonmilitary technologies, however, there is more skepticism about U.S. motivation and the efficacy of CoCom. None of the foreign countries investigated has an explicit technology transfer policy toward Communist nations; in none is East-West technology transfer a matter of public debate. There is a general feeling that no Western country has a monopoly on any form of technology; there are always alternative suppliers for Communist nations. Officials in each country argue that if they are not allowed to sell a technology to the U.S.S.R., it will purchase the equivalent elsewhere. Furthermore, given the lack of a coordinated Western technology transfer policy, the perceived impossibility of arriving at a general agreement in the West on which nonmilitary technologies should be controlled, and CoCom's inability to invoke sanctions against violators, the overall inclination is to export most nonmilitary technologies to the East.

This, according to U.S. businessmen, is the point at which official U.S. policy diverges from that of its allies—to the detriment of the interests of the U.S. economy. The United States has always maintained a wider interpretation than its allies of military significance, and has at times attempted to control a large range of items

[6]While France only reluctantly concedes its membership, several European officials reported that Italy is even more reticent about admitting that it is in CoCom—although the President of CoCom is Italian. The fact that the Italian Communist Party received just under one-third of the vote in the last general election may not be unconnected to this reticence.

because of what might seem very indirect military implications. (For a complete discussion of the gamut of military relevance, see chapter V.) So long as these differences continue, U.S. firms will be prohibited from exporting items that, in Europe or Japan, would raise no national security problems.

Further, most European officials agree that there is no satisfactory all-purpose definition of technology that can be applied to export controls in a realistic operational way. U.S. attempts to shift the emphasis of control from end products to design and manufacturing know-how have led other CoCom members to object that their Governments lack the legislative authority to control the sale of "know-how," as opposed to equipment. The general view is that one cannot separate technology from the product and that attempting to switch from end product to technology control will aggravate CoCom's problem.

Some businessmen in Europe remain skeptical about the value of the end-use statements and safeguards demanded by CoCom. It is difficult to take action against a Communist end user who violates the signed consignee statement, and even harder to detect possible diversions of equipment from civil to military use. Many businessmen would prefer a less elaborate system with fewer safeguard obligations. An American official discussing these problems admitted that end-use statements may have limited effectiveness, but said "it makes us feel good" to receive the signed end-use letters. Moreover, there is a general agreement that there are various ways for corporations to evade CoCom through American multinational subsidiaries in Europe and through third countries, although it is obviously impossible to fully document these evasions.

Both the Europeans and Americans accuse each other of hypocrisy in CoCom. The Europeans claim that, on the one hand, the United States upbraids them for not being vigilant enough in their export control, for being too crassly commercial, and for not considering political factors enough in their export policies; on the other hand, America, which possesses the most advanced technology, submits more exception applications to CoCom than any other country. Moreover, cases such as that of the TASS computer inevitably lead to charges of U.S. commercial motivation in first holding up an export license and then granting it after corporations in other countries have been deterred from pursuing it. American officials point out that while the Europeans criticize the United States for its overly political attitude towards East-West technology transfer, they profit commercially from its more restrictive East-West trade policy by securing orders that might otherwise have gone to U.S. firms.

The skepticism about the degree to which one can or should control the transfer of technology to Communist nations is reinforced by other lingering suspicions of American motivation in Western Europe. A brief discussion of two separate cases—the 1962-63 North Atlantic Treaty Organization (NATO) pipe embargo and a West German nuclear powerplant deal—highlights these concerns.

In November 1962, the United States attempted to impede the completion of the Friendship oil pipeline from the U.S.S.R. to Eastern Europe by preventing its allies from exporting large-diameter steel pipe to the Soviet bloc. The pipe embargo order was passed in NATO because the United States knew that the British would not agree to it in CoCom. Several instructive points emerged from this attempt to prevent export to the U.S.S.R. Washington was able to prevail on the West German Government to force several German corporations to cancel already concluded deals for the sale of pipe. This naturally caused an outcry in the German business community. However, since the United States did not have as much political leverage over its other allies as it did over Bonn, it was unable to prevent Great Britain, Italy, or Japan (which, of course, was not in NATO) from selling similar pipe to the U.S.S.R. Britain, in particular, chose to ignore the NATO directive. Ultimately, not only did the U.S.S.R. find alternative sources of supply for pipe, but the embargo induced it to step up the development of its own pipemaking capacity. The completion of the Friendship pipeline was delayed by only a year as a result of the embargo. Moreover, there was a strong suspicion in most European capitals that the real American motivation for the embargo was not fear of enhancing Soviet military capabilities, but rather the wariness of U.S. oil companies about the U.S.S.R. dumping cheap oil on the West European market. Whatever the truth of this allegation, and it may well have been utterly unfounded, it made a powerful argument. The net result of the pipe embargo was damage to U.S.-West European relations and only a marginal effect on the Soviet ability to complete an oil pipeline that supplied Red Army troops in Eastern Europe.

The pipe embargo has not been forgotten in Western Europe and the issues it raised remain important to the perceptions of Western Europeans of the utility of export controls and of the U.S. role in them. The basic themes that recur are the suspected hypocrisy of the United States, its overzealousness in enforcing stricter controls than necessary, and the basic ineffectiveness of such controls in preventing the acquisition of technology or technical capacity in the Communist bloc.

Similar issues resurfaced in 1973, when West Germany and the U.S.S.R. settled the economic aspects of a major project in which the West German Kraftwerkunion would build a $600 million, 12,000 megawatt nuclear powerplant in Kaliningrad. The Soviets undertook to supply West Germany with electric current from this plant. Both the United States and Britain raised objections to the deal in CoCom in January 1975, claiming that the issue at stake was inspection of nuclear facilities by the International Atomic Energy Agency: the Soviets had never agreed to onsite inspections as part of the nuclear nonproliferation treaty. The United States also raised doubts about the

security aspects of selling nuclear technology to the U.S.S.R. West Germany, however, claimed that the Soviets already possessed nuclear power technology. The situation was complicated because West Germany imports enriched uranium from the United States, and it feared that Washington might embargo exports of uranium to West Germany as a means of preventing the deal.

In March 1976, the West German Government announced that the Kaliningrad project had been abandoned because of insuperable difficulties. The ostensible reason was the failure of West Germany and the U.S.S.R. to agree over electricity supplies from the plant to West Berlin. However, there was much speculation that the real reason for the cancellation of the deal was American objections in CoCom.[7] Moreover—and here similarities to the 1962 pipe embargo case are clear—there was speculation that the real reason for U.S. objection was not fear of enhancing Soviet nuclear capabilities, but commercial rivalry. Westinghouse had outbid the German Kraftwerkunion (jointly owned by Siemens and AEG-Telefunken) in reactor sales to Spain and Yugoslavia, and was reportedly also interested in the Soviet deal. It is equally obvious, of course, that if CoCom prevented the sale, Westinghouse stood less chance of selling nuclear reactors to the U.S.S.R. in the future. Nevertheless, whatever the truth of the matter, it is significant that American motives in CoCom were still perceived as dominated by an interest in controlling commercial competition rather than retarding Soviet technological development and protecting Western security. Some have even argued that the CoCom procedure consti-

---

[7]Hanns-Dieter Jacobsen, "Die Entwicklung der Wirtschaftlichen Ost-West Beziehungen als Problem der Westeuropaeischen und Atlantischen Gemeinschaft," *Stiftung Wissenschaft und Politik* (Ebenhausen, 1975), p. 70; Simone Courteix, "Le Comite de Coordination Des Echanges Est-Quest (CoCom)," *Annuaire de l'URSS et de Pays Socialistes Europeens*, pp. 1-7. German Government and business officials, in interviews, disagreed about whether CoCom had given final approval for this project.

tutes a forum for intra-Western commercial espionage by providing market opportunity information.

These charges may have been politically motivated and entirely unfounded. It is significant, however, that U.S. behavior has permitted this kind of interpretation, although recently Europeans seem increasingly willing to attribute U.S. recalcitrance and delays to the complex and time-consuming U.S. export control procedures rather than to sinister commercial reasons.[8] This may or may not be regarded as a positive step in U.S. relations with its CoCom partners.

In a larger context, many European officials question the utility of embargoes in general, and of the CoCom embargo in particular. Some argue that the most effective embargo against the U.S.S.R. would be an embargo of wheat as opposed to technology. The Soviets cannot produce wheat if there is a bad harvest, but they can always manufacture equipment, however primitive.[9] Disagreements about the effectiveness of embargoes depend not only on one's evaluation of the U.S.S.R.'s ability to absorb and diffuse technology (see chapters IV and X). They once again invoke the problem of defining national security and determining how directly useful to the military sector a technology must be before it is considered "significant." In addition, some argue that, were it not for CoCom, the U.S.S.R. today would be a more formidable antagonist militarily, and that it has made most of its technological breakthroughs by obtaining Western technologies. A counterargument holds that it is better to create interdependencies through the export of technology to the U.S.S.R. than to force it to become more technologically self-sufficient through a strategic embargo. According to this view, technology transfer to the East is a factor for stability to be balanced against the potentially destabilizing economic effects of a continued complementary trade structure between East and West.[10]

If there were no CoCom, would things be different? In Europe and Japan, the general answer appears to be "yes, but not much." Government officials in West Germany, France, and Japan insist that they would have their own national controls on the export of military technology. Granted, the absence of multinational controls might mean that some military hardware would leak through to the Communist countries. In addition, European governments would probably lessen controls on industrial exports that contain dual-use technology. Most important, the value of regular and continuous channels of communication between exporting nations would be lost. However, given the problems of trading with Communist nations and their lack of suitable exports to pay for Western imports, the economic limits to East-West trade would act as a barrier to any great increase in technology exports to Communist nations (see chapter III). Businessmen in all countries agree with this assessment. Even if export restrictions were removed, they say, the problems of selling to the U.S.S.R. would limit the amount of technology exported. One computer executive gave as an example of this problem a Soviet offer to have his firm build a computer in Sverdlovsk. Even if there was no CoCom, he claims, the logistical problems of dealing with the Soviet bureaucracy and installing machines in the provinces would deter him. Thus, there are economic deterrents to East-West technology transfer that exist regardless of the political environment.

In sum, so long as the Soviet Union is perceived as presenting a threat to Western security, there will be general agreement in Western Europe about the need to embargo exports of directly military technology. However, the question of the transfer of

---

[8]Comptroller General, *Export Controls: Need to Simplify Administration*, op. cit., p. 11.

[9]For a discussion of this theory, see Marie Lavigne, *Les Relations Economiques Est-Quest* (Paris, 1979), pp. 73-77.

[10]Friedemann Mueller, *Sicherheitspolitische Aspekte der Ost-West Beziehungen* (Security Aspects of East-West Trade), *Stiftung Wissenschaft und Politik* (Ebenhausen, March 1977).

dual-use technology may become more controversial. The crucial issue to the future of controlling technology for national security purposes will therefore center on the question of the ultimate contribution of such technologies to Communist military and strategic capabilities. It appears that America is at present more seriously concerned than its CoCom allies over this problem. Certainly, the German approach to detente is somewhat different from that of the United States, as the recent U.S.-West German disagreements at the Belgrade followup to the Conference on Security and Cooperation in Europe showed. In the current international climate, exports of technology are seen as good business in West Germany, beneficial to the economy, and helpful to West German foreign policy inasmuch as they promote a better atmosphere in relations with Communist nations. They are not portrayed as a potential threat to national security; indeed uncertain U.S. leadership is perceived by some as a greater security danger than the transfer of technology. The French and Japanese take a similar view, and go even further in divorcing technology transfer from political considerations. The fact remains, however, that CoCom has remained a viable, albeit imperfect, institution for some 30 years, weathering disagreements among its members over both practice and policy. The United States may continue to depart from its partners, but in the absence of attempts to drastically alter the organization or of events that dramatically affect allied foreign policy, there is little reason to expect much change in CoCom—if it is unlikely to adopt more stringent restrictions, so too is it unlikely to disappear.

CHAPTER IX

The East-West Trade Policies of America's CoCom Allies

CONTENTS

The East-West Trade Policies of America's CoCom Allies

East-West trade has always been more important to Western Europe and Japan than to the United States. Economic and political imperatives in other Organization for Economic Cooperation and Development (OECD) nations have combined to create a generally favorable attitude toward trading with the Communist world, an atmosphere that prevails today in both government and business circles. It has also carried over to Western European and Japanese views on export controls and technology transfer. Indeed, while the issue of technology transfer to Communist nations is a matter of controversy in the United States, such debate is virtually nonexistent in Japan and most West European countries; they accept trading with Communist countries as a more or less normal part of foreign economic policy. For a variety of reasons, America's European and Japanese allies do not necessarily share its concern over the economic and security problem raised by trade and technology transfer to Eastern Europe, the U.S.S.R., and the People's Republic of China (PRC). This fact has relevance to the debate within the United States over East-West trade and technology transfer policy.

WEST GERMANY

INTRODUCTION

Current West German policies towards trading with the East are best understood in their historical context.

Germany has been transferring technology to Russia for well over a century. The long historical tradition of Russo-German economic interdependence is characterized by the German export of machinery in return for imports of Russian raw materials. Table 28 demonstrates the degree to which the two economies were oriented towards each other prior to World War I.

Bilateral trade was always more important for Russia than for Germany, and after the Bolshevik Revolution, the Soviet Government continued to seek German machinery imports. In fact, there was considerable

Table 28.—Russo-German Trade Percent of Total Russian Imports and Exports

	Russian imports	Russian exports
1858-62	28%	16%
1868-72	44	24
1914	47	29

SOURCE: Juergen Kuczynski and Grete Wittkowski, *Die deutsch-russischen Handelsbeziehungen in den letzten 150 Jahren* (Berlin, 1947). pp. 24, 25.

German-Soviet clandestine military cooperation.[1] Although the volume of trade between the two countries declined after the Nazis came to power, German-Soviet economic cooperation continued until the 1941 Nazi invasion of the U.S.S.R. After the War, for political reasons, West German trade was

---

[1]Adam B. Ulam, *Expansion and Coexistence* (New York, 1969), pp. 152-153.

largely reoriented away from the East, but the historical legacy of close economic ties has profoundly affected present German attitudes towards trading with Russia. Despite the cold war, many West German businessmen look on the U.S.S.R. and Eastern Europe as natural and desirable markets for their manufactures.

The German business community has long favored trading with Communist countries and separating political from economic relations with the U.S.S.R., but early postwar West German Governments did not take this view.[2] Prior to the election of Chancellor Willy Brandt in 1969 and the inauguration of a new *Ostpolitik*, German Governments led by the Christian Democrats (CDU) looked upon trade with the East as a primarily *political* problem: since trade with the East was relatively unimportant economically, it should be controlled and used to promote West Germany's political goals. The most important of these was the reunification of Germany under a Western system or, failing that, a strengthening of East Berlin's ties with Bonn. Governments of West Germany under Chancellors Adenauer, Erhard, and Kiesinger attempted to implement linkage strategies, making trade conditional on Soviet political concessions on the German question. Both negative linkage or trade denial (the predominant form of leverage prior to 1969) and positive linkage or trade inducement were attempted, but these produced only marginal compromises, and no substantial Soviet political concessions.[3]

As early as 1952, however, an organization of businessmen was created to promote and coordinate East-West trade. Known as the *Ostauschuss der deutschen Wirtschaft*, or Eastern Committee, the group was sponsored by the powerful Federation of German Industry, and charged both with furthering trade with the Communist countries and with representing West German business on that subject before the Government. The *Ostauschuss* has always had a semiofficial character; it has nevertheless disagreed publicly with some Government attempts to restrict trade. It remains today an important organization for furthering East-West trade, although some German firms claim it has lost momentum in recent years. The *Ostauschuss* favors technology exports to the East, but it also supports strict German adherence to the Coordinating Committee for Multilateral Export Controls (CoCom).

Adenauer in particular felt that since the U.S.S.R. was West Germany's political antagonist, it was wrong in principle to sell it any products that could enhance its military capacities. Until 1963, the various Adenauer administrations used political and security arguments against trading with Communist nations, claiming that such trade was dangerous and economically unimportant. That part of the business community engaged in trading with Eastern countries countered with economic arguments in favor of trade, holding that it was economically beneficial for German industry and could even have positive political consequences, by creating desirable interdependencies and giving the U.S.S.R. a stake in the stability of the West.

The clash between Government and industry over exporting to Communist nations peaked in the late 1950's, although as late as 1963 Chancellor Adenauer upbraided those German businessmen exporting to the Russians for their disloyalty to the German reunification cause. Soon after erection of the Berlin wall, he also criticized the United States for selling wheat to the U.S.S.R.: "I can't stand any more of this wretched talk of detente."

---

[2]In 1954, a survey by Gabriel Almond found that, on the question of trade with the East,

> It remains interesting that the business community in Germany is relatively unaware of the priority of political factors in Communist policy-making. Only government officials and some leaders of business pressure-groups seem to see this point. Few of the industrialists, even in the largest establishments, are aware of it. Their thinking about the possibilities of the Communist market is dominated by simple, apolitical economic calculation.

See Gabriel Almond, "The Politics of German Business," in Hans Speier and W. Phillips Davison, eds., *West German Leadership and Foreign Policy* (Evanston, Ill., 1957), p. 237.

[3]For a fuller discussion of the policy of linkage in West German-Soviet relations, see Angela Stent Yergin, "The Political Economy of West-German-Soviet Relations, 1955-1973" (unpublished Harvard University Ph.D. dissertation, 1977).

In the mid-1960's, the Erhard and Kiesinger administrations retained a basically restrictive attitude towards trade with the Communists, but attempted to use it to induce greater polycentrism in Eastern Europe. Thus, they offered more favorable credit terms to Eastern Europe than to the Soviet Union. The business community was largely opposed to this policy, favoring the total separation of trade and politics.

After Willy Brandt's election, the Social Democrats (SPD) took over the formulation of *Osthandel* (trade with the East) policy. In general, they favored a depoliticization of trade and eschewed negative linkage, although they were not averse to employing positive linkage strategies and offering economic incentives in return for relatively minor political concessions. Since 1969, the German Government and the business community have converged in their desire to separate trade and politics. Whereas previous Governments intervened to hinder East-West trade (e.g., by preventing the granting of credits, or by canceling specific orders), present ones frequently act to facilitate it. In several of the biggest deals involving the export of technology to the U.S.S.R. (for instance, the Kursk deal, described below) the Government has pressured the Soviet Government to accept terms favorable to German corporations. Since 1952, the *Ostauschuss* has consistently favored trade with Communist nations while warning against exaggerated expectations. During the last Brezhnev visit to Bonn, the business community was noticeably cooler than the Government towards the 25-year economic agreement signed by Brezhnev and Schmidt. Ironically, the business community is sometimes now more skeptical than the Government about the value of increasing trade with Communist nations, including the benefits to German employment.

As the West German Government has relaxed its political restrictions on trade with the Eastern bloc, the U.S. Government has been moving in the opposite direction. As a result, the United States and West Germany disagree with increasing frequency over the politics of East-West trade. The reaction of a German Government spokesman to the Jackson-Vanik amendment illustrates the prevailing German attitude toward U.S. efforts to use trade for political purposes in dealing with Communist countries:

> A policy like the one Congress thought was right or like what our own opposition occasionally recommends, cannot only fail to achieve the desired goal, but can even make it more difficult.[4]

ECONOMIC FACTORS

West Germany is heavily dependent on oreign trade, which accounts for 30 percent of its gross national product (GNP). A healthy export sector is a vital component of its economic viability. Since West Germany is far more heavily trade dependent than is the United States, it tends to favor exports regardless of the destination. West Germany's postwar economy was built on Chancellor Erhard's implementation of the *Soziale Marktwirtschaft* (social market economy) theory, designed to create a truly competitive market. In the 1950's, exports of manufacturers and engineering goods led the way to remarkable economic growth, and German officials and businessmen generally assume that exports of technology are necessary for continued economic growth. This applies to exports to the East.

In 1978, total West German trade with Communist countries (excluding East Germany and including China, North Korea, and Mongolia) amounted to over 30 billion Deutschmarks (DM) (roughly $15 billion, see table 29). This was 5.7 percent of total West German foreign trade. Trade with East Germany, which West Germany considers "inner-German trade," rather than foreign trade, came to 8 billion DM. The U.S.S.R.

---

[4]*New York Times,* Jan. 18, 1975. The West Germans have succeeded, behind the scenes, in securing the emigration of about 60,000 ethnic Germans per year from the U.S.S.R. and Eastern Europe. In 1978, 58,000 emigrated: 36,000 from Poland, 12,000 from Romania, 8,500 from the U.S.S.R., 900 from Czechoslovakia, and 500 from elsewhere. After the Jews, ethnic Germans are the largest group allowed to emigrate from the U.S.S.R.

Table 29.—West German Trade With Communist Nations, 1970 and 1978
(million DM, figures rounded off)

	1970				1978			
	Total	Import	Export	Balance	Total	Import	Export	Balance
U.S.S.R.	2,800	1,254	1,546	+ 292	11,707	5,406	6,301	+ 895
Poland	1,402	744	658	− 86	4,732	2,086	2,646	+ 560
Czechoslovakia	1,785	727	1,058	+ 331	3,178	1,357	1,821	+ 464
Hungary	1,012	490	522	+ 32	3,487	1,293	2,194	+ 901
Romania	1,302	580	722	+ 142	2,983	1,214	1,769	+ 551
Bulgaria	477	237	240	+ 3	1,032	314	718	+ 404
East Germany	4,548	2,064	2,484	+ 420	8,820	4,066	4,754	+ 688
China	1,001	389	612	+ 223	2,723	734	1,989	+ 1,255
Total	14,327	6,485	7,842	+ 1,357	38,662	16,470	22,192	+ 5,722

SOURCE: Statistisches Bundesamt, *Statistiches Jahrbuch fuer die Bundesrepublik Deutschland.*
Der Bundesminister fuer Wirtschraft, *Der Deutsche Osthandel zu Beginn des Jahres 1979.*
Der Bundesminister fuer Innerdeutsche Beziehungen, *Die Entwicklung der Beziehungen zwischen der Bundesrepublik Deutschland und der Deutschen Demokratischen Republik, 1969-1976.*

was Germany's single most important Communist trading partner ($5½ billion total turnover), representing 39 percent of its *Osthandel.* Poland followed at 16 percent, Hungary and Czechoslovakia at 11 percent, Romania and China at 9 percent[5] (see figure 11). West Germany is also the most important Western trading partner of the Communist countries and represents 21 percent of all OECD exports to Communist nations.[6] Today West Germany is the leading capitalist trading partner of the U.S.S.R., Bulgaria, Poland, and Hungary. Indeed, it is Hungary's second largest trading partner, after the U.S.S.R.[7] Manufactured goods constitute 89 percent of Germany's exports to the East, while 50 percent of its imports are of raw materials. Trade with all of these countries is conducted under bilateral trading agreements, and West Germany grants most-favored-nation status to all Communist countries, within limits imposed by the European Economic Community (EEC). From the point of view of the German economy, overall dependence on trade with Communist countries is small. Nevertheless, certain sectors of the West German economy (e.g., the steel industry) are quite dependent on exports to the East, and technology

transfer to the East also provides substantial employment in certain industries. The Soviet Union and Eastern Europe are the largest export market for the West German machine-tool industry; approximately one-third of machine-tool exports from West

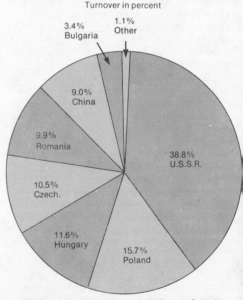

Figure 11.—Regional Distribution of West Germany's Trade With the East, 1978

Turnover in percent

3.4% Bulgaria
1.1% Other
9.0% China
9.9% Romania
10.5% Czech.
11.6% Hungary
38.8% U.S.S.R.
15.7% Poland

SOURCE: Der Bundesminister fuer Wirtschaft, *Der Deutsche Osthandel zu Beginn de Jahres 1979.*

---

[5]Der Bundesminister fuer Wirtschaft, *Der Deutsche Osthandel zu Beginn des Jahres 1979* (Bonn, 1979), p. 26.
[6]Ibid., p. 16.
[7]Ibid., pp. 4-10.

Germany go to Communist nations, and East-West technology trade is a significant employment guarantor for medium-sized machine-tool firms. Ten percent of German investment goods exports go to Eastern Europe and the U.S.S.R., and some large firms, Salzgitter for example, are fairly dependent on trade with Communist nations to keep their workers employed. Indeed, to the extent that *Osthandel* is still debated in West Germany, Government statements emphasize its positive economic aspects. During the 1976 election campaign, the opposition CDU criticized the Government for granting overly generous credit to Communist nations. The Government's reply stressed the importance of trade with the East to domestic employment. It claimed that *Osthandel* provided 300,000 jobs, and was vital for West Germany's continuing economic health. Interestingly, it was left to the business community, and particularly companies engaged in trade with Communist countries, to point out that imports from the East could have a negative effect on employment.[8]

In opposition to this historic trade dependency and regardless of the political climate, a number of economic factors act as deterrents to greater technology transfer from West Germany to the Communist nations. The central problem limiting trade with the Communist countries, as one German official put it, is simply that the Russians and Eastern Europeans cannot pay. The chronic shortage of hard currency in the East, combined with the Soviet and East European insistence that trade be bilateral, means that Eastern trading partners must find noncurrency means of paying for their West German imports. Although Eastern-bloc countries would like to export machinery to Germany, the West Germans are still unwilling to buy such manufactures in large quantities: they are of inferior quality and

pose a potential competitive threat to domestic products. The lack of suitable East European and Soviet imports therefore places substantial limitations on West Germany's exports of technology. Additionally, since the German Government does not subsidize interest rates on credits, the Soviets prefer to deal with nations from whom more favorable credit terms are available. Such credits do, however, provide one means of freeing hard-currency to pay for additional German imports. In some Eastern European nations, German corporations are able to enter into joint ventures, giving them 49-percent ownership, but the U.S.S.R. does not permit any foreign investment in Soviet enterprises.

Given these problems, the most viable form of West German-East European economic relations has been the use of compensation agreements. Under these the Soviets and East Europeans pay for their imports of German technology with products produced from German machinery and equipment. The West Germans see limited benefits in such countertrade, however, and are reluctant to respond to Soviet and East European initiatives to expand this kind of business.[9]

Together, these built-in economic constraints inhibit a massive expansion in West German trade with the East. Moreover, the need to protect German domestic production has given rise to import restrictions. These further limit the amount and kind of East European goods that can be imported to pay for West German exports.

POLITICAL FACTORS

A variety of political factors, some of them contradictory, also affect West German trade with the East. Most important is West Germany's desire to maintain and im-

---

[8]The head of the West German Eastern Committee (Ostauschuss) of the Federation of German Industry replied to Government claims by arguing, "If you calculate that way, you must compare the jobs which are secured through trade with the East with those that are threatened through imports from the Eastern bloc." See *Writschaftswoche*, Oct. 1, 1976.

[9]For a German discussion of the limits of compensation deals and the advantages of cooperation, see Matthias Schmitt, *Industrielle Ost-West Kooperation* (Stuttgart, 1974).

prove relations with East Germany. Inner-German relations are a key factor determining West Germany's differentiated technology transfer policies toward Eastern Europe, the U.S.S.R., and the PRC. Prior to 1969, West Germany predicated its economic relations with East Germany on the declared aim of achieving German reunification. This goal has shifted somewhat since the Brandt *Ostpolitik* and the 1972 basic treaty between West and East Germany. In this treaty, Bonn gave *de facto*, but not *de jure*, recognition to East Berlin. Although the East Germans count their trade with West Germany as foreign trade, Bonn does not. To do so would be to imply recognition of East Germany as a legitimate foreign country. Inner-German trade comes under a special category; imports from East Germany are called "supplies," and exports to it are called "deliveries." Inner-German trade is today more highly politicized than West German trade with other Communist countries; its chief stated aims are to improve political contacts between the two countries and to facilitate greater family reunification and a more favorable environment in Berlin. West Germany uses technology transfer as well as other forms of trade to facilitate greater inner-German contacts. Thus, inner-German economic relations are a special and unique category in West German export control policies towards the Communist countries.

Inner-German relations have largely determined West Germany's political and economic relations with other Communist countries, and at various times Bonn has implemented different trade policies toward Eastern Europe. In the mid-1960's, a "bridge-building" policy was designed to encourage East European independence from Moscow by offering Eastern Europe (excluding East Germany) special economic incentives, including easier credit terms than those available for the U.S.S.R. This policy proved politically counterproductive when the Soviets began to resist Western encouragement of polycentrism in Eastern Europe, and after the invasion of Czechoslovakia, Bonn altered its policy so as not to alarm Moscow. Eastern Europe remains an important market for West Germany, and differences remain in economic policies toward Eastern Europe and the U.S.S.R., but Bonn no longer views its trade with Eastern Europe as an instrument with the express political purpose of creating greater independence from Moscow.[10]

Today, the economic and political aspects of *Osthandel* are generally perceived as mutually reinforcing rather than as a specific source of leverage. West German policies regarding technology transfer to Communist countries reflect Bonn's desire to promote better relations with East Germany and the rest of Eastern Europe, while ensuring that the U.S.S.R. complies with the 1971 Four-Power Agreement on Berlin. Bonn prefers to deal with the U.S.S.R. only on matters of bilateral foreign policy; it eschews public criticism of Soviet domestic policies. The German goal of encouraging greater civil rights in the U.S.S.R. and Eastern Europe is pursued by carefully offering the U.S.S.R. certain political and economic incentives out of the public eye. This policy has led to U.S.-West German disagreements over President Carter's human rights policy. In Chancellor Schmidt's words, "As regards human rights, we on this side of the Atlantic—and that includes my Government—are on the whole more reserved in our approach than the United States."[11] The German approach is basically that of Egon Bahr, special assistant to former Chancellor Willy Brandt, whose policy for dealing with the East was "change through rapprochement," as opposed to change through confrontation. Bonn views technology transfer to the East as one of several means of improving relations with the U.S.S.R., which, it hopes, will eventually ease the inner-German situation.

[10]West Germany does differentiate between East European countries in its arrangements for the emigration of ethnic Germans. It gives large-scale credits to Poland for the 35,000 or so Polish Germans who emigrate every year, but does not have this arrangement with other East European countries.

[11]Interview in *Die Zeit*, July 21, 1978.

According to one German official, it is advantageous to Germany to assist through the export of technology in the construction of a Soviet industrial infrastructure. A more industrialized Soviet Union, so the argument goes, is less likely to be interested in war. Some in Germany feel that purchases of West German technology give Soviet leaders an increasing stake in peace and the economic status quo, thereby discouraging aggression. Proponents of this view favor increasing technology exports to the U.S.S.R. and ensuring its development as a highly technological society. Not all German officials agree, however.

Germany continues to place the effectiveness of the North Atlantic Treaty Organization (NATO) at the center of its foreign policy.[12] Bonn's concern with the security of the Western alliance therefore tempers its inclination to use technology transfer as a reward and incentive for good Soviet behavior; West Germany recognizes the need to comply with American security goals and to avoid sales of militarily useful technologies. Most German firms dealing with the East also accept the necessity of export controls and the need to comply with CoCom. Thus, there is a fairly comfortable *modus vivendi* between business and Government on questions of technology exports to the East. But different concepts of national security within the Atlantic Alliance create contradictory pressures influencing West Germany's technology transfer policy towards the Communist nations. Its *Wespolitik* demands a restrictive stance toward technology transfer, while its *Ostpolitik* suggests a more flexible use of trade to promote political moderation.

West Germany has been wary of any significant rapprochement—either political or economic—with the PRC. Although West Germany has sold technology to China, Eastern Europe and the U.S.S.R. remain far more important both politically and economically, and West Germany is wary of playing the "China card" for fear of repercussions in Europe. The recent heightened Western interest in the PRC has led to concern in Germany that other Western nations may begin to differentiate among Eastern trading partners, giving China preferential treatment. West German Government officials stress their even-handedness in dealings with all Communist countries. The guiding German principle is that one should not export to Czechoslovakia (or even Yugoslavia) any technology that one would not export to the Soviet Union; there is no guarantee that technology sold to any Eastern European country will not find its way to the U.S.S.R. The German Government applies the same logic to China. It favors a consistent East-West export control policy. Moreover, since trade with both China and the U.S.S.R. remains a marginal portion of West Germany's foreign trade from the point of view of the *Ostauschuss*, German exporters could well increase sales to both China and the Soviet Union. Discriminating against or in favor of either makes no economic sense.

TRADE IN TECHNOLOGY

Most officials claim that West Germany does not have a technology transfer policy towards the Communist nations. It does cooperate with other Western nations in CoCom, but apart from these multilateral export controls, the German administration has not developed clear, national guidelines on West-East technology transfer. Rather, a series of German laws and procedures comprise in aggregate an operating system for regulating the export of technology to Communist nations, and the technology transfer system therefore defines, rather than reflects, policy.

Interviews with Government and business officials in Bonn and Frankfurt reveal the absence of a generally accepted defini-

---

[12]For a general survey of Germany's position, see Angela Stent Yergin, "Soviet-West German Relations: Finlandization or Normalization?" in George Ginsburgs and Alvin Z. Rubinstein, eds., *Soviet Foreign Policy Toward Western Europe* (New York: Praeger Publishers, 1978), pp. 102-133.

tion of technology; as in the United States, most agree that the concept lacks precise definition. On the whole, international technology transfer is considered "part of the general process of diffusion by which the knowledge of, and use of, new products and processes passes from one production unit to another."[13] Most officials differentiate between technology and machinery in theory, but many have found it impossible to distinguish between software and hardware on a practical level, since technology is often embedded in equipment. Those officials who had read the Bucy report were somewhat skeptical of the feasibility of implementing it in CoCom.

West Germany is the largest single Western supplier of advanced technology to the Soviet Union. In 1977, for instance, 34 percent of the Soviet imports of high technology came from West Germany, as did 29 percent of its imports of manufactured goods. Japan was the next largest supplier, providing 17 percent of the U.S.S.R.'s high-technology imports and 20 percent of her imported manufactures. Germany is the second largest supplier of high technology to China, following Japan. In 1977, 15 percent of Chinese high-technology imports came from West Germany, as did 16 percent of its imported manufactures.[14]

West Germany exports a variety of technologies to the U.S.S.R., with machine tools and petrochemical plants leading the field. One-third of West Germany's machine-tool exports go to Communist countries.[15] Other larger scale exports are of mechanical engineering, electrotechnical, optical, and capital construction goods.

The most notable example of recent West German technology transfer to the U.S.S.R. is a giant steel complex being constructed at

Kursk by a consortium of German firms (among them, Slazgitter, Krupp, Korf Stahl, Siemens, AEG, and DEMAG), originally slated to cost more than 5 billion DM. Although recently reduced in size, the Kursk plant represents the largest single deal in the history of East-West trade and will be the world's largest steel complex based on the direct reduction method. Under the terms of the agreement, the Soviets will sell back to West Germany iron ore pellets made at the plant.

Other large-scale German-Soviet deals involve the building of the new Sheremetyevo airport at Moscow, construction of several petrochemical plants and automobile factories, exports of energy technology, and collaboration on nuclear energy research. In Eastern Europe, among other projects, West Germany is supplying passenger jets to Romania, and constructing coal gasification plants in Poland and petrochemical plants in Hungary.

The Germans operate most often through turnkey projects, although the sale of licenses is growing.[16] West Germany has sold licenses for waste incineration to Czechoslovakia; know-how to make bus engines, color TVs, washing machines, chemicals, and windows to Hungary; production technology for high-pressure safety-valves, concrete mixers, and shoe products to Poland; and know-how for axis-blowers for nuclear power stations, furnaces for sulfur burning, herbicides, electrical equipment, and wedge presses to the U.S.S.R.

Private agreements between German firms and the Soviet State Committee on Science and Technology (SCST) are also becoming increasingly common. Siemens, Germany's largest computer manufacturer, has concluded an agreement with SCST to estab-

[13]Philip Hanson and Heinrich Vogel, "Technology Transfer Between East and West: A Review of the Issue," *Osteuropa Wirtschaft*, February 1978, p. 97.

[14]John P. Young, "Quantification of Western Exports of High Technology Products to Communist Countries," U.S. Department of Commerce, pp. 15-16

[15]John Dornberg, "Trade With the East Bloc is Causing Some Concern," *International Herald Tribune*, Apr. 19, 1977.

[16]According to E.C.E. Secretariat, "General Aspects of East-West Licensing," *Eastern Business Magazine*, vol. 3, No. 1, the United States and the United Kingdom are the world's two greatest technology-exporting countries, with 30 to 40 percent of world trade in licenses.

lish a Center for Automation in Moscow. This is the first joint East-West scientific institute in the U.S.S.R., and Siemens hopes to facilitate its computer exports through it. While Siemens does not compete with American corporations in the large computer field, its smaller computers and microprocessors are competitive with American products. Yet computers represent less than 1 percent of German exports to the East, and Germany exports only a fifth as many computers to the Communist world as does the United States. About 6 percent of Siemens' computer exports currently go to the U.S.S.R.; Hungary and Czechoslovakia are more important markets at present. Despite the desire to expand computer sales to the East, when President Carter vetoed the Sperry-Univac computer sale to TASS in July 1978, the German Government discouraged Siemens from bidding for the contract.

German officials point out that Soviet intransigence on the Berlin issue continues to inhibit their exports of high technology to the U.S.S.R. For some years, the Soviets have delayed signing of a West German-U.S.S.R. agreement on scientific cooperation because of West Germany's insistence that West Berlin be included in its provisions. The absence of an agreement now means that Germany has less State-financed technological cooperation with the Communist world than do other Western countries. Although German officials believe that the Berlin issue will eventually be resolved and the agreement signed, they point out that today France is more likely than Germany to export high technology to Communist nations. German exports will remain largely composed of machinery.

In the last year, German trade with the PRC has grown faster than its trade with any other Communist country, and the question of technology exports to China is becoming more important. Germany today is China's third largest trading partner after Japan and Hong Kong. German imports

from China are mainly textiles and primary products. Two-way Sino-German trade rose by 49 percent in 1978, to $1½ billion, and West German exports to China rose by a spectacular 72 percent over 1977. German businesses are now discussing cooperation with the Chinese in the energy and nonferrous metal areas where, in exchange for technology, China will export raw materials. German exports to China now include large-diameter pipe, machinery, electrotechnical equipment, trucks, ships, and chemical products.[17] A $4 billion deal involving the sale of West German coal mining equipment and the training of Chinese personnel was recently concluded,[18] and another $14 billion contract has been discussed. If it is concluded the German *Metalgesellschaft* will supply the Chinese Ministry of Metallurgy with 22 plants for nonferrous metal industry, jointly explore iron ore deposits, and market ores and processed materials.[19] But while many German businessmen welcome the potential of the vast Chinese market, Government officials and the *Ostauschuss* warn of exaggerated expectations for trade with China. The Chinese, they argue, are reluctant to incur a crippling debt; moreover, they prefer to accept credits in dollars rather than in Deutschmarks.[20]

EXPORT CONTROL POLICY

The legal framework under which exports of technology to the Communist world take place in West Germany differs markedly from that of the United States. Exports are unrestricted under German law, although the administration may recommend exceptions to the *Bundestag* (Parliament).[21] German exports are governed by the 1961 law on foreign trade and payments (*Aussenwirtschaftsgesetz* or AWG), supplemented by a

---

[17]Der Bundesminister fuer Wirtschaft, op. cit., pp. 11-12.
[18]*Financial Times*, Sept. 27, 1978.
[19]Ibid., Nov. 15, 1978.
[20]*Handelsblatt*, Dec. 29/30, 1978.
[21]This section is based on Russell Baker and Robert Bohlig, "The Control of Exports: A Comparison of the Laws of the United States, Canada, Japan, and the Federal Republic of Germany," *International Lawyer*, vol. 1, No. 2, 1967, pp. 163-191.

foreign trade ordinance. Although the Federal Government has authorization to control exports of commodities and technical data (sec. 5-7, AWG) in the interests of national security, the *Bundestag* may cancel these restrictions within 4 months of their promulgation. This gives the legislature immediate and compulsory review of all foreign trade regulations issued by the West German Government. The working presumption of the law is that exports should remain uncontrolled except where special circumstances require use of standby statutory controls. This approach runs directly counter to U.S. export control policies, which require special approval for virtually all exports to Communist nations (see chapter VII).

Section 7 of AWG cites national security and foreign policy as criteria for restricting trade, but the AWG regulations encourage the granting of licenses for restricted goods unless it can be conclusively proven that the exports will injure Germany's economic or political security. The details of license requirements are found in periodic supplements to Annex AL of the Foreign Trade Ordinance of 1961. (The latest full list dates from December 1976, and the latest extra supplement from December 1978.)

The first three control lists—covering munitions, atomic energy, and "other strategic goods"—are essentially the CoCom lists, or the International Strategic Lists. The German Government recently added three items not covered under CoCom to the national atomic energy and industrial list—heavy-water containers, installations of fuel elements for nuclear powerplants (which the German Government tried unsuccessfully to add to the CoCom international atomic energy list), and rocket installations. The fourth list largely covers nonindustrial goods, like botanical plants, alcohol, and raw materials, controlled either because of Common Market regulations or because they are in short supply in Germany.[22] AWG also restricts the export of certain kinds of technical data, and of

documents concerning the manufacture of goods on the International Strategic Lists.

The German system of export licensing for technology sales to Communist nations is well-organized, and information on the procedures is readily available to businessmen. A company wishing to export a controlled item to the East applies to the *Bundesamt fuer gewerbliche Wirtschaft* (BGW or Ministry of Industrial Economy) in Frankfurt. BGW is empowered to grant licenses that do not need CoCom permission, and 30 to 40 percent of all applications are decided in the Ministry. Once the application is complete, it takes only about 3 weeks for the Ministry to grant a license. Applicants must promise in writing that the products will be used only for the stated purposes. For instance, when the Soviet Academy of Sciences purchased a Siemens computer, it had to sign an end-user statement regarding its use. German corporations must also include assurances that the products will remain in the country to which they are exported. Officials admit, however, that technology may reach Communist nations via neutral non-CoCom countries like Austria, Switzerland, and Sweden. Bonn has few means of preventing this.

BGW refers about 60 percent of all license applications to the Ministry of Economics for further consideration. The Ministry of Economics cooperates with the Ministries of Foreign Affairs and Defense in deciding whether the application should be brought up in CoCom. The criteria for deciding whether a license should be granted internally or should go to CoCom relate to possible strategic use, end use in general, whether other corporations or countries have already sold the item to Communist countries, and the character of the technology embodied in the product. German cases in CoCom require about 3 months for a decision, unless American reexport licenses are required. In this case, they may take up to 2 years.

The Foreign Ministry's most important licensing criteria are political, but do not in-

---

[22]For the latest list of goods subject to export licensing, see *Beilage zum Bundesanzeiger*, No. 246, Dec. 30, 1976.

clude the use of export control as a reaction to short-term political developments as in the United States. President Carter's denial of a license to Sperry-Univac in response to the Soviet dissident trials in 1978 prompted the West German Foreign Ministry to publicly state that Germany would never link exports to domestic developments within the Soviet Union. There is, in fact, no legal way for Bonn to deny export licenses for such reasons. The Government is bound by AWG to grant licenses unless it can show that there is a real national security danger involved in the export. However, the Foreign Ministry may deny export licenses if it judges that the export of a particular product could exacerbate international conflict and threaten German security. For instance, Germany once exported small handguns to a Communist country. These were intended for sports use only and were permitted under CoCom regulations. When very similar guns later appeared in terrorist hands in the Middle East, the German Government refused, despite repeated requests, to grant any more export licenses for this type of weapon. German firms can sue the Government if they are not satisfied that their export license denial was based on a threat to national security. Another political criterion for export license decisions is the German Government's desire to maintain equivalent technology export policies toward both the PRC and the Soviet Union. Germany does not want to exacerbate tensions with the U.S.S.R. by adopting a more lenient stance toward China.

Because West Germany's trade with East Germany is considered domestic inner-German trade, it is not covered by AWG. Rather, it is regulated by a special legal system. The highly sensitive nature of inner-German relations makes it extremely difficult to elicit information on technology transfer between East and West Germany. Here, export control more closely resembles American law than the West German foreign trade law; exports are forbidden unless expressly permitted, and CoCom regulations apply to East Germany. Thus the system of granting export licenses is the same for East Germany as it is for the rest of the Communist world. There have periodically been proposals to differentiate in export license-granting procedures as between East Germany and other Communist countries, but these suggestions have so far been rejected.

Inner-German technology transfer is sometimes controversial among West European nations because, according to the 1957 Treaty of Rome, East Germany receives the same tariff privileges as does West Germany, making it the *de facto* 10th member of the Common Market. East Germany is considered part of the West German domestic market; it is not treated as a foreign country, and therefore the EEC common external tariff does not apply to West German imports from East Germany.[23] Other EEC members periodically complain about East Germany's privileged treatment, but Bonn has thus far succeeded in retaining this special status. Inner-German trade, like Germany's trade with other Communist countries, consists largely of the exchange of West German manufactures for East German primary products.

There are periodic charges that high technology finds its way to the Communist nations via inner-German trade because of loopholes in the intricate system. German Government officials deny these allegations. They assert that technology is unlikely to slip through to East Germany because there is less technological cooperation with East Germany than with any other Communist nation. Moreover, East Germany is the most highly developed Communist society, well able to produce its own technology. The question of inner-German technology transfer remains a matter of debate, and hard data on the issue are difficult to find. The East German press ignores the subject of technology transfer, but defends trade with West Germany in general.

---

[23]For a detailed discussion of East Germany's status in the Common Market, see Sighart Nehring, "Der Sonderstatus des Innerdeutschen Handels" (The Special Status of Inner-German Trade), *Wirtschaftsdienst*, 1977/XLL, pp. 631-637.

CREDITS AND TARIFFS

West Germany's credit regulations differ from those of many Western countries. Although commercial interest rates have tended to be lower than other countries, the West German Government does not subsidize interest rates on official credits. An Economics Ministry spokesman states, "West Germany is unlikely to set up an institution similar to the Eximbank [Export-Import Bank] unless competition from other Western countries forces us to."

In general, therefore, West German official credits are no more competitive than those of other NATO countries. This gives rise to frequent East European complaints. Nevertheless, Government-guaranteed, long-term commercial bank credits are readily available to Communist countries. An interagency Government committee meets bimonthly with representatives of banking and industry to discuss export credit decisions. West German banks sometimes charge lower than the current market rate, but in such cases companies charge the purchaser higher prices and reimburse the banks for the difference between the market rate and the rate of interest charged. Given the healthy state of the German export economy, and West Germany's continual trade surplus with Eastern nations, there is little pressure from business to facilitate more East European purchases.

In 1977, German banks and firms extended a total of $11 billion in credits to Communist nations; of these, two-thirds were bank-to-bank credits that did not qualify for Federal insurance. The Soviet debt to West Germany is currently about $2.8 billion, and the total East European debt is about $8 billion, or a quarter of the Communist countries' total debt to the West.[24] In addition West Germany has granted East Germany an annual interest-free "swing" credit—meaning that either side can use it—of 850 million DM.

Compensation deals now constitute about 15 percent of West German trade with Communist nations. In May 1978, Brezhnev and Schmidt signed a 25-year agreement on economic cooperation which envisages a broadening of trade relations and cooperation.[25] Given Communist lack of hard currency, it is likely that the number of compensation deals will increase in the future.

While West Germany still restricts some imports, particularly textiles, to protect domestic industries, it has gradually liberalized its import restrictions on Communist goods. Only 7 percent of imports from the East are restricted—mainly textiles, steel, glass, ceramics, and leather goods. Import restrictions do not apply to the minimal technology imports from Communist nations. The Soviets export only a tenth as much technology to West Germany as the West Germans export to Russia. The U.S.S.R. has sold West Germany some steel technology and Hungary has sold pharmaceutical technology. Another example is Salzgitter's purchase from the U.S.S.R. and East Germany of a licensed process to produce low-density polyethylene.[26] The U.S.S.R. has also sold at least 18 licenses to West Germany for metallurgical, chemical, and electronic products.[27] Despite these very limited technology imports from Communist countries, West German-COMECON technology transfer is, according to one German official, essentially a "one-way street."

[24]*Le Monde*, Aug. 1/2, 1976.
[25]For the text of the "Agreement on Developing and Deepening Economic Cooperation," see *Soviet News*, May 9, 1978.
[26]Leslie Colitt, "East-West Licensing Turnaround," *Financial Times*, Apr. 30, 1976.
[27]John W. Kiser III, "Report on the Potential for Technology Transfer From the Soviet Union to the United States" (unpublished), October 1977.

FRANCE

INTRODUCTION

France's approach towards East-West trade and technology transfer is determined by its overall foreign policy stance: a strong desire to be independent, sovereign, and free from American hegemony; a preference for diversifying international links as much as possible, irrespective of the ideological and political character of other nations; and an aversion to mixing politics and economics, particularly where doing so would interfere with France's goal of maximizing foreign policy options. Technology transfer to Communist nations is not a political issue in France; indeed, relations with the U.S.S.R. arouse little controversy within the country, while relations with the United States are a far more delicate and contentious subject, particularly given the strength of the French Communist Party. The question of adhering to American-based export control policy through CoCom arouses much more disagreement in Paris than the issue of what should be sold to Communist countries. The general French approach is to support and increase trade with Communist nations in any goods and by all possible means; to eschew the use of economic levers in the pursuit of political goals; and to maximize the economic and political benefits to France which can be gained from East-West economic exchanges.

ECONOMIC AND POLITICAL FACTORS

France, like West Germany and Britain, enjoys an export-oriented economy, and its high level of trade dependence influences its attitudes towards trading with Communist nations. The French Government views trade with the U.S.S.R. and Eastern Europe as "normal"—"just like our trade with the United States," according to a Foreign Ministry official. Trade links with the PRC are newer and less well-established. The French are primarily concerned with diversifying their exports as much as possible, and wel-come Communist markets for their trade expansion potential. They also view trade with the East as an important employment source. Although trade with the U.S.S.R. and Eastern Europe forms only 4 percent of total French trade, officials point out that trade with Communist countries has kept alive some branches of French industry. Six years ago, the medium-sized machine-tool industry was struggling to survive; since then exports to the U.S.S.R. have revived machine-tool production and enabled the industry to reorganize more efficiently.

One official also claims that trade with Communist countries is a stabilizing factor for the French economy, since the Eastern countries with their countercyclical markets are secure even in times of economic crisis in the West. (This is a debatable point. See chapter III.) In global terms, France's comparative trade advantage lies in its consumer products, but there is relatively little demand for these in Communist nations. French exports to the East are primarily industrial goods. The main drawbacks of East-West trade for France lie in the lack of suitable Eastern imports and the difficulties raised by compensation deals.

France, therefore, faces many of the same economic problems in its relations with the U.S.S.R. and Eastern Europe as do its European allies and the United States. It is the political aspects of French trade with Communist countries, not the economic factors, which distinguish French policy towards trade and export control.

The political determinants of French export control policy reinforce the economic determinants, and both point towards encouraging East-West trade. Since de Gaulle's 1966 overtures towards the Soviet Union, France has sought to establish a "special relationship" with Russia, and has consistently tried to improve Franco-Soviet relations. According to official spokesmen, French national security is enhanced by improving economic and political relations with the U.S.S.R., and to this extent economics and politics are linked. France has historically accorded top priority to its relations with the U.S.S.R. (as opposed to the rest of Eastern Europe), and it is more concerned with Franco-Soviet trade than with economic relations with other Communist nations. Poland comes next in economic and political importance.

TRADE IN TECHNOLOGY

France has been losing relative importance as a trading partner for the Communist countries, moving from third largest capitalist trading partner of Communist countries in 1970 to fourth in 1976, after Germany, Japan, and the United States.[28] French trade with the Communist world amounted to $5.3 billion in 1976. Nearly 90 percent of this was with the Council for Mutual Economic Assistance (CMEA) nations, the most important of which are the U.S.S.R. and Poland (see table 30).

France is the Soviet Union's third most important source of Western technology, supplying 11 percent of Soviet high-technology imports, and 14 percent of its manufactured imports in 1977. It also ranks third among Western exporters of technology to

China, supplying 14 percent of China's high-technology imports in 1977, but only 2 percent of its manufactured imports.[29]

Technology plays a relatively important role in French exports to the East, although the French Government does not have an official definition of technology. Some officials define technology as know-how, exclusive of machinery; others say one cannot separate technology and equipment; still others seem to apply different definitions of technology to different circumstances. The share of machinery in French exports to the U.S.S.R. is significantly higher than in exports to other countries. Fifty percent of French exports to the U.S.S.R. are of machinery; 30 percent are of semifinished products. The main technologies sold to the Communist nations are turnkey plants for chemicals and gas-lift equipment (which uses computers); computers; and metallurgical, industrial, and petrochemical equipment.

In April 1979, Giscard d'Estaing and Brezhnev signed a 10-year economic accord designed to invigorate Franco-Soviet trade during the 1980's. The treaty provides for tripled bilateral trade and emphasizes industrial cooperation agreements and long-term deals. These cooperative projects in-

---

[28]Le Courier des Pays de L'Est, Mensuel D'Informations Economiques, *Le Commerce de la France Avec les Pays de L'Est en 1970-1976*, pp. 7-8.

[29]John P. Young, op. cit., pp. 15-16.

Table 30.—French Trade With CMEA Members, 1970-76
(in millions of dollars)

	1970			1976		
	Export	Import	Balance	Export	Import	Balance
CMEA	647,640	452,640	+ 195,000	2,735,400	1,995,840	+ 739,560
Bulgaria	47,760	18,960	+ 28,800	102,240	49,920	+ 52,320
Hungary	46,680	27,120	+ 19,560	129,360	98,520	+ 30,840
Poland	81,240	67,920	+ 13,220	749,880	429,840	+ 320,040
East Germany	59,640	42,240	+ 17,400	214,680	187,680	+ 27,000
Romania	82,080	53,280	+ 28,800	258,360	206,040	+ 52,320
Czechoslovakia	57,120	39,720	+ 17,400	161,640	108,480	+ 53,160
U.S.S.R.	273,120	203,400	+ 69,720	1,119,240	915,360	+ 203,880
Total Communist countries	$734,520	$535,440	+ $199,080	$3,143,760	$2,210,160	+ $933,600

CMEA = Council for Mutual Economic Assistance.

SOURCE: Organization for Economic Cooperation and Development Statistics.

clude new sources of energy, energy-saving equipment, electronic products (including computers), machinery and machine tools, and metals, chemicals, and petrochemical products. The agreement further specifies that French companies and banks will take part in building industrial complexes in the U.S.S.R. and that similar Soviet institutions will participate in projects in France. The French companies providing equipment, licenses, and credit for factory construction in the U.S.S.R. will be partially or totally reimbursed in products manufactured by these Soviet industries.[30]

In the first 6 months of 1979, $340 million worth of new Franco-Soviet contracts were signed, the key ones in high-technology areas. Meanwhile, in 1978 the French company Technip won one of the biggest single Soviet orders placed in a Western country, a $213 million contract for gas-lift installations to improve oil recovery levels in Western Siberia.[31] Another major new contract won by Thompson-CSF will supply $100 million of telephone equipment to the U.S.S.R.[32]

Under the most publicized Franco-Soviet high-technology deal, a group of French companies will sell a computer and ancillary equipment to the Soviet news agency TASS. The $20 million contract involves an Iris 80 computer from CII-Honeywell Bull SA, ancillary equipment from three subsidiaries of Thompson-CSF SA, and programing and software from Steria (Societe de Realizations en Informatique et Automatique). Although the 1980 Olympics will be a major user of the system, it will not be fully operational until 1981.[33] This order originally went to the American firm Sperry-Univac, but in July 1978, President Carter vetoed the license application as a sign of U.S. displeasure with Soviet dissident trials. In the wake of this action, and following U.S. guidelines, both the German and British Governments discouraged firms from bidding for the deal.[34]

[30]Financial Times, Apr. 30, 1979; Le Monde, May 2, 1979.
[31]Financial Times, Apr. 27, 1979.
[32]Financial Times, Apr. 4, 1979.
[33]East-West Trade News, Apr. 4, 1979.
[34]Les Echos, Aug. 18, 1978.

The French corporation won the contract before President Carter decided to reverse his decision. The French Government opposes the practice of subjecting technology exports to short-term political interference.

EXPORT CONTROL, CREDIT, AND TARIFF POLICY

France enjoys good cooperation between business and Government on questions of Soviet and East European trade. The Government encourages industry through subsidized credit and other policies, and the Department of Industry maintains close contacts with businessmen on East-West issues. Government and business also appear to have worked out a viable modus vivendi for licensing technology exports. According to a computer industry spokesman, firms have learned to write applications for export licenses that are virtually certain to be approved. More skeptical observers claim that the French Government turns a blind eye to violations of export license application procedures, particularly where end-use statements from Soviet organizations are concerned. In any event, industry and Government appear harmonious over these questions. The French Communist Party, working through various companies, encourages trade with Communist nations, and several prominent Communist businessmen are engaged in East-West trade.

France's ambivalent attitude towards the United States in general, and towards U.S. attempts to limit technology exports through CoCom in particular, complicates its export control policy. As one spokesman put it, "We can't always align ourselves with Washington—we would not have a foreign policy if we did that." This gives rise to paradoxes. Officials point out that France is even more concerned about Soviet military strength than is the United States: Paris is geographically more vulnerable to Moscow than is Washington. On the other hand, French export control policy is greatly affected by France's general aversion to complying with U.S. demands. Unlike Britain and Germany, staunch supporters of NATO, France is only a marginal member. In 1966, after resisting American efforts to integrate NATO forces, France removed its troops from NATO. Presently it participates only selectively in military and related activities. The French are also sensitive about admitting that they belong to CoCom; official spokesmen even claim it is not known publicly that France is a member. The great secrecy surrounding France's relationship to CoCom testifies to the extreme sensitivity of the whole subject of allied cooperation on technology exports to Communist countries.[35] Officials indicate that France alone is the best judge of its security interests, an attitude that applies equally to relationships in both NATO and CoCom. While France shares with Germany and England a basically favorable predisposition toward East-West trade, it resists U.S. attempts to control technology exports.

The domestic legal framework governing technology transfer is elusive. While no laws state the rules on export licensing, various pieces of information suggest how the system functions. Products for which export licenses are needed appear on export control lists published periodically by the Ministry of Economics and Finance in the *Journal Officiel de la Republique Francaise.*[36] An interministerial committee establishes the criteria for items requiring export controls. The lists are essentially the three CoCom lists. France has few, if any, unilaterally controlled items, and only about 8 percent of exports to Communist nations need licenses.[37]

---

[35] Although it is housed in a section of the U.S. Embassy in Paris, CoCom is not listed in the Paris telephone directory—neither under its name nor in the street directory.

[36] See Ministere de L'Economie et de Finances, "Avis aux Importateurs et aux Exportateurs relatif aux produits soumis, au control de la destination finale," *Journal Officiel de la Republique Francaise,* July 14, 1977, for the latest list. Curiously, the first export list was published in this journal on Dec. 31, 1961. The question of how French businessmen knew prior to 1961 for which goods they required licenses remains unanswered.

[37] *Le Monde Diplomatique,* September 1978. However, Professor Marie Lavigne, author of this article, cites a U.S. source, explaining that it is impossible to obtain French information on these figures.

Unlike the other systems under review, the French process of export licensing begins with customs officials, to whom firms needing licenses submit applications and copies of contracts. If customs officials decide that the license application requires CoCom approval, they send it to the Ministry of Industry, which sends it in turn to the Ministry of Foreign Affairs. On more sensitive items, the Ministry of Defense may also become involved. No regular interministerial committee comparable to those in Britain and West Germany exists to deal with sensitive technology licensing applications. Nevertheless, if the CoCom representative in the Ministry of Foreign Affairs requires further consultation, an unofficial committee of intergovernmental advisors can be summoned to discuss the case. The CoCom representative from the Ministry of Foreign Affairs also presents the French position before CoCom, signs the license application after approval, and sends it through the Ministry of Industry to customs, which notifies the firm. From the companies' point of view, therefore, the export-licensing system begins and ends with customs.

France appears not to require formal third-country statements (promising that the technology will remain in the country to which it is sold and not be exported to third parties) prior to licensing, as do Britain and Germany. Soviet and East European end-user statements are required, however.

The French Government facilitates the export of technology to the East by providing generous credit supports and other financial facilities. Medium-term credit insurance (up to 3 years) is available in France from the *Compagnie Francaise d'Assurance pour le Commerce Exterieur (COFACE),* a quasi-public agency under the supervision of the French Government. COFACE provides East-West trade credit insurance with both commercial and political risk coverage for 8 to 95 percent of the credit. A few large commercial and investment banks provide the bulk of export credits for East-West trade. The most prominent of these are the *Credit Lyonnais* and the *Societe Generale,* both large nationalized banks, and the *Banque de Paris et des Pays Bas.* A corporation seeking to finance trade with Communist countries usually deals directly with one of the contracting French banks and then with COFACE. After it has secured COFACE credit insurance, it has access to French Government-supported refinancing facilities through the *Banque Francaise du Commerce Exterieur,* a publicly chartered bank whose capital is held by the *Banque de France* (the Central French bank) and various other banks.

France charges all-inclusive rates on Government-supported export credits.[38] In this it differs from other major Western trading partners of Communist countries. For the 1974-79 period, for example, the Franco-Soviet intergovernmental agreement stipulates that France grant the U.S.S.R. a $3 billion credit at interest rates ranging from 7.20 percent to 7.55 percent (depending on the value of the projects) to facilitate Soviet purchases of French machinery. Similar agreements exist with other Communist countries.[39] There is, therefore, ample credit support available for exports of French technology to the East.

Like Germany, France has felt the lack of sufficient and suitable East European imports to balance its exports. Only a few

---

[38]For a more detailed discussion of French export credit support, see Suzanne F. Porter, *East-West Trade Financing: An Introductory Guide,* U.S. Department of Commerce (Washington, D.C.: Government Printing Office, 1976).

[39]See *Le Courier des Pays de L'Est,* October 1978, No. 222, pp. 18-21, for details of credit agreements with other socialist countries.

French import controls cover Eastern goods, and these are mainly on textiles and shoes. No import controls exist on Communist technology, perhaps because very little Soviet or East European technology is presently imported. In 1971, after the U.S.S.R. criticized France for not buying enough of its finished goods, a French company called *Gisofrac* was established to promote Soviet manufactured imports. Supported by the Government and the three nationalized banks, *Gisofrac* deals exclusively with Soviet, not French, exporters. The director of the company admits that the results so far have been disappointing. On some occasions, the Soviets have been unable to supply machines in the quantities ordered. France imports a small number of Soviet Lada automobiles and it has also acquired some Soviet technology, including a recooling system (purchased on license), a press (the price of which was considered by some to be too high), and some petrochemicals. The problems of inferior Soviet quality limit the attraction of these products for the French market. France and the U.S.S.R. have a mutual credit agreement, but until now very few Soviet credits have been used to finance Soviet exports to France.[40] A Franco-Soviet intergovernmental commission has formed working groups to resolve some of these problems, but for the moment Soviet technology exports are only a marginal part of Franco-Soviet trade.

---

[40]Interview with Paul Nouailhac, Gisofrac, May 30, 1979, Paris.

THE UNITED KINGDOM

INTRODUCTION

The British attitude toward technology transfer to Communist nations more closely resembles the West German than the French approach, although the British share the French view that trade and politics are two separate activities which should be linked tenuously, if at all. While the Germans have revised their views of the political dimensions of *Osthandel,* the British have fairly consistently separated their economic from their political relations with Communist countries. This stance has caused occasional friction in British-U.S. relations. But the United Kingdom not only values its membership in NATO; it also prizes its close relations with the United States. As a result,

Britain is not averse to cooperating with the American-inspired export control policy toward Communist nations.

ECONOMIC AND POLITICAL FACTORS

The economic and political determinants of British technology transfer policy to Communist countries are somewhat different from those of West Germany and France. Britain, too, is an export-oriented economy with a high trade dependence, and trade with Communist countries is viewed as a guarantor of employment. Consequently, the United Kingdom favors technology transfer to Communist nations for domestic economic reasons. Nevertheless, there is also some concern about the negative employment effects of imports from the East and about the problems of countertrade. British-Soviet trade suffers from the same economic restraints as do German-Soviet and Franco-Soviet trade; the predisposition to trade with the East is modified by the difficulties of the Communist nations in paying for these imports.

Because there is no problem equivalent to that of the two Germanies in British foreign policy, London's trade with the Eastern countries is less influenced by political goals than is that of Bonn. Moreover, the constraints limiting Germany's China policy do not apply to the United Kingdom. For some years, in fact, the United Kingdom has been actively engaged in the transfer of technology to China, and it intends to continue this policy. National security considerations, therefore, place only limited restraints on British technology transfer. However, given Britain's perception of its "special relationship" with America, its economic interest in expanding all forms of trade can conflict with its political goal of maintaining a relationship with the United States.

TRADE IN TECHNOLOGY, EXPORT CONTROLS, AND CREDITS

Total United Kingdom-Soviet trade in 1978 amounted to $2.2 billion (see table 31). Major exports were of machinery, chemicals, and nonferrous metals, and the primary imports were petroleum products and nonmetallic minerals.

Britain's technology exports to the East are primarily petrochemical plants, machine tools, transport equipment, gas pipeline, polyethylene plants, methanyl plants, secondary recovery for oil, and glass fibers. Exports of energy technology are expected to increase as the United Kingdom develops its North Sea oil reserves. Some British energy technology is also being exported via U.S. multinational subsidiaries in Britain, and British credits are used for these exports. Sale of turnkey plants is the predominant form of technology transfer, although license sales are also important. Britain is not as large a supplier of high technology to Communist nations as West Germany, however. In 1977, the United Kingdom supplied the U.S.S.R. with 2.2 percent of its high-technology imports and 6 percent of its manufactured imports.[41]

Table 31.—United Kingdom Trade With the U.S.S.R. (in million pounds sterling)

	1972	1978
Imports	218.7	688.2
Exports	90.3	423.1
Total	390.0	1,111.3
Balance	− 128.4	− 265.1

SOURCE: Department of Trade, London.

---

[41]John P. Young, op. cit., p. 15.

The British definition of technology stresses the software concept: it includes technical data, expertise, information, and patents, and not merely equipment, machinery, and other hardware. However, controls are regarded as most practical in the case of hardware. Indeed, the British generally regard the Bucy report recommendations on control of "know-how" as too restrictive and unworkable. One former British delegate to CoCom argues that end-use statements on software do have value. This official expresses the hope that the Bucy recommendation should not "herald a change in Western policy and practice."[42]

Technology transfer to Communist nations is regulated by the export of goods (control) order, supplemented by a Consolidated List of Goods Subject to Security Export Control, which includes all those goods requiring export licenses when sold to Communist nations. The lists contain three sections: the Munitions List, the Atomic Energy List, and the Industrial List. These closely parallel the German lists, and contain the items on the CoCom lists. In addition, Britain prohibits the export of certain goods for domestic reasons.[43] The licensing system functions on the exceptions principle. License applications are handled by the Department of Trade, which discusses the license requests with an interdepartmental committee. About 1,000 license applications are processed every year, and the average time required for a decision is 1 month. The major criteria affecting these decisions are CoCom considerations, national security, and the possibility of technology diversion

to Communist nations via third countries. These are balanced against the effect on domestic employment.

The British Government encourages industry to consult with the Department of Trade before submitting license requests, so that by the time the applications are made the outcome is usually assured. The British Government has, however, been known to turn down licenses for which CoCom approval has been granted. There is an appeals procedure for licenses that are denied. After the British Government has approved the license, it goes to CoCom and, if the product embodies American-originated technology, to the United States. In 1978, a series of interdepartmental meetings investigated the effectiveness of controls on the export of technology to third countries, and decided that they were effective.

The British Government subsidizes interest rates for credits to Communist countries, and guarantees credits granted by commercial banks through the Export Credit Guarantee Department. In an attempt to boost United Kingdom-Soviet trade, in 1975 Britain offered the U.S.S.R. a $2 billion credit line for the purchase of British technology over a 5-year period at an interest rate of about 7 percent—a rate lower than that paid by Britain itself for money borrowed overseas.[44] The U.S.S.R. has been very slow to take up these credits, and Britain still maintains a large trade deficit with the Soviet Union. Given Britain's economic problems, its major concern regarding East-West trade is on the import, rather than the export, side.

---

[42]R. J. Carrick, *East-West Technology Transfer in Perspective* (Berkeley, Calif.: University of California Policy Papers in International Affairs, 1978), pp. 42-43.

[43]See Consolidated List of Goods Subject to Security Export Control, *Trade and Industry* (London, Apr. 30, 1976).

[44]Christopher S. Wren, "Britain to Offer Soviet $2 Billion in Trade Credits," *New York Times*, Feb. 18, 1975; Melvyn Westlake, "Where Critics of Russian Trade Credits Go Wrong," *Times*, Mar. 4, 1975.

JAPAN

INTRODUCTION

The volume and policy framework of Japan's trade with the Communist world derive more from commercial than from political factors. Though small in terms of Japan's overall foreign trade, business with the East operates under liberal Government policies. Export controls reflect purely economic concerns: balance of payments, the stability of the yen, the growth of the Japanese economy, and the development of its foreign trade. Singularly absent from the theory and practice of Japanese foreign trade and technology transfer are strategic and national security, or political concerns (e.g., human rights). Nonetheless, Japan has traditionally cooperated with the United States through CoCom on matters of export control.

ECONOMIC FACTORS

Japan's dramatic postwar development was facilitated both by cheap labor and by the prudent use of imported technology to boost productivity. As Japan's economic miracle emerged and domestic industry developed its own technologies, the country's dependency on imported raw materials and energy grew. To balance such imports, Japan has looked increasingly to foreign markets for finished goods and consumer products, machinery, and technology. Given the limited markets for consumer goods in China, the Soviet Union, and—to a lesser extent—Eastern Europe, the Japanese are increasingly likely to emphasize machinery exports to those areas.

Japan is the world's second largest petroleum consumer (after the United States) and the largest importer of crude oil. In the mid-1970's, its purchases accounted for 16.7 percent of the international oil market, compared with 13.4 percent for the United States. Furthermore, Japan is heavily dependent on both OPEC and what the Japanese call the "umbrella of the majors:" 72 percent of its oil imports, 43 percent of its refining operations, and 47 percent of its

distribution network is in the hands of the large multinational oil companies. Japanese oil imports from the Middle East require a month in transit in supertankers of 100,000 tons, and cost 1,000 yen per ton to transport.

These considerations provide a powerful incentive to seek alternative suppliers of petroleum. China and the Soviet Union both offer such sources to Japan, albeit with a number of unresolved questions about the Communist nations' own future energy needs and their ability to quickly and economically bring new oil reserves into production. If such questions could be resolved, oil imported from the U.S.S.R. would require only 2 days in transit aboard smaller tankers in the 25,000- to 50,000-ton range, and would cost only an estimated 200 yen per ton to transport, exclusive of transportation from the Soviet oilfields to port.[45] Soviet oil would also provide the Japanese with some protection against supply interruptions caused by unrest in the Middle East, sparing them the traumas they endured during the 1973-74 embargo.

In the now-moribund Tiumen oil development project, the U.S.S.R. promised Japan a maximum of 25 million tons of oil a year at a cost of $1 billion, with delivery scheduled to commence in 1980. Soviet behavior, however, indicated that any such deal would be fraught with political and strategic difficulties—as when, in March 1974, the Soviet side suddenly shifted its plans from a Tiumen-Nakhodka pipeline (to be built with Japanese assistance) to a request for Japanese aid in building the BAM (a second Trans-Siberian railroad) in order to ship the oil by train.[46]

[45]John P. Hardt, George D. Holliday, and Young C. Kim, *Western Investment in Communist Economies* (Washington, D.C.: Government Printing Office, 1974), p. 45; Roger Swearingen, *The Soviet Union and Post-War Japan: Escalating Challenge and Response* (Stanford, Calif.: Hoover Institution, 1978), pp. 121-128.

[46]Gerald L. Curtis, "The Tyumen Oil Development Project and Japanese Foreign Policy Decision-Making," in Robert A. Scalapino, ed., *The Foreign Policy of Modern Japan* (Berkeley, Calif.: University of California Press, 1977), pp. 157-158.

Meanwhile, the Chinese have persistently offered "oil without strings" to Japan. Shipments have grown in relative terms but remained small in absolute quantity in comparison to the Soviet proposal: 1 million tons in 1973, 4.9 million tons in 1974, and about 8 million tons in 1975. The Chinese apparently feel that the provision of "oil for the lamps of Japan" (and Japanese factories) is advantageous to them politically as well as economically. Among other things, it can serve to mute Japanese enthusiasm for an "energy alliance" with the U.S.S.R. Thus, Chinese oil exports to Japan have risen steadily even as PRC exports in other areas have diminished, and the 1975 quoted price per barrel, worldwide inflation in oil prices notwithstanding, was 70 cents lower than in the previous year.[47] But the fact remains that Chinese oil is heavy, with a high wax and sulfur content, and is generally difficult and expensive to refine, especially by comparison with the lighter crude the Soviet Union can provide. Coal, which provided roughly one-sixth of Japan's energy needs in 1972, is almost two-thirds imported. In the mid-1970's, Japan could import coal from Siberia for $3 per ton, compared with $18 a ton for U.S. coal.[48] Lower costs, transportation savings, and the guarantee of long-term, stable supplies through imports from either the U.S.S.R. or the PRC will continue to make such nations attractive as trading partners. Japan extended $150 million in bank credits to the U.S.S.R. in 1974, in return for which Japan will receive 104 million tons of coal between 1979 and 1998. Approximately one-seventh of these credits will be returned to Japan for the purchase of consumer and manufactured goods.

Japan has also recognized the Soviet Union as an important source of the other raw materials required by Japanese industry. In 1975, the U.S.S.R. provided 20 percent of Japan's lumber and cotton imports, 21 percent of its potassium salt imports, 26 percent of its nickel imports, 29 percent of its asbestos, and between 40 and 80 percent of all precious metal imports. The Soviet Union ranked third in importance as a source of iron ore, chromium, and copper respectively, and fourth in terms of coking coal. For Japan, each of these commodities falls in an area of high import dependency (in many cases, 85 to 100 percent).[49]

Given the critical nature of these raw materials for Japanese industry, it is safe to say that the U.S.S.R. is more important as a supplier to Japan than Japan is to the Soviet Union, despite the recent achievement of a trade balance favorable to Japan. In fact, the volume of Japanese trade with the Communist world is relatively small. In 1977, China and the U.S.S.R. ranked 10th and 11th, respectively, among Japan's export markets, and 11th and 13th among the sources of Japanese imports. This trade is growing, however. Trade figures for 1978 (see table 32) reveal that exports to the PRC rose by 47 percent and those to the U.S.S.R. by 26 percent over the previous year. Similarly, imports from China rose by 32 percent from 1977 to 1978, although imports from the U.S.S.R. remained virtually unchanged.

Table 32.—Japanese Trade With the U.S.S.R., East Europe, and the PRC—1978 (in thousands of dollars)

	Dollars	Percent
Exports		
U.S.S.R.	$2,253,840	(+26.2%, 1977)
PRC.	2,613,736	(+47.2%, 1977)
East Europe	616,318	(−17.0%, 1977)
Total	$5,837,935	(+30.9%, 1977)
Imports		
U.S.S.R.	$1,318,765	(+0.6%, 1977)
PRC.	1,809,000	(+32.2%, 1977)
East Europe	212,291	(+9.1%, 1977)
Total	$3,473,793	(+15.7%, 1977)

SOURCE: *Summary Report: Trade of Japan*, no. 11, November 1978, p. 68. Data for January-November 1978, period only.

[47]Ibid., pp. 169-170.
[48]Sir John Crawford and Saburo Okita, eds., *Raw Materials and Pacific Economic Integration* (Vancouver: University of British Columbia Press, 1978), p. 218.

[49]Hardt, Holliday, and Kim, op. cit., p. 44.

Seen in their historical context, these figures indicate that while the *relative* importance of Japanese-Soviet trade has not increased markedly since World War II, the increase in absolute terms has been phenomenal, reflecting the remarkable overall growth of the Japanese economy and foreign trade. Table 33 summarizes Japan's postwar trade with the Soviet Union.

Japan's trade with Eastern Europe has remained small and largely stagnant. Poland and Romania are Japan's two major East European trading partners, but trade with them between 1977 and 1978 either increased only marginally (Poland 1.1 percent) or actually declined (Romania, minus 16.6 percent). Japan reportedly hopes to remedy this situation with a major breakthrough in computer exports to the Eastern European telecommunications market. Its major competitor here would be Britain.

Japan appears to be in a better position than Germany, Britain, and the United States to circumvent the problem of severe limits in the Communist nations' ability to pay for imports. Given the nonmarket countries' preference for bilateral deals involving counterpurchase, barter, or buy-back arrangements, the Japanese *sogo shosha* (all-round trading companies) have an unmatched natural affinity for compensation negotiations that stems from a long history of multifaceted, multilateral business and trading arrangements.[50] Japanese trading companies possess worldwide marketing networks, and have the financial ability to engage in triangular or "switch" trading (whereby Japanese firms sell the ruble credits they have earned to a third country or company at a markup), as well as the ability to dispose of a wide variety of unrelated products. The Japanese have also shown themselves sensitive to the importance of structural arrangements in East-West trade. In recent times, major Japanese traders have modified their organizational frameworks (especially where the sales function was geared primarily to handle a single item) in order to better manage the kinds of multidimensional projects demanded by East-West trade.[51]

POLITICAL AND FOREIGN POLICY FACTORS

Taken by themselves, a number of political, strategic, and foreign policy aspects of the Soviet-Japanese relationship might be expected to affect Japanese policies on trade with and technology transfer to the Communist world. That they are not the major determinants of Japanese policies and practices in these areas can be attributed to the primacy of economic factors and to Japanese recognition of the fact that "linkage" between its trade and foreign policy objectives cannot be carried out successfully. Japan simply does not have the cards to play in a trade-and-foreign-policy poker game with the Soviet Union. Economic and political relations with the U.S.S.R. are therefore kept clearly separated.

The Soviet Union presents Japan with several irksome diplomatic problems. Most important, perhaps, is the insistence of the Soviets on retaining the Northern Territories, a group of Japanese islands captured in the final days of World War II. In fact, the U.S.S.R. and Japan have never signed a formal peace treaty, because neither side has been willing to yield on the Northern Territories issue. Fishing rights have been

Table 33.—Postwar Development of
Japanese-Soviet Trade
(in thousands of U.S. dollars)

Year	Exports	Imports	Total	Balance
1946...	$ 24	$ 0	$ 24	+ $ 24
1950...	723	738	1,461	− 15
1960...	59,976	87,025	147,001	− 27,049
1970...	340,932	481,038	821,970	− 140,106
1975...	1,626,200	1,169,618	2,795,818	+ 456,582
1978...	2,502,195	1,441,723	3,943,918	+ 1,060,472

SOURCE: Japanese-Soviet-East European Trade Association, Tokyo, Japan, 1979.

[50]Raymond Mathieson, *Japan's Role in Soviet Economic Growth: Transfer of Technology Since 1965* (New York: Praeger Publishers, 1979), p. 29, pp. 236-237.

[51]Japan International Trade Organization (JETRO), *Japan's Plant Exports*, No. 11, (Tokyo: JETRO, 1977), pp. 13-14.

another source of Japanese-Soviet dispute. In this area the Soviets have clearly and consistently held the upper hand. Annual negotiations over catch quotas, and the recent imposition of Soviet sovereignty over waters within 200 miles of its coast, have slowly eroded Japanese competitiveness with the giant and technologically advanced Soviet fishing industry.

Numerous public opinion surveys have indicated that the U.S.S.R. is "the most disliked" country among the Japanese public. Indeed, in 1978, anti-Soviet feeling ran as strong as at anytime during the past 15 years, with 40 percent of all respondents listing the U.S.S.R. at the top of the list of nations they most disliked.[52]

China, by contrast, has fared much better, with only about 10 percent of the 1978 sample listing it as "the most disliked" (and with 15 percent describing it as the "most liked," in contrast to less than 5 percent for the U.S.S.R.).[53] Moreover, with the exception of a brief reversal during the height of the Great Cultural Revolution, Japanese public opinion has tilted increasingly towards China. For whatever it is worth in terms of its actual influence over the making of Japanese foreign policy, public opinion does not seem to regard the Communist bloc homogeneously as a "security threat," and is unlikely to be sympathetic to a campaign to restrict exports on those terms.

Unfortunately, there is no comparable information on the general attitudes of Government and foreign policy elites. Nevertheless, American Japan-watchers and Japanese scholars agree that the Japanese defense and foreign policy establishments view the U.S.S.R. as the chief military threat to Japanese security. But if Japanese strategic and diplomatic vulnerability has affected trade and technology transfer policies, the effect appears to be more to encourage than discourage trade. As a nation defenseless on

all sides, Japan pursues a strategy of "being friends with everybody" and maintaining an evenhanded stance in the Sino-Soviet conflict. Reflecting the great importance of economic needs, a 1974 White Paper on Foreign Trade prepared by the Ministry of International Trade and Industry (MITI) noted that to ensure stable supplies of essential resources, Japan would be required to pursue a policy of "orderly imports" and "diversification of import markets." Among other things, this meant deepening interchanges with the Communist bloc as well as with Latin America and Africa.[54] Similar sentiments were expressed by Mr. Hatoyama, Minister of Foreign Affairs, in a speech to the Diet in October 1977. According to the strategy he outlined for Japanese-Soviet relations in the coming year, Japan would (a) strive to develop relations across a number of fronts (economic, cultural, political) simultaneously, but (b) would insist that any long-term relations depend on a peace treaty and a settlement of the territorial issue.[55]

Practical experience effectively precludes the Japanese from accepting the idea of controls or embargoes on technology transfer on political or general foreign policy grounds. Such a stance is reinforced by the fact that the "leverage-through-linkage" strategy has its critics on the Japanese scene. Japanese scholars and officials alike have argued variously that the Japanese need to take the initiative in increasing Soviet trust in their intentions, that issues such as the return of the Northern Territories should be subordinated to pragmatic trade considerations, and that the overwhelming military might of the U.S.S.R. makes linkage an impractical ploy. Others see increased trade and other exchanges as one way to halt the "spiral model of insecurity and conflict" that has been at the root of Japanese-Soviet relations since at least the end of the 19th century. From these perspectives, Japanese demonstrations of "good faith" and "reliability"

[52]*Foreign Opinion Notes: U.S. International Communication Agency,* May 7, 1979, p. 4 (figure 2).
[53]Ibid., p. 2 (figure 1).

[54]Crawford and Okita, pp. 170-171.
[55]*Speech by Itchiro Hatoyama, Minister of Foreign Affairs* (Tokyo: Foreign Press Center, Oct. 3, 1977).

through trade, they feel, can serve to alter prevalent attitudes and to create a more fruitful atmosphere for political negotiations.

On occasion, controversy internal to the Japanese Government has arisen over the issue of linking or not linking trade and economic cooperation with foreign policy considerations. A case study of the now defunct (for economic, not foreign policy, reasons) Tiumen oil project reveals that certain officials in the Foreign Ministry did favor a linkage strategy vis-a-vis Japanese participation in Siberian development, while MITI and the Ministry of Finance were adamantly opposed to the idea. In the end, the opponents of linkage (including then-Foreign Minister Ohira) carried the day.[56]

Nor is there any evidence that officials in the Japan Defense Agency (JDA) see a threat to Japan's security as a result of the transfer of Japanese technology to the Soviet Union. Unlike the U.S. Department of Defense, JDA appears to be a captive of other bureaucratic interests, centered chiefly in MITI, Foreign Affairs, and Finance. Its position papers, at least until quite recently, failed to reflect an independent agency viewpoint. In fact, according to some sources, they were drafted outside the confines of JDA, though bearing its imprimatur. In any event, though a recent JDA paper called for increased R&D expenditures for domestic production of defense equipment, there is no mention of the strategic ramifications of technology transfer, economic competition with the Communist bloc, or the military implications of foreign trade.[57]

Japanese defense and foreign policy specialists who are concerned about the Soviet military threat view a continuing defense relationship with the United States as their best defense. They see a substantially lower level of security threat from China. Indeed, Japanese officials have suggested that Japanese technical assistance to aid in China's

modernization and industrialization makes a positive contribution to Japanese security by reducing the possibility of domestic upheavals and foreign policy radicalism.

This view would also appear to coincide with Japanese public opinion. According to a December 1978, poll by the Japanese Public Survey Opinion Organization, most Japanese see "domestic political order" (47 percent) and the "state of the economy" (33 percent) as more critical to Japan's security than military measures or defense *per se* (14 percent).[58]

TRADE IN TECHNOLOGY

Determining the share of Japanese trade that can be categorized as "technology transfer" is difficult. In 1976, machinery accounted for 40 percent of Japan's exports to the Soviet Union and 11 percent of its exports to China, in terms of dollar value. Among exports to all nations, Japanese machinery represented slightly under a third of total value. But Japanese machinery exports appear to be rapidly expanding. By 1978, machinery represented 64 percent of all Japanese exports, and 35 percent of all exports to Communist-bloc countries (including 47 percent of exports to the Soviet Union and 20 percent of exports to the PRC) (see table 34). This expansion can be at least partly explained by the fact that Japanese industrial output and export product lines have been affected by increasing competition from abroad, as other Asian nations (such as Korea and Taiwan) have gained the advantage of cheap labor in labor-intensive industries like ceramics and textiles. These changes have led the Japanese to develop new export lines in machinery, technology-intensive goods, and metals.

However, by using the U.S. Commerce Department's definition of "high-technology items" (see chapter VI) Japan is a relatively

---

[56]Curtis, pp. 163-164.
[57]Japan Defense Agency, *The Defense of Japan* (Tokyo: JDA, 1976), p. 128.

[58]*Research Memorandum: U.S. International Communication Agency*, Apr. 27, 1979, p. 7.

Table 34.—Japanese Machinery Exports,
1978, by Region
(in thousands of U.S. dollars)

	Machinery exports	All exports	Machinery as % of total
IW nations	$27,144,870	$37,268,971	72.8
LDC.	22,158,104	37,017,595	59.9
Communist bloc	1,817,140	5,181,217	35.1
U.S.S.R.	952,194	2,012,288	47.3
PRC.	460,546	2,311,332	19.9
East Europe	310,269	554,815	55.9
Total	$51,120,220	$79,467,933	64.3

SOURCE: *Summary Report: Trade of Japan*, no. 10, October 1978, pp. 128-29 (table 10). Data available for January-October 1978, only.

Table 35.—Soviet Imports of Machinery and
Equipment From Japan, the United States,
and West Germany, 1977

	M&E imports	% of all M&E imports	% of all M&E imports from West
Japan	684.9	6.0	18.3
United States. . .	350.8	3.1	9.4
West Germany. .	1,041.6	9.1	27.8

SOURCE: *Summary Report: Trade of Japan*, no. 10, October 1978. pp. 128-29 (table 10). Data available for January-October 1978. only.

insignificant source of technology to the U.S.S.R., which purchased only about 18 percent of its Western machinery and equipment imports from Japan in 1977 (as opposed to 28 percent from West Germany and 9.4 percent from the United States, see table 35). The Japanese led the competition (West Germany, France, Britain, and the United States) only in Soviet imports of calculating machines (including computers), special-purpose vessels, and optical instruments between 1972 and 1977; and ranked second as suppliers of valves; batteries and cells; tubes, transistors, and photocells; optical elements; and image projectors. Japan supplied the Soviet Union with none of the following high-technology items, which could be viewed as strategically or militarily sensitive: aircraft turbines; nuclear reactors; telecommunications equipment; electron and proton accelerators; aircraft; and aircraft parts. In the area of oil-refining equipment, however, Japan far outstripped all competition. In this category it supplied 87 percent of the Soviet Union's imports from Western sources, although this only amounts to 36 percent of total Soviet oil-refining equipment imports. Other categories of technology-intensive imports in which Japan was an important Western supplier include power-generating and electrical equipment (with 40.6 percent of imports from Western sources); chemical industry equipment (21 percent); excavation equipment (36 percent); and compressor equipment (19 percent).

Plant exports comprise the dominant share of technology transfer by the Japanese, with license and patent exports playing only a minor role. Japan's plant exports to the rest of the world totaled $6.5 million in 1976, an increase of about one-third over the previous year, even though total exports grew by only one-fifth. Chemical plants represented about 40 percent of the value of transferred technology, and plants to manufacture heavy electrical equipment also figured prominently. The Communist bloc ranked first among regional customers for Japanese plants (purchasing almost 30 percent of the total), with the Middle East second, and Latin America third.

In many cases, the Japanese have proven particularly adept at importing technologies, improving on them, and exporting the new generation to both Communist and free world countries. For instance, in 1968, Toyo Engineering obtained basic patents for an ammonia production process. In a year the firm's alterations led to a 30-percent increase in output. Mitsui and Co. then sold the process to the Soviet Union.[59]

Two recent Japanese economic forecasts, one developed by the prestigious Japanese Economic Research Center (JERC) and one by MITI, predict that technology will become an increasingly important component of Japanese exports in the next 5 years.[60] Specifically, these studies foresee the following changes in Japanese output and exports:

- 1985 machinery exports, especially products related to new technology and

---

[59]Japan International Trade Organization, op. cit.
[60]Kiyoshi Kojima, *Japan and a New World Economic Order* (Boulder, Colo.: Westview Press, 1977), pp. 130-136.

commodities incorporating electronics, will be 22 percent above 1975 levels;
- exports of plant construction material will increase;
- exports of finished chemicals, petrochemical products, plastics, iron and steel will decline;
- the share of value-added, "knowledge-intensive" products in the chemical industry's output will rise; and
- 1985 exports of precision instruments will increase 15.7 percent above 1975 levels.

In general, then, the Japanese economy is expected to continue its current move away from labor-intensive, low-productivity industries—which will emerge increasingly in the developing countries and will find a growing Japanese import market—towards knowledge-intensive industries dependent on high value-added per unit of raw material and labor. Since the Soviet capacity to absorb imports of finished products and consumer goods is somewhat limited, Japan can be expected to continue seeking Soviet markets for other Japanese exports—particularly those in which Japanese technology is embedded—to counterbalance imports of Soviet raw materials.

At the same time, according to JERC, China will pursue a course of economic development that will closely parallel the earlier Japanese experience. China, like early postwar Japan, will rely heavily on low-paid but plentiful and highly motivated labor, a high degree of capital formation, and extensive imports of technology. This pattern would make China an increasingly important market for Japanese technology exports. This attractive market can be expected to reinforce Japan's already-liberal stance toward technology transfer to Communist nations.

EXPORT CONTROL POLICY

The basis for Japanese export controls is provided by the foreign exchange and foreign trade control law enacted on December 1, 1949. It presumes that trade development is desirable and that trade with all nations, including Communist countries, should be permitted without controls except under certain circumstances relating primarily to fiscal considerations. This principle of "export freedom" is secured by article 47 of the law, as modified by the provisions of article 48. Licenses may be required for export at the discretion of the Government, depending on the goods involved, the designated recipients, or the mode of payment. The latter provision allows the Government to exercise authority over sales involving payments other than "standard measures of financial settlement" or cash.[61] It is under this provision that the executive branch—or, more precisely, MITI—is authorized to regulate the transfer of technology to other nations.

The law itself, however, contains no mention of the control of goods or technology for either military or political reasons. Nor is it specific regarding either the areas or the commodities for which controls are to be invoked. Substantive limitations are contained only in a Government export trade control order, of which 89 variants have been promulgated between June 1950, and June 1977.[62] The order is altered from one to eight times a year, depending on changes in Japan's domestic economic situation, its balance-of-payments positions, or shifts of Japan's national list. The latest available varient of the order contains 204 items. The few items that are not derived from CoCom are included for reasons of domestic short supply, to prevent dumping, or to improve quality control.

"Area restrictions" figure only marginally in the order's control provisions. They are not applied to Communist states *per se*, for the latter are lumped together with market economies under a broad designation "Area A." "Area B" restrictions apply only to those countries for which there are special, non-CoCom, embargo provisions (e.g., Rhodesia) and to others (e.g., Iran, Iraq, Nigeria)

---

[61]Baker and Bohlig, op. cit., pp. 163-191.
[62]Ibid., pp. 174-176; *Export Control: Export Trade Control Order* (Tokyo: MITI, 1978), n.p. (pt. II-B).

with whom Japan has special balance-of-payments problems.

Both the law and the various orders operate in an atmosphere best characterized as a "presumption of license" rather than a "presumption of denial." In contrast to the United States, Japan has never introduced a so-called "blanket clause" whereby restrictions pertain unless or until a general license is established or validated.[63] Indeed, article 2 and article 47 of the law foresee the eventual removal of export restrictions entirely by means of a periodic review of each order and through administration of both the law and orders according to minimal rather than maximal standards.[64] Currently (1979), certain additional steps in this direction appear to be under consideration. One involves a redrafting of the 1949 law by MITI so as to further reinforce the presumption-of-license provisions and to eliminate any vestiges of an atmosphere of restriction.

Explicit provisions for the regulation of technology *per se* are conspicuous by their absence in the law, the orders, and Japanese discussions of export controls. As noted above, such transfers can be regulated legally according to the provisions pertaining to methods of payment. But all interviews strongly suggest that they are not. It appears, rather, that Government officials strongly believe (and businessmen concur) that restrictions on the flow of technology are best left to the normal operation of commercial forces, i.e., to the desire of firms to retain a competitive advantage. They also suggest that the economic rather than security aspects of export regulation are firmly fixed in the minds of both those administering the law and those subject to it.

It should be noted here that the restrictions contained in the law and the orders apply only to Japanese firms located on Japanese territory and not to foreign subsidiaries, branches, or Japanese-based multinationals. There is no concept of extraterritoriality in Japanese trade law—or in Japanese law in general for that matter, except for serious criminal cases. Violation of export controls is a criminal offense, with sanctions of up to 3 years in prison and a minimum fine of 300,000 yen (about $1,500 at the current exchange rate). If, however, the price of the item involved "times three" exceeds the value of the minimum fine, the penalty is automatically trebled. Although the law contains no formal provisions for the Government to revoke a firm's export privileges, MITI's legal authority to exercise "administrative guidance" in granting licenses means that it can *de facto* indefinitely delay an offending firm's export privileges.

The licensing process in Japan operates on consensus between business and Government. It is not an adversary procedure, and it provokes few complaints, if any, from Japanese businessmen. Most Japanese-Soviet and Sino-Japanese trade passes through the hands of the 14 or 15 "all round trading companies" (the *sogo shosha*).[65] Thus, in comparison to the United States where large and complex deals involving the exchange of multiple products by trading conglomerates are more common in validated license decision cases than individual contracts signed by relatively small firms, there are probably few licensing instances that must be handled by the Japanese Government.

The setting in which the Japanese licensing process operates differs from that in the United States in other ways. Since its resumption following the 1956 agreements, Japanese trade with the Soviet Union has always taken place within the framework of intergovernmental trade compacts. Since 1966, these have provided comprehensive, 5-year projections of all aspects of trade exchanges. Not coincidentally, these also correspond to the time frames provided by Soviet 5-year plans. The original initiative for these

---

[63]Ibid., p. 174.

[64]Ibid., p. 171; *Export Control,* op. cit.

[65]Yataro Terada, "System of Trade Between Japan and Eastern Europe, Including the Soviet Union," *Law and Contemporary Problems,* 37, 3 (summer 1972): 434-435; Alexander K. Young, *The Sogo Shosha: Japan's Multi-National Trading Companies* (Boulder, Colo.: Westview Press, 1979), pp. 195-221.

agreements came from the Soviets and, since 1968, the U.S.S.R. has tried unsuccessfully to get Japanese negotiators to commit themselves to trade agreements stretching over 15 to 20 years. Until the 1971-75 trade agreement, each trade "plan" also included a list of products to be exchanged and estimates of the volume or monetary amounts involved. For subsequent agreements, no estimates of amounts have been supplied, and the *annual* breakdowns of both products and amounts have been suspended. The list of items, however, has been retained. The 1971-75 trade agreement listed some 300 separate items for export and import, with a supplemental schedule of items provided to cover the "coastal trade" of the Soviet Far East.

The Japanese do not regard these agreements as legally binding, but they nonetheless affect licensing operations in important ways, as summarized below:

1. *A priori* agreement on what can and cannot be traded exists. Hence, instances of conflict between Government and business (and instances of denial) are extraordinarily rare.
2. The lists are useful to Japanese companies seeking trade as a means of identifying favorable market opportunities in the U.S.S.R. They also shape the long-term planning of production and immediate production decisions of firms involved in import-export exchanges with the U.S.S.R. In point of fact, the quantities actually traded usually surpass the levels provided for in the agreements. Both sides use the item lists and designated quantities in their annual trade reviews to determine if "trade on a regular basis," the catch phrase of every agreement, is in fact occurring.
3. The lists are credited with "greasing the wheels" of the licensing bureaucracy.
4. Japanese business is protected by the agreements against Soviet dumping.[66]

[66]Terada, op. cit., pp. 432, 433.

The licensing process itself is characterized by extensive informal consultation between MITI and the exporting firm even before negotiations with a Communist foreign trade organization commence. Companies brief MITI on the trade and payment provisions, returning for further consultation at successive stages if the package alters. For exports to non-Communist nations, such a review process apparently occurs only when Japan Export-Import Bank credits are involved.

In effect, therefore, a system of thorough preliminary clearance operates. The licensing process is expedited by the flexibility of the Soviets. If they do not think that a license will be forthcoming, they do not seek trade agreements. There is no evidence of Soviet pressure in cases where a license has been denied. At a second stage, the Japanese firm brings the negotiated package back to MITI for approval of credit and payments provisions, which will also likely involve the Ministry of Finance and the Export-Import Bank. At the present time, about half of all Japanese-Soviet deals involve exporter credits. The other half involve buyer's credits that depend on loans from the Export-Import Bank to the Foreign Trade Bank of the U.S.S.R.

By law as well as practice, export licensing remains largely the prerogative of MITI. No interagency boards or committees are involved. The Ministry of Foreign Affairs apparently plays an occasional consultative role, being contacted to "hear its views," while the Ministry of Finance is involved on a more regular basis.

Issues of conflict rarely surface in the licensing process. When there are differences of opinion within MITI or strong communications by one of the other ministries, then MITI convenes an informal "committee," which is usually made up of members from the economic agencies (Ministry of Finance, Export-Import Bank, Economic Planning) and Foreign Affairs. Only when bank-to-bank loans are at stake is the conflict likely to be referred to the Cabinet and Prime Minister.

There are no provisions for public accountability in the licensing process. The Government is not required to bring instances of approval or denial to the attention of the Diet (Japanese Parliament) or the public at large. Although notification of completed contracts is published in an official gazette, information about license approvals or denials is not. There has been no discussion of the issue of export controls, technology transfer, or licensing procedures in the Diet during the past few years.

SUMMARY

A number of factors, both economic and political, serve to create differences between the United States and its CoCom allies in East-West trade and export control policy. Our major Western trading partners—West Germany, France, Britain, and Japan—are all far more heavily dependent on foreign trade; West Germany, for example, derives nearly one-third of its GNP from international commerce. Similarly, trade with the Soviet Union, China, and Eastern Europe provides economic benefits of far greater relative importance to our CoCom allies than it does to the United States.

While the value of technology exports to the U.S.S.R., Eastern Europe, and the PRC is not large as a percentage of total German, British, French, or Japanese foreign trade, such exports do provide critical support of key sectors within each nation's economy. Jobs, markets, foreign exchange, and balance of payments are all at stake, and these economic factors clearly influence the policies and practices of our CoCom partners with respect to the transfer of technology to Eastern-bloc nations. In general, it can be said that Germany, France, and Britain tend either to see their economic and political interests as harmonious or to separate economic interests from any conflicting political factors, and in either case to base their East-West technology transfer policies primarily on economic factors. To the limited extent that trade with the East is used for political leverage, the linkage tends to be positive rather than negative—that is, trade is used as an inducement to political accommodation, and not as a weapon for punishment.

Japan's economic circumstances differ somewhat from those of the European nations, but the Japanese situation also encourages East-West trade and technology transfer. Highly dependent on imports of raw materials and energy from both the Soviet Union and China, Japan relies heavily on export markets that are shifting increasingly away from consumer goods and toward technological items. The Communist nations provide attractive markets for technology exports. Furthermore, problems in Soviet-Japanese relations are submerged by these economic interests, partly because Japan lacks the strategic and diplomatic strength to use foreign trade as a diplomatic playing card, and partly because the Japanese Government sees trade with the U.S.S.R. as a tool for lessening tensions between the two nations.

The CoCom nations' generally favorable stance regarding trade with and technology transfer to the East is reflected in the ease with which export licenses are granted. The export control systems employed by West Germany, France, Britain, and Japan all operate on the presumption that exports should be permitted in all cases except those involving items with clear and exclusive military value. A cooperative relationship between business and Government appears to exist in each of our allies' export control programs, making it possible for licenses to be granted swiftly and easily. In most cases, a time-consuming scrutiny by Government officials is not considered necessary before permission to export technology is granted.

Nonetheless, all four CoCom allies adhere to CoCom's policies and regulations with respect to exports that might jeopardize Western security. Even France, a nation that in recent years has pursued a foreign policy pointedly independent of United States interests, has found it expedient to follow CoCom guidance in regulating exports to the East. Each nation maintains a list of embargoed exports or items requiring special permission for export, and in most cases the national lists largely coincide with CoCom lists. In at least one instance involving nuclear powerplant components, West Germany has taken an even stricter stance than CoCom in adding items to its list.

The major constraint on West European and Japanese transfers of technology to Communist nations stems from the inability of the purchasing nations, particularly the U.S.S.R., to arrange for payment. The Soviet Union suffers both from a shortage of hard currency and from a lack of exports attractive in Western markets. Consequently, the sale of Western technology must frequently be based on buy-back agreements that involve future payment in imports of products produced with the exported technology. The CoCom nations generally provide the U.S.S.R. with favorable credit terms through subsidized interest rates, and even Germany which does not offer lower official rates is generous in the amount of official credit availabe to the East.

In summary, our West European and Japanese allies and trading partners perceive East-West technology transfer as part of a larger picture involving trade in general, rather than as an issue in its own right. Like other aspects of export policy and other features of diplomatic relations with Communist nations, the sale of technology to the East is dealt with through a relatively routine weighing of national economic and political costs and benefits. Germany, France, Britain, and Japan have weighed their interests and determined that they are best served by a technology transfer policy that is generally liberal, yet remains within the boundaries of the strategic requirements of Western security.

CHAPTER X

Western Technology in the Soviet Union

CONTENTS

Western Technology in the Soviet Union

THE HISTORY OF WESTERN TECHNOLOGY IN THE SOVIET UNION

Debates in the United States over the national security implications of bolstering the Soviet economy through the sale of advanced technology are of relatively recent origin, but the desire to profit from Western technological advances vastly predates both the cold war and the creation of the Soviet State. In this sense, Western technology transfer to the U.S.S.R. has had ample precedent; foreign technology and capital infusion have played a relatively large role in Russian economic growth for the past 300 years. From Petrine times until the present, Russian statesmen have attempted to compensate for domestic inability to generate competitive innovation by importing know-how from abroad. The motivation for this interest in technical and economic progress has varied, and technical advance, economic growth, and military power have all been closely intertwined. Successive heads of both the Russian and Soviet Governments have emphasized the necessity of competing with the advanced states of Europe not only in terms of domestic standard of living, but also in terms of national security.

The first systematic and nationwide attempts at modernizing the Russian State through Government edict occurred during the reign of Peter the Great (1682-1725). During his tenure, the number of manufacturers and mining enterprises quadrupled. Western impact in this period was felt more through the transfer of know-how, ideas, and people than through the transfer of hardware. The main thrust of the Petrine economic reforms directed toward the development of an efficient, modernized Russian armed force that could match those of Poland and Sweden, Russia's major European adversaries. The almost continuous state of war, punctuated by periodic invasions of the Russian homeland, made the development of a modern navy and munitions industry seem crucial to the survival of the Tsarist State.

The State bureaucracy under Peter I, remolded along Western lines, was the prime mover in the development of key military-related sectors of the economy. This established a pattern which was to persist until the October revolution. Growth in the new armaments, metallurgy, shipbuilding, and textiles industries was encouraged by guaranteed demand for their products from the Government sector. The State, in turn, strictly regulated the quality of the product, demanding standards comparable to German and Dutch industry. In 1702, Peter initiated a drive to induce foreigners to settle in Russia. This was intended to be a spur to innovation; Russia was importing both the necessary know-how and what the Tsar regarded as superior Western cultural traits.

Peter attempted a deep and comprehensive Westernization of Russia, but it was based on narrow premises. While a relatively competitive military sector was established by the middle of the 18th century, the structure of the new industries precluded ongoing growth and innovation in the absence of State influence. Manufacturing was based on serf labor, and there was no impetus to discover labor-saving and capital-intensive modes of production. Product quality and quantity in those industries wholly dependent on the State were in many cases determined by administrative decree, but low quality in the private sector was tolerated in the absence of alternatives. Finally, raising the proportion of foreigners in the intelligentsia was an insufficient first step toward the comprehensive educational system necessary for permanent increases in worker productivity and domestic innovation.

The 18th and early 19th century modernization drives depended for the most part on State resources as their motive force, but, by the end of the Crimean War, it was clear that this technique could not support industrial development on a par with that in Western Europe. It was not until Count Sergius Witte became Minister of Finance in 1892 that Government financial policy deliberately focused on industrial development. Witte stabilized State finances, returned the ruble to the gold standard and borrowed extensively abroad. At the same time, he channeled a great deal of foreign capital into the expansion of the railway system, thus lending an added impetus to growth.

Witte was a disciple of Frederick List, whose ideas on tariff protection for developing industries had helped to industrialize Prussia. The result of these policies was a surge of industrial growth unprecedented in Russian history, and based essentially on private initiative. Between 1892 and 1903, when Witte left office, the annual rate of industrial growth consistently exceeded 8 percent.

During this period, the major vehicle of technology transfer was the import of foreign machinery. Its role in the modernization process was considerable: in 1912, only 55 percent of the ruble value of all machinery sold in Russia was of domestic origin, and

imports of agricultural machinery increased from 6 million to 50 million rubles from 1895 to 1914. Foreign investment in the last Tsarist period was particularly important in the mining, metallurgy, textile, and chemical industries.

The success of the October revolution ended the period during which economic growth was nurtured by private initiative. According to Lenin, the industrial growth of the prerevolutionary era was based on the exploitation of the masses by the capitalist class and had fostered backwardness in the Russian worker. The Soviet task was to rebuild through State control a society as productive as the most advanced Western nation. Western assistance remained vital to this enterprise.

Lenin's New Economic Policy, which came into effect after 1921, had as its central mechanism of technology transfer the granting of concessions to Western entrepreneurs. Technical assistance contracts, the employment of foreign engineers and experts in the U.S.S.R., and the dispatch of Soviet experts to training positions in the West were also utilized.

Over 200 concessions were made to foreign firms between Lenin's death and the first 5-year plan. While Soviet literature downgrades the contributions of foreign technology transfers accomplished through this medium, it is clear in retrospect that much of the rapid growth of the 1920's was dependent on foreign operative and technical skills. The Soviets at this time made little or no attempt to develop completely new mechanisms of domestic productions; even experimentation was limited and soon abandoned. They concentrated on acquiring new productive processes from the West, training politically reliable engineers, and establishing basic and applied research institutes.

By the end of the 1920's the Soviets were convinced that they had found a more effective mode than the pure concession or the joint venture for the transfer of Western skills and technology. After 1928, technical-assistance agreements and individual work contracts with foreign companies, engineers, skilled workers, and consultants replaced

"Model A" Fords roll off the Gorki assembly lines in 1929

the pure and mixed concessions. Under these arrangements the capitalist firms could no longer claim a share of ownership. In addition, the control of technology transfer operations lay totally in the hands of the Soviets. Existing concessions were closed out through taxation, breach of contract, harassment and, in some cases, physical force.[1]

In their place, in the summer of 1929, many wide-ranging technical-assistance agreements were concluded with foreign firms. These were to be of specific, limited duration. The units designed and begun between 1929 and 1932 were some of the largest in the world, so large in fact that in many cases contracting Western firms had not previously dealt on a similar scale. Design and layout of these complexes came mostly from America, with Ford, General Motors, Packard, General Electric, and U.S. Steel contributing heavily. And although nearly a half of the installed equipment was German, it was very often manufactured in Germany to American specifications.

For 2 years there was an unparalleled infusion of foreign technology in the form of skilled labor, technical data, and equipment. Although most of the engineers were gone by 1932, they left behind designs based on Western models which contributed to a large increase in manufacturing capacity. Until 1941, production increases in most Soviet industrial sectors were the result of the installation and expansion of the Western plants acquired in the massive transfers which took place during this brief period.

Stalin had used the threat of war to initiate the new era of industrialization and collectivization in 1919. First priority was therefore given to the military departments of the new works, and many plants built in this period simultaneously produced civilian and military equipment. After World War II, the most significant vehicle of technology transfer was the stripping of German industry. It has been estimated that at least two-

thirds of the German aircraft and electrical industries, most of the rocket production industry, several automobile plants, several hundred ships, and a host of military equipment were transferred *en masse* to the U.S.S.R.

In the late-1950's, the Soviets turned their attention to technology transfer in industries where the German acquisitions had been slight—the chemical, computer, shipbuilding, and consumer industries. During this period, the U.S.S.R. began a massive complete plant-purchasing drive. Between 1959 and 1963, at least 50 complete chemical plants were bought for chemicals not previously produced in the U.S.S.R. In addition, a large ship-purchasing program was initiated in order to expand the Soviet merchant fleet.

In sum, whatever the role of technology transfer in the contemporary Soviet economy, it is clear that Western technology has long been looked on as a way to overcome domestic economic shortcomings. These imports have played a major—and continuous—role in both the Russian and Soviet States. In this sense, Soviet efforts to obtain imported technology are neither surprising nor new. In addition, throughout both Russian and Soviet history such transfers of know-how and capital from the West have been conscious tools of State economic and military policy. The centralization of economic decisionmaking, particularly as it relates to the selection and use of foreign technology, has been practiced in Russia for at least 300 years.

Equally normal, however, has been great vascillation in the ways in which foreign exporters and technicians have been treated. While Western-Soviet trade has had a long history, this history has been characterized by periodic State-imposed deteriorations of trading conditions and by a conspicuous lack of predictability in commercial contacts. On the basis of the historical evidence, at least, there is no reason to expect that increased sales of technology to the U.S.S.R. will much enhance the opportunities for Western exports of manufactured goods.

---

[1]Anthony Sutton, *Western Technology and Soviet Economic Development, 1930 to 1945* (Stanford, Calif., 1977), pp. 20-26.

THE NATURE OF THE SOVIET ECONOMY

THE COMMUNIST PARTY

No market mechanism officially operates in the U.S.S.R. Instead, economic decisions concerning allocation of resources and rates of expansion of different sectors are made administratively, and basic economic policy formulation is one of the principal functions of the Communist Party.

The Party exercises control and supervision over the economy in a number of ways. Many branches of the Government report directly to Party organs. The State Planning Committee (Gosplan), for example, reports directly to the Politburo (Executive Committee) of the Party.[2] At lower levels, building projects are first submitted to the Party before being submitted to the appropriate Government office. At the enterprise level, the Party organization both mobilizes workers to fulfill the plans and monitors the activities of enterprise managers.

The most potent tool used by the Party to direct the economy is the *nomenklatura* system. The *nomenklatura* is a comprehensive list of appointments under Party control. It nominates individuals to all important posts in the State, industry, and army. As a result, although only about 6 percent of the Soviet population belongs to the Party, nearly all agricultural or industrial managers are Party members.

THE GOVERNMENT APPARATUS

The State apparatus administers the detailed planning and organization of the econ-omy. The Soviet economy operates under a ministerial system in which individual enterprises belonging to a particular branch of the economy (petrochemicals, metallurgy, etc.) are subordinated to a single ministry. There are three types of ministries: the all-union ministries run the enterprises under their control directly from Moscow, and these enterprises are not answerable to regional authorities; the union-republic ministries have offices both in Moscow and the republics; and the republic ministries direct enterprises in their own republics. The heads of these ministries are either members of the Council of Ministers of the U.S.S.R. or of the other republic Councils of Ministers.

ECONOMIC PLANNING

Coordination of ministry activities is done primarily by Gosplan, the principal planning agency.[3] While only a limited number of commodities are centrally planned and distributed by Gosplan, the planning process is extremely complex.

The first step in this process is for the Party to establish priorities, in the form of output targets, for the upcoming plan period. These targets are sent to Gosplan, which tentatively formulates a detailed set of output goals and determines the resources required to produce them. These goals or "control figures" are sent down through the planning hierarchy to the individual enterprises. At this point, enterprises and ministries formulate their own input estimates for Gosplan's output targets. Gosplan must reconcile the two. Should demand for a particular commodity input exceed supply over the

---

[2]P. Gregory and R. Stuart, *Soviet Economic Structure and Performance* (New York, 1975), p. 118.

[3]Ibid., p. 119.

economy as a whole, Gosplan may decide to reduce demand, to draw on stocks, or to import. After it has arrived at this "material balance," Gosplan submits the plan to the Council of Ministers for approval and/or modification. The finalized targets are then communicated down the hierarchy to individual firms.

This system of material balance planning is cumbersome and slow; it stresses quantitative output goals and requires the maintenance of a vast bureaucracy. While it strives for consistency (equating outputs to inputs), it has proven incapable of achieving optimality, i.e., the most productive resource mix for desired production levels. On the positive side, material balance does permit the Government to channel growth in high-priority sectors while maintaining strict control over the economy.

The monetary counterpart of each enterprise's input and output plans are financial plans. These facilitate planner control over enterprise operations to the extent that deviations from the financial plan signal deviations from the physical plan. This control is reinforced by the fact that all legal interfirm transactions, with the exception of certain investment allocations and foreign trade, are handled by the State Bank (Gosbank), which is the sole center for settling of accounts.

Each year, Gosplan formulates and the Council of Ministries approves an investment plan for the entire economy. The plan is carried out by "project-making" organizations in charge of investment planning at the enterprise level, and its implementation is supervised by Gosplan and the ministries. Thus, decisions to expand enterprise capacity are made outside the enterprise itself. Investment choice in the U.S.S.R. is hampered by the inefficiencies that arise from the reluctance of planners to rely exclusively on profitability criteria, and from overly taut investment planning.

Soviet enterprises operate on an independent "economic accounting" system. This is often taken to mean that they operate to maximize profits. The system guarantees,

however, only that enterprises have financial relations with external organs such as Gosbank and that their operations are evaluated in terms of value indicators using official prices. Under this system, future production targets bear no relation to profits.[4]

The plan, formal and informal constraints, and the managerial incentive structure have made gross output the most important indicator of enterprise performance. A manager is rewarded primarily for rapid expansion in physical output in a given planning period, irrespective of poor performance in other areas. Managers therefore tend to avoid change, expecting negative impacts from innovation in process or products.

These factors, which inhibit incentives and may result in misallocation of investment funds, are endemic to the Soviet system of economic organization. Even where a measure of local decisionmaking power exists, such decisions must conform to the wishes of the central planners, who perform without necessarily according priority to issues such as prices or profits.

The declining rate of economic growth in recent years has lent impetus to attempts to reform the Soviet economy. In 1965, Premier Kosygin submitted a plan designed to reduce the number of enterprise targets set from above and, most important, to replace gross output by "realized output" (sales) as the primary indicator of the success of an enterprise. Further, profits were to be an important source of funds for decentralized investment by enterprise managers and were to be used as a source of funds for bonus payments to workers. These changes were to be phased over 5 years.

The period since 1971 has witnessed a reversal of official attitudes toward the solution of basic economic problems. Rather than relying on economic "levers" at the enterprise level—the basis of the Kosygin reforms—attention is being increasingly directed toward improving planning methods and increasing control over enterprises

---

[4]Ibid., pp. 179-230.

to improve economic performance. Emphasis is now on new planning methods such as perspective planning, automated plan calculations, automated information retrieval systems, and new organization methods.

One of the major motivations of the 1965 reforms and of the later modifications to them was the continuing reluctance of managers to introduce new technology and raise product quality. The subsequent recourse to more centralized administrative techniques means that those features of the economy that deterred innovation in the past continue to exist.[5] Major innovations in Soviet tech-

nology are therefore more likely to come from technological infusions from the West than from domestic R&D.

In conclusion, while modest attempts at reform have been undertaken in the Brezhnev era, the basic problems of economic incentive in the innovation process have not, in the final analysis, been seriously addressed. The economic reforms of 1965 have been so modified as to dilute their effect. In lieu of emphasis on economic "levers" as spurs to innovation, reorganizational and administrative solutions have met with little success. Western technology continues to be important to future Soviet economic growth.

---

[5]See Gertrude Schroeder, "Recent Development in Soviet Planning and Incentives," in *Soviet Economic Prospects for the Seventies*, Joint Economic Committee, 1973.

DECISIONMAKING ON FOREIGN TECHNOLOGY

INSTITUTIONS INVOLVED IN THE ACQUISITION OF FOREIGN TECHNOLOGY

Decisions concerning the purchase of foreign technology, like any other economic decision in the Soviet Union, take place within the framework of a system of central economic planning. A brief catalog of the major institutions involved in this process suggests the variety of the interests involved in such purchases and the complexity of the process itself. These actors fall into two major categories, the State and the Communist Party apparatus.

State Apparatus and Technology Acquisition

The Council of Ministers.—At the top of the Soviet Government organization is the Council of Ministers of the U.S.S.R. This body is the formal repository of all State authority. As such, it is the theoretical locus of administrative responsibility for trade matters. In practice, however, decisions are usually taken in ministries and agencies that operate under the Council and are rubber-stamped at the highest level. To administer

the massive Soviet economy, the Government relies on the operation of a variety of general and specialized bodies.

Gosplan.—Gosplan, the State Planning Commission, is the central Government's chief agency for conducting the work of general economic planning. Part of its work consists of import planning, which is conducted by Gosplan's own Department of Foreign Trade. The primary responsibility of this Department is to integrate foreign trade into the national economic plans. In addition, Gosplan is responsible for planning R&D and innovation. This work is carried out in a separate Department for the Comprehensive Planning of the Introduction of New Technology into the National Economy.[6]

Gostekhnika.—Known in the West as the State Committee on Science and Technology (SCST), this organization bears primary responsibility for the coordination of R&D work throughout the economy.[7] It is chief advisor to the central Government on national technological policy. Part of the latter

---

[6]See Joseph S. Berliner, *The Innovation Decision in Soviet Industry* (Cambridge, Mass.: MIT Press, 1976).
[7]Ibid.

function consists of developing strategies to acquire Western technology and integrate it with domestic R&D capabilities. SCST participates in negotiation for the acquisition of sophisticated technology from the West, often providing technical expertise.

U.S.S.R. Academy of Sciences.—This body consists of about 600 members who bear the responsibility for supervising the greater part of scientific research work in the U.S.S.R. The Academy's jurisdiction includes about 200 scientific establishments employing some 30,000 scientists. Through its Administration of Foreign Affairs, the Academy not only monitors scientific developments in the West, but plays an active role in scientific exchanges. While the Academy's primary concern is basic research, it is obliged to submit proposals to SCST concerning applied R&D leading to innovation.

Military Industrial Committee of the U.S.S.R. Council of Ministers.—The existence of this committee has never been officially confirmed, but it is probable that it holds primary responsibility in the State structure for the coordination of all activities in the area of armaments production. While the role of the Military Industrial Committee in technology acquisition is unclear, it undoubtedly participates in decisionmaking on technology purchases.

Ministries.—The central administration could not possibly directly supervise each of the 43,788 industrial enterprises that fall under the Soviet system of central planning. An intermediate level of administration is therefore provided by ministries, which are interposed between enterprises (or production associations) and the central and republic Governments.

Ministries are organized by branch and those dealing with the economy are differentiated by product (e.g., Petroleum Ministry) (see figure 12). A major function of the economic ministries is to formulate and implement technical policies in relevant sectors. This function is accomplished through the Main Technical Administration of each ministry. Each ministry also includes a Department of Foreign Affairs.

Different economic ministries are involved in acquisition decisions to the extent that foreign technology can be incorporated into their sectors. At present, the extractive industries, chemicals, and machine tools are especially active in foreign trade.

The Ministry of Foreign Trade administers all Soviet trade; no foreign trade operations can be processed outside of its structure. The Ministry encompasses dozens of import-export foreign trade associations organized according to product category. These associations act as intermediaries between relevant Soviet ministries and foreign firms and are empowered to sign contracts. They are governed by boards which are composed of specialists of the associations and representatives of the relevant ministries.

Administrative decisions in the Ministry of Foreign Trade are made through the cooperation of three internal divisions:

1. The trade-political administrations. These are divided by region. A separate trade-political administration exists for trade with the United States, while a second administration deals with all other capitalist countries.
2. Functional administrations for planning, currency, legal matters, etc.
3. Administrations for single commodity groups. A separate administration of this type exists for machinery and equipment imports from capitalist countries. The relationship of these administrations to their import-export associations is shown in figure 13.

Other Agencies.—There are many other Government agencies involved in some portion of the process of technology acquisition. The Ministry of Finance participates in the development of hard-currency plans and administers their implementation. The Vneshtorgbank, or Bank of Foreign Trade, is subordinated to the State Bank. It gives credit to all Soviet organizations for foreign trade

Figure 12.—U.S.S.R. Council of Ministers

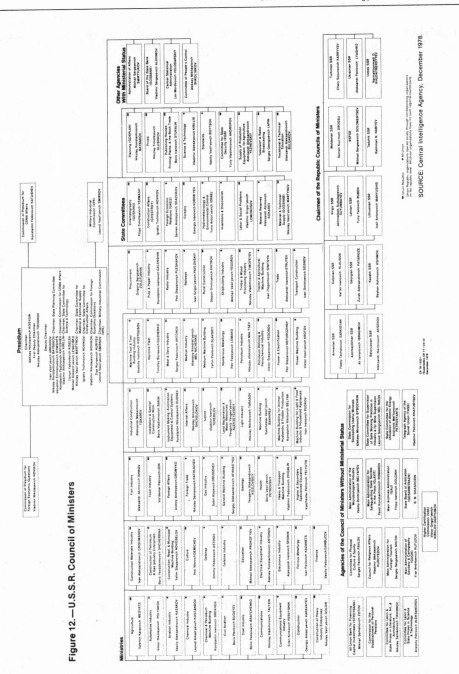

SOURCE: Central Intelligence Agency, December 1978.

Figure 13.—Organizational Structure of Operational Management of Foreign Trade

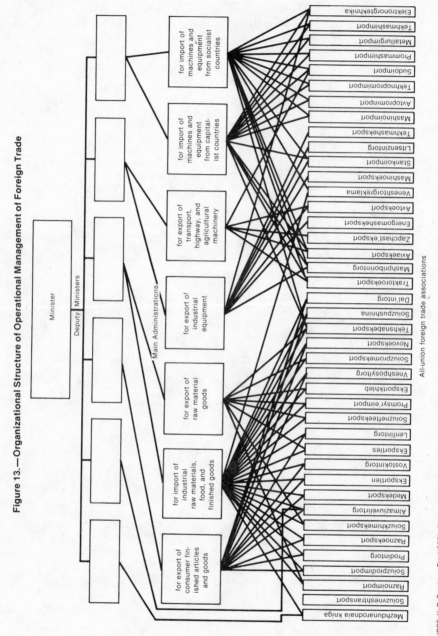

All-union foreign trade associations

SOURCE: V. P. Gruzinov, *The USSR's Management of Foreign Trade* (Moscow, 1975).

in rubles and deals with clients in hard currency. The All-Union Chamber of Commerce primarily arranges exhibitions and contacts between foreign firms and Soviet organizations.

Within the Soviet R&D establishment, organizations exist which in many cases adapt Western technology purchases and advances in applied research to domestic production. Research and development institutes (under individual ministries) specialize in applied research in a specific technological area. Once a new product or process has been developed to a point where it is thought ready for commercial application, it is handed over to engineering-design organizations, which mark out the details of materials, grades, sizes, shapes, and other technical specifications of the final product and the precise machinery, assembly quality control, and other production arrangements for manufacturing it. If reverse engineering of a Western product is possible, these organizations will have the expertise to accomplish it. There are over 2,000 such organizations in the U.S.S.R. subordinated to various ministries.

In addition to the organizations listed, a number of other segments of the State structure intercede in the process of technology acquisition. In particular it is clear that the Ministry of Defense is not only concerned with Western technological achievements, but may have a deciding voice in individual import decisions. The precise structure of the relationship between the Ministry of Defense and the negotiations conducted by the Ministry of Foreign Trade is not, however, known in detail.

The Communist Party and Technology Acquisition

All levels of Soviet administration—including that of the Communist Party—may provide inputs in the process of foreign technology acquisition. In the most general sense the Government, including the planning bodies, exercises detailed control over planning and purchase of technology, and the Party bureaucracy avoids direct involvement in practical decisions once broad policy goals have been met. The relationship between State and Party in technology acquisition is, however, ambiguous and varies not only with time but with the political sensitivity of a given purchase. Under the present regime, Party organs ordinarily exercise a veto over initiatives made by the state bodies while eschewing contact with representatives of Western firms.

In addition to its functions of policy formulation and monitoring of administrative operations, the Party bureaucracy exercises ultimate control over technology acquisition as it would over any other Government function, through its absolute control of personnel in the State structure. All officials concerned with technology acquisition are carefully screened not only by State but also by relevant Party organs.

The influence of Party organs is not confined to the national level. Party structures on the republic and provincial levels often have considerable input in technology acquisition. This is particularly true in the case of construction of facilities to house new equipment and machinery. Local Party organizations are directly responsible for monitoring construction of all plants in their regions. Their inability to organize such efforts has often proved to be a major barrier to swift implementation of Western technology purchases.

TECHNOLOGY ACQUISITION AND PLANNING

There are indications that the role of foreign technology transfer in the foreign trade planning system as a whole is being reevaluated. At present, import decisions are made as part of annual planning cycles, and foreign trade is often utilized as a means of filling short-term planning shortfalls. The result of this is that, as figure 14 demonstrates, the vast majority of Soviet imports from the industrialized West have consisted of non-technology-intensive manufactured goods, and agricultural and other primary products. Attempts to more fully integrate current and prospective foreign trade plans into national economic plans are likely to result in a greater proportion of hard-currency expenditure devoted to more productive high-technology imports.

The acquisition of technology from the West is accomplished in two general stages. First, hard currency is allocated among sectors. Secondly, individual purchases are determined through the participation of ministries, their production associations, research institutes, and engineering-design bureaus. In sensitive cases, detailed decisions are formally made by higher levels of administration.

Both the distribution of hard currency and concrete purchases are accomplished either in the framework of the 1- and 5-year plans or through irregular (ad hoc) decrees of relevance to single industrial branches or enterprises. From year to year the allocation of hard currency—the primary quantitative determinant of imports—is basically preserved across sectors. Changes in particular priority targets or drastic reductions in the hard-currency stock do, however, periodically alter these proportions. World market prices quoted in hard currencies are utilized for export-import operations with the West.

As is true of all aspects of Soviet planning, hard-currency allocations are determined on the basis of level achieved—every year the allocation is marginally increased as compared to the preceding year (subject to high-level changes in national economic priorities).[8]

Engineering-design bureaus, which project new construction or modernization needs, determine which particular types of Western technologies can be used. They present to their ministries specifications of needed equipment and know-how. Ministries then send drafts of their requests to Gosplan and the Ministry of Foreign Trade.

If these requests are within the limits of the hard-currency plan, they are routinely approved and included in the trade plan. But this is usually not the case. Ministry requests often exceed the hard-currency allocation. Such discrepancies are resolved through bureaucratic negotiation between

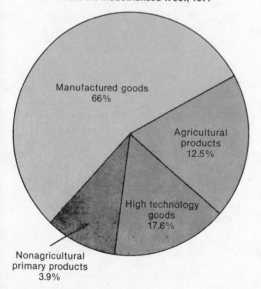

Figure 14.—Composition of Soviet Imports From the Industrialized West, 1977

Manufactured goods 66%

Agricultural products 12.5%

High technology goods 17.6%

Nonagricultural primary products 3.9%

SOURCE: Office of Technology Assessment.

[8]Igor Birman, "From the Achieved Level," *Soviet Studies*, XXX (2).

ministries, SCST, local Party organs, etc. If Gosplan cannot resolve the discord, it is usually settled by a Deputy Chairman or Chairman of the Council of Ministers and, in the most crucial cases, by the Politburo of the Party.

National economic plans specify only large purchases of Western technology. In addition, ministries are allocated limited amounts of hard currency with which to deal directly with the associations of the Ministry of Foreign Trade. In recent years some large enterprises engaged in the production of goods for export have similarly been permitted relatively small amounts of hard currency to be used at their own discretion for the purchase of capital goods.

Flexibility in planning is achieved through irregular decrees, issued every 3 to 7 years by the Central Committee of the Party and the Council of Ministers for each branch of the economy. Such decrees often plan shifts in the distribution of hard currency among sectors and are very concrete in nature, exactly itemizing equipment and technology to be imported. They are incorporated into subsequent national plans.

Decisionmaking on individual technology purchases is based on a coordinated system of collecting and processing Western scientific and technical information. This is supervised by SCST. Nearly all R&D bodies—in particular the engineering-design bureaus—and many large enterprises collect relevant information. In addition, each ministry includes at least one Institute of Scientific and Technical Information, one of the functions of which is to process available Western scientific and technical data

Under this system, Western technical literature is translated, published, and made available to relevant specialists in a relatively short time. Specialists who are sent abroad are required to report on Western technological achievements. Soviet intelligence services also engage in scientific and technical espionage.

As a buyer of Western technology, the U.S.S.R. actively encourages trade fairs and other exhibitions in which foreign firms may bring their most advanced and salable products to Moscow. While such exhibitions are accepted practice in overseas marketing, it is common Soviet practice to attempt to obtain as much detailed technical and operational data as possible on desired products without actually concluding purchase agreements.

The Soviet system of information gathering on Western technological developments, while not ideal, guarantees that the Soviet negotiator is relatively well-informed and cognizant of both the technical specifications and availability of a given product in different Western markets.

CRITERIA AND PRIORITIES FOR TECHNOLOGY PURCHASES

Writing in 1941, Soviet economist D. Mishustin summarized the basic aims of Soviet technology acquisition from the framework of import policy:

> The basic task of Soviet importation is to use foreign goods, and first of all machinery, for the rapid accomplishment of the plans of socialist construction and for the technical and economic independence of the U.S.S.R.[9]

Then, as now, one of the fundamental goals of Soviet import policy in general was to improve the technological base of production with the help of foreign technology while at the same time carefully avoiding dependence on those imports.

In the Brezhnev era the concept of comparative advantage has been added to the dominant theme of technical and economic independence:

> In the final analysis, the purpose of foreign trade is the procurement of imported goods and services a) which are not produced

---

[9]D. Mishustin, *Vneshnaya Torgovlya SSSR* (Moscow, 1941), p. 6.

within the country at all, b) are produced, as a result of whatever temporary reasons, in insufficient quantity and c) whose production within the country is more expensive than their purchase on the foreign market.[10]

After hard currency is distributed sectorally in the planning process, a number of criteria based on these general policies of technical independence and economic advantage are utilized to make individual purchase decisions. These criteria are not entirely economic. Since the middle of the 1960's, Soviet economists have attempted to determine the economic benefits of import choices through the use of foreign trade efficiency indices. These are formulae that provide a measure of the cost to the national economy of producing a good for export relative to the foreign exchange received abroad, or of the foreign exchange expended abroad in purchasing a good or technology relative to what it would have cost to produce the good domestically.[11] Thus far, such attempts have been singularly unsuccessful, and there is at present no reliable method of measuring foreign trade efficiency at the disposal of decisionmakers. At the heart of the problem lies the insulation of the Soviet price system from world markets and the failure of internal prices to reflect relative scarcity. In lieu of reliable economic evaluation of technology purchases, Soviet buyers simply attempt to minimize hard-currency cost within the context of a shifting set of preferences and priorities.

The first of these is military. All other factors being equal, those types of technology that directly or indirectly enhance military capabilities are given first priority. While many Soviet purchases do not, in fact, embody any military potential whatsoever, it is true that some transfers of an ostensibly purely civilian nature have been given higher allocation priority due to their potential contribution or convertibility to military use.

A second factor involves a general preference for disembodied as opposed to embodied technology, i.e., know-how as opposed to products. The transfer of disembodied technology may require a relatively high domestic contribution of R&D, but buying large amounts of hardware generally raises hard-currency costs. Since an individual ministry is allocated a fixed sum of hard currency, whenever possible it will attempt to minimize the cost of Western inputs while maximizing relative domestic inputs in the development of a given innovation.

Third, purchases of technological complexes are preferred to purchases of single items or processes, so long as the hard-currency cost is not prohibitive. Such systematic transfers ensure the swiftest and most productive utilization of foreign technology purchases.

Another element in setting technology purchase priorities is that preference be afforded those products and processes that can be easily duplicated for production in the U.S.S.R. This tendency stems from the desire to minimize increasing dependence on Western technology.

Finally, an increasingly important criteria for technology purchases involves their use in export industries. Since the generation of hard currency (and further imports) is directly dependent on export potential to the West, increasing priority has been given to projects producing goods for Western markets. It is impossible to definitely rank these criteria in order of their importance in the decisionmaking process. The factors influencing the choice of an individual technology purchase are often ambiguous and priorities vary according to situation. It is clear, however, that a lack of definitive economic formulae for import decisions allows for the influence of noneconomic—e.g., military—factors in the decisionmaking process.

---

[10]G. Smirnov, "K Voprosu Ob Otsenki Economicheskoi Effektivnosti Vneshnei Torgovlyi SSSR," in *Voprosy Ekonomiki*, No. 12 (1965), p. 94.

[11]See Lawrence J. Brainard, "Soviet Foreign Trade Planning" in *Soviet Economy in a New Perspective*, Joint Economic Committee, 1976.

THE ROLE OF WESTERN TECHNOLOGY IN THE SOVIET ECONOMY

ABSORPTION AND DIFFUSION OF WESTERN TECHNOLOGY

A recent study of Soviet technological levels done under the auspices of the University of Birmingham (England),[12] found that in most of the industries it examined—armaments, nuclear, electric power, metallurgy, machine tools, computers, and chemicals—the technology gap between the U.S.S.R. and the West has not diminished substantially over the past 15 to 20 years, either at the prototype/commercial application stages or in diffusion of advanced technology. The Birmingham study further concluded that Soviet growth has been largely based on output using traditional technology. For example, the study points out that even though the higher technology sector of petrochemicals has grown relative to the chemical industry overall and dominates the total industry output, petrochemical products are manufactured with older, proven technologies.

This pattern of growth, a result of slow absorption and diffusion of new technology, can be seen in the areas in which Soviet industry has performed best. The U.S.S.R.'s most productive technological developments came in industries that were based on well-established technology, with advances coming primarily from successful scaling-up of existing technology. Advances in the metallurgy, power generation, and power transmission industries, for example, are the result more of engineering than of innovation in processes.

The pattern of better performance of those industries that are not based on rapidly changing technology is part of the reason behind the apparent shift in the technology import policies of the U.S.S.R. over the past decade. As the Soviet economy expanded during the early postrevolutionary years, and again following World War II, the industries most needed to support growth were traditional ones such as metallurgy, machine-building, machine tools, and the energy sector. The growth was produced by relying on slowly changing technologies, limited but sometimes essential imports of foreign technology, and massive increases in the supply of labor and capital.

This chapter has already shown that to bring about the rapid industrialization of the economy envisioned by the first 5-year plan which began in 1928, the U.S.S.R. turned to large imports of machinery and equipment. Prior to this period, such imports had averaged only about 0.3 billion rubles per year; during the next 5 years, they rose to an average of 1.4 billion rubles per year. Following the end of the first plan, imports of machinery and equipment dropped back to an average of 0.3 billion rubles per year.[13] Relations with Western firms supplying technology were designed to be short-lived, with the aim of minimizing Soviet dependence. This aim also guided the country's overall import and export policy.

The fear of relying on a potential adversary was one reason for the Soviets' strong desire to minimize dependence on the West for technology and products. In time, such technology transfers were also limited by constraints imposed by Western export controls.

The Soviets began to copy prototypes of equipment that they had been able to obtain from the West. The ultimate failure of this practice, coupled with an inability to rely on domestic innovation, has led to several changes in import policies during the past decade: an apparent shift toward greater reliance on Western technology; a willingness

---

[12]R. Amman, J. M. Cooper, and R. W. Davies, ed., *The Technological Level of Soviet Industry* (New Haven, Conn.: Yale University Press, 1977).

[13]George Holliday, "The Role of Western Technology in the Soviet Economy," in *Issues in East-West Commercial Relations,* Joint Economic Committee, January 1979, p. 47.

to permit—in fact encourage—long-term agreements involving large volumes of foreign exchange with Western firms supplying technology; and in some cases, limited changes in Soviet management practices.

Although the Soviet Union is not willing to open its economy to full interdependence with the West, more extensive use of Western technology is no longer feared. At the same time, the West has also liberalized its constraints on export control. In the United States, for instance, the Export Administration Act of 1969 as amended has reduced the list and raised the permissible performance characteristics of controlled items (see chapter VII).

Soviet technology transfer policy is also strongly affected by the country's growth policy, which has been revised. The Soviet Union no longer enjoys vast pools of underutilized labor that could be mobilized for economic growth by transfers from the agricultural sector to industry or by increased labor participation rates of women. According to the Central Intelligence Agency's (CIA) projections, the Soviet Union will experience a sharp decline in the rate of expansion of its labor force in the 1980's to less than 1 percent per annum by 1982.[14] Other factors, such as the distribution of population, will further strain the amount of growth that can be obtained through larger labor inputs.

This labor constraint also comes at a time of decreasing productivity of capital inputs. Furthermore, an increasing share of Soviet capital investment is now going by necessity to agriculture, the consumer goods industries, and other sectors that do not directly increase the productive capacity of the economy. As a result, increases in labor productivity are expected to account for up to 90 percent of all growth in industrial output, and virtually all growth in the agricultural sector.

Technological improvement is to be the basis of planned increases in Soviet labor productivity. Industries based on traditional technology are becoming less important relative to those based on sophisticated, rapidly changing technology. These include the organic chemical, electronics, and computer industries. More traditional sectors, such as oil and gas and machine tools, are being modernized with technology from the electronics industry: computer numerical control for precision machine tools, computer analysis of seismic data, and automatic control of hydrocarbon production.

The growing importance of new technology, and the increasing importance of industries that are experiencing rapid advances in technology, coincides with a continuing weakness in the Soviet economic system's capacity to absorb and diffuse technology. The U.S.S.R.'s problems in this area increase the need for importing technology from the West, since the transformation of domestic innovation into new technology is often slow. On the other hand, the same problem reduces the effectiveness of imported technology. The problem lies not so much in the quality of Soviet basic research, nor in the level of theoretical knowledge, but rather in the system's inability to turn theoretical knowledge into prototypes, and even more importantly, to move rapidly from prototypes to large-scale industrial production. The reasons for this lie in such factors as insufficient incentive, poor organization, and the rigidities that result from central planning.[15]

In the West, innovation and new technology development are encouraged by a desire to beat the competition and thereby maximize profits and cut costs. Rewards for firms that innovate successfully, and competitive pressures felt by those firms that do not innovate or at least duplicate new technology a

---

[14]Central Intelligence Agency, *Soviet Economic Problems and Prospects: 1977*, p. ii.

[15]David Granick, *Soviet Introduction of New Technology: A Depiction of the Process*, Stanford Research Institute, January 1975.

short time after a competitor does so, provide sufficient incentive to ensure Western capabilities for development, absorption, and diffusion of new technology.

Large research institutes are responsible for R&D in the U.S.S.R., but such institutes lack incentives to consider adequately the practical application of their work. Planners determine the direction that scientific inquiry will follow in research institutions. They are encouraged to develop ideas that qualify as innovations, but not to apply the ideas to the production process. Separate institutions, called Engineering Design Organizations, have responsibility for applying new technologies, but their successes do not reflect as favorably on the research institutes as does the propagation of additional new ideas, whether practicable or not. Research and development receive the greatest emphasis, followed by application engineering and—with the least concentration of funds and effort—product development. In the West, the emphasis is reversed.

A number of other factors at the enterprise level also inhibit the introduction of new technology. The use of new processes and the development of new products involve risks; since not all attempts are successful in the West, the risks must be minimized and the rewards for success maximized in order to promote such efforts. The Soviet system works in reverse, maximizing risk and minimizing reward. Success for a production enterprise in the U.S.S.R. is measured primarily in ability to exceed the output quota set for the year, although recent reforms permit limited consideration of other factors in determining bonuses for workers and plant managers. The risk of trying a new technology or product is great, since an unsuccessful effort is bound to result in failure to meet the plant's output goal for the year. Even if the innovation is moderately successful, the increased output once the new equipment is operating may not be sufficient to offset the loss of output during conversion. The benefits of very successful efforts are short-lived, because while a jump in output due to the introduction of

new technology will surely result in bonuses for managers during the first year, they will just as surely result in a jump in the plant's output quota the following year.

In the West, managers often try several new technologies before finding one that provides sufficient long-run benefits to justify the cost of all the experimental efforts. Similarly, managers find that new processes often require several years to work the "bugs" out of the system. The ability to judge a new technology on its return over a long period of time, and the willingness to accept the fact that most innovations will probably not prove to be successful, are major advantages of the Western competitive system over the Soviet model.

Cost reduction is a major incentive for innovation in the West. In the U.S.S.R., on the other hand, even otherwise successful innovations often produce unacceptably high overall costs due to the rigidities of central planning. A Soviet plant seeking to employ a new process often must rely on other plants to supply related new equipment, and may find the necessary equipment unavailable. In the latter case, the plant may be forced to develop the equipment on its own at relatively high cost. Similarly, a Soviet plant beginning to produce a new product does not have the right to determine the price at which the new product will be sold. This means that the centrally determined price may not cover the plant's product development and production costs.

The rigidities of central planning lead to another problem which tends to inhibit the diffusion of technology. In the U.S.S.R. extensive use is made of vertical integration of production facilities. This minimizes the enterprise's dependence on outside suppliers. While this structure does encourage a firm to develop equipment and technology to meet its own needs, it also leads to a lack of standardization and to the inefficient production of equipment in small quantities. Furthermore, this horizontal independence means that new technology is less likely to be transferred to other plants that might benefit from it.[16]

In theory, the Soviet Union should have a distinct advantage over the West in the diffusion of technology, since new technologies developed in the U.S.S.R. are State property rather than trade secrets, and should be freely available to any enterprise able to use them. In practice, however, communication among Soviet production enterprises is poor, and there may be long delays in the publication and dissemination of information about new developments. The slowness of journals to publish research papers and report other developments may result in a great deal of duplicated effort. It has been reported that the average time between submission of a paper and its publication in the important Soviet journal "Electrochemistry" was as much as 2½ years, and individual articles have been delayed as long as 4 years.[17] Cooperation and communication among plants within the same industry, and between organizations in different industries, are seriously inadequate.

The factors mentioned above all inhibit the introduction and diffusion of new technology throughout the Soviet economy; as a result, the share of total output in the U.S.S.R. due to the introduction of new methods or new products is lower than the comparable share for other industrialized countries. When performance is measured solely in terms of increased output, there is little incentive to change the form of the output. This leads to production of unchanged equipment over a long period of time, often even after better equipment has been developed.

Other forces also encourage U.S.S.R. enterprises to continue using outdated production equipment. Because of shortages of

equipment and the lack of direct connection between the cost of production and the cost of the final product, depreciation rates for Soviet equipment are set very low by Western standards. In the West, out-dated machinery is taken quickly out of production and replaced by newer equipment that will lower production costs. High depreciation rates for equipment encourage these shifts, as does a strong market for secondhand equipment. No such secondhand market exists in the U.S.S.R., so the equipment continues to be used at the plant. As a result, the replacement rate for Soviet equipment in most industries is much lower than in corresponding industries in the West.

Many of these same problems reduce the effective introduction of new technology imported into the U.S.S.R. from abroad. This seems to be particularly true for those technology transfers that require application of Soviet design and manufacturing engineering to the imported technologies and can best be seen in U.S.S.R. attempts to duplicate Western technology based on trade publications, product literature, plant tours, personal conversations, etc., or the import of a small number of product units to serve as models or prototypes for Soviet production. As previously noted, the U.S.S.R. also maintains an extensive collection of Western journals, some of which are systematically translated into Russian.

While the use of Western equipment as prototypes for production of new technology reduces the need for Soviet R&D, it still requires significant application of domestic effort, particularly in terms of developing production methods. For rare pieces of equipment or products that can be dismantled so as to uncover production techniques by examination, "reverse engineering" is relatively simple. With more sophisticated products, such as integrated circuits or petrochemicals, however, reverse engineering is much more difficult and impractical.

One of the highest cost components in innovation in the West, and one of the great advantages of the competitive market sys-

[16]John Hardt and George Holliday, "Technology Transfer and Change in the Soviet Economic System," *Issues in East-West Commercial Relations,* Joint Economic Committee, January 1979, p. 74.

[17]M. Perakh, "Utilization of Western Technological Advances in Soviet Industry," in *East-West Technological Cooperation* (Geneva: NATO Colloquium, 1976), p. 179.

tem over central planning, is the determination of which products and processes will eventually prove to be economically and technically viable. In the West, the market system makes this selection based on efficiency and profitability, thus screening out innovations that are not worth further development. The work done on products and processes that never reach the final stage of commercial introduction and acceptance is as much a cost of technological advancement as the work done on successful innovations. Marx considered this process to be a major flaw in the capitalist economy—a wasteful misallocation of resources. The innovation engendered by this method of selection, however, tends in the long run to more than compensate for its real costs. Thus, by concentrating their efforts only on those new products or processes that have already been screened by the Western market mechanism, the U.S.S.R. is able to avoid the cost of following infeasible or uneconomical ideas.

According to East European officials, an average of 5 to 7 years elapses between the beginning of efforts to copy a Western product and successfully readying it for production in worthwhile quantities.[18] This time-lag means that the copied equipment is often outdated, at least in Western terms, by the time it is used. This period may, however, be shorter than the time it would have taken for the U.S.S.R. to develop the product completely on its own.

But the U.S.S.R. is increasingly interested in obtaining from the West technology of the type that is difficult to copy without assistance. The Soviet desire and willingness to seek more active forms of technology transfer is enhanced by the rate at which Western technology is advancing in such leading industrial fields as petrochemicals, electronics, and precision instruments. In the past, if the technology embodied in a piece of equipment could be duplicated within a few years, the U.S.S.R. could remain only slightly behind

the level of technology being used in the West. Now, where technology is advancing rapidly, keeping up with the West is more difficult, and there is pressure to increase the speed with which technology is imported from abroad, assimilated, and diffused. These circumstances also encourage greater emphasis on more efficient selection of technology imports.

Studies by Western specialists have noted several factors that make Soviet technology acquisition less efficient than it could be. One frequently voiced criticism is the length of time it takes for the Soviet Union, once a decision has been made to import certain equipment or technology, to decide which nation and firm will supply it, and then to accept delivery and get the equipment set up and into working order. One study has compared the time required for the U.S.S.R. to accomplish this with the average time required in the West in the chemical and machine-tool sectors. It found first, that the U.S.S.R. required about twice as long to sign a contract for a particular need as would have been the case in the West. This was due to several factors. Initial inquiries from the Soviet Union were frequently vague, and the form of the final order was often different from the original specifications. The study concluded that vagueness at the initial stage probably results from a genuine lack of Soviet knowledge or decision on what will be the best choice, rather than from any deliberate attempt to make the process more difficult.

Second, the Soviets require much more extensive documentation than other countries. While some of this may be attributed to a lack of trust on the part of Soviet trade officials, there is no doubt that the additional documents ultimately make it easier for the U.S.S.R. to assimilate and possibly duplicate the technology being provided.

Third, there is no direct contact between the supplier of technology and the final user. This is an important source of delay in the acquisition process. The supplier must work with the Foreign Trade Organization which

---

[18]Business International S. A., *Selling Technology and Know-How to Eastern Europe: Practices and Problems* (Geneva, Switzerland, November 1978),p. 6.

handles that type of equipment, and communication between the supplier and user takes much longer and is subject to greater possibilities of misunderstanding than would be the case in the West. Inexperience in the installation, operation, and maintenance of complex equipment, along with poor management and planning, frequent shortages of adequately trained personnel to learn to operate and repair equipment, and problems with the quality of raw materials or other inputs to be processed with the new equipment, are additional factors that lengthen the time between equipment delivery and proper startup.[19]

Delays, the desire to reach certain production levels within a set period of time, and the inability of Soviet industry to supply sufficient equipment to meet those goals, are also important factors in the Soviet decision to import machine tools and chemical equipment. Once a decision to import is made, the user typically seeks the best equipment available. This leads to the purchase of equipment with performance characteristics that exceed anything the U.S.S.R. is itself capable of producing.

The Soviet Union seems to be increasingly aware of the need to use foreign trade and technology acquisition to improve its economic performance. Foreign trade is no longer viewed as a necessary evil; in fact, there is a growing awareness that the demands of Western markets can have a positive effect on the quality of goods produced for export, and thus on the level of quality in the entire economy. Some of the hard currency earned by an enterprise's exports is being returned to the enterprise, providing an incentive to improve the technology used by the plant, as well as the means by which the enterprise can afford to import additional Western technology.[20] Importing Western technology has become more attractive during the 1970's, as Western suppliers have competed for the Soviet market, as long-term credits have been made increasingly

available, and as Western firms have become increasingly willing to accept product buyback provisions as a means of financing technology imports.

The increased attractiveness of technology transfer, coupled with the U.S.S.R.'s growing need, has resulted in more purchases of machinery and equipment from the West, together with the use of cooperation agreements and other arrangements with Western firms and countries to promote technology transfer. The CIA and other sources have estimated that as much as 10 to 12 percent of total Soviet investment in machinery and equipment has come from abroad during the 1970's. While purchases of equipment from the West have increased rapidly, purchases of licenses have been growing quickly as well; according to Soviet officials, license purchases are expected to increase even faster than equipment purchases.[21]

The U.S.S.R. has signed Government-to-Government cooperation agreements with most countries of the West since its initial agreement with the United States in 1972. Some 150 authorized projects are either underway or planned on the basis of these technology agreements with the United States. A number of American, West European, and Japanese firms have also signed private cooperation agreements with the State Committee for Science and Technology. These cooperation agreements have been concentrated in high-technology areas such as electronics, computers, instruments, and various types of engineering.

But while the U.S.S.R. has expanded the number and variety of technology transfer mechanisms available to it, the most effective form of technology transfer, the joint venture, has not been permitted. Joint ventures with Western interests have been used in Yugoslavia since 1967, in Romania since 1971, and in Hungary since 1972.

---

[19]Philip Hanson and M. R. Hill, unpublished manuscript.
[20]Holliday, op. cit., p. 55.

[21]Z. Zeman, "East-West Technology Transfers and Their Impact in Eastern Europe," in *East-West Technological Cooperation*, op. cit., p. 171.

It is very difficult to estimate the actual impact of Western equipment and technology on the performance of the Soviet economy. The available information does not even permit an accurate determination of the share of Soviet capital equipment that comes from the West, although most specialists who have studied this problem estimate the share to be between 4 and 6 percent. This low level is the combined result of the shortage of hard currency in the U.S.S.R., the Soviet policy of wanting to avoid excessive dependence on the West, and Western export controls.

In sum, however, a general picture of Soviet import policies and their effectiveness may be drawn. The U.S.S.R. has had a long history of systematically utilizing Western technology to compensate for domestic economic shortcomings. The present system through which decisions regarding imported technology are made is incompletely understood in the West, and is characterized by its complexity and slowness. The prioritization of technology for import seems to be dominated by the availability of hard currency and the potential economic and military impacts of the technology, but no consistent and universally applicable set of criteria has emerged. The Soviets are well-informed, however, about Western technologies under consideration and their selections usually reflect careful evaluation of the properties of the technology relative to their specific needs.

The absorption and diffusion of Western technology in the U.S.S.R. have been retarded by structural features of the Soviet economy and the rigidities inherent in central planning. The Soviets appear to be aware of these defects and may attempt to correct them with further purchases of Western management and other know-how. Meanwhile, the economic impact of imported technology is not as great as it might have been on a Western nation purchasing on a similar scale.

THE ECONOMIC IMPACTS OF WESTERN TECHNOLOGY

Several factors affect the degree of economic benefit to be derived from the purchase of any technology. Obviously the initial selection is important. In situations where the availability of hard currency poses restraints on the amount of technology that can be acquired, a Communist country can ill afford to make a poor choice—either in terms of the industry or sector singled out as liable to benefit from Western technology, or in the selection of a particular machine or process from all those available in the West. The criteria that ideally govern this choice include fundamental investment decisions (the choice of capital versus labor-intensive technologies); the sophistication of available domestic technology relative to the imported technology; the indigenous capabilities of the country's R&D sector; and the available infrastructure.

A Western technology may prove economically beneficial in several ways. First, assuming the existence of the necessary infrastructure, including trained manpower, the productive capacity of an industry may be enhanced. Even if no diffusion of the technology occurs, this may be a net gain to the economy. Of course, in the absence of infrastructure, the new technology may produce a net loss in macroeconomic terms. This is the case with "resource-demanding" technologies, i.e., those that require substantial capital or labor inputs before they become operative.

Second, the economic benefits of the technology may be enhanced if it can be used to increase productivity in other industrial sectors, or if the technology embodied in imported equipment can be replicated in equipment produced by the domestic economy. Such diffusion requires certain capabilities in the domestic R&D sector, yet this alone is not sufficient to close a technology lag or gap. The true test of the effectiveness of technology transfer is not only whether imported technology can be diffused at a technological level comparable to that of the West, but if it can also be the basis of domestic R&D efforts to upgrade it. It is only when imported technology can be fully absorbed in the economy—and improved on—that technology gaps can be reduced.

It is generally true that innovations in the Soviet economy have followed their introduction in the West. This impression is supported by a number of studies, some of them concentrating on a single industry, others taking a broader perspective and attempting to measure the effects of technology transfer on Soviet productivity, income, and technological level.

The impact of Western capital equipment on Soviet economic performance appears to be much larger than the small Western share of total capital stock would suggest. The decision to import technology and equipment is based on the judgment that it will produce better results than if that money were spent on domestic equipment. Thus, theoretically at least, the worth of a given unit of imported equipment has a greater effect on economic performance than the same unit's worth of the domestic equipment for which it is being substituted.

Whether all import decisions are made with net productivity as the deciding factor is, however, open to question. Import decisions are based partly on noneconomic criteria, and the Foreign Trade Organization negotiating a purchase often does not know the grounds for the decision or the net effect of the purchase on the industrial sector. This is not only due to a lack of communication between organizations responsible for putting any new process into production, but also to the administered price system which does not reflect relative scarcities. Within the Council for Mutual Economic Assistance

(CMEA), a growing body of literature on the use of foreign trade indexes, which would address this problem, has appeared. But such indexes are not used extensively and provide only one of many kinds of information on which import decisions are based. It must be noted, however, that in spite of the Soviets' inability to determine precisely the profitability of proposed technology imports, they have rarely had to make decisions on projects of marginal value. Owing to the relatively small volume of trade, the Soviets have had their choice of transactions in which productivity gains were clearly high.

An econometric study conducted jointly at the Stanford Research Institute and Wharton School constructed an input-output model of the Soviet economy (SovMod), which attempted to determine the effect of the growth in equipment and technology imports from the West between 1968 and 1972 on Soviet overall economic performance.[22] The study concluded that if Western exports during this period had stayed at 1968 levels, the Soviet Union would have had an installed stock of Western equipment that was 20 percent below the actual 1973 level, and that Soviet growth during this period would have dropped from 32.1 to 29.6 percent. This conclusion implies that Western equipment accounted for approximately 2.5 percent of the U.S.S.R.'s rate of growth during this period, or several times the share of this equipment in Soviet capital investment.

Studies like this are controversial, however. The assumptions on which the model is based have been questioned and other researchers have reached significantly different conclusions using the same data. It has been contended, for instance, that the existence of significant differences between the productivity of Western and Soviet capital equipment is not supported by statistical analysis.[23] This finding implies that the contribution of Western equipment to the per-

[22]See Herbert Levine and Donald W. Green, "Implications of Technology Transfer for the U.S.S.R.," in East-West Technological Cooperation, op. cit.

[23]Philip Hanson and M. R. Hill, unpublished manuscript.

formance of the Soviet economy is not significantly different from the contribution of Soviet equipment. More than anything else, the conflicting results obtained from these macroeconomic approaches point to the wisdom of reverting to the study of the actual effect of Western equipment and technology on the capacity of individual sectors of the Soviet economy. A disaggregated approach in which each industry is examined individually to determine what equipment and technology has been transferred, how well and how quickly it has been absorbed and diffused, and what changes there have been in comparative levels of technology, may be more productive and accurate. It must be noted that such information is very difficult to obtain even in the West, where access to information is relatively free. The details assembled here must necessarily be taken as partial and impressionistic. This report will concentrate on two in-depth examinations of the industries in which Western technology is most important—the oil and gas equipment industry, and the computer industry. This will be preceded by brief discussions of the other Soviet industries that have received significant attention from researchers concerned with Western technology transfer; they are chemicals, machine tools, and motor vehicles.

Chemicals

The Soviet chemical industry has been long and heavily dependent on the West as a source of both technology and productive capacity. The subsectors of the chemical industry in which technology has remained fairly traditional—basic inorganic chemicals and the production of phosphates and potash fertilizers—have performed relatively well. But performance in petrochemicals and nitrogenous fertilizers has lagged considerably. In the latter two areas, modern technology in the West has changed rapidly in ways that have allowed a significant expansion of plant size at reduced production costs. The Soviet chemical industry has been unable both to duplicate the technology and to keep up with the constant development of new processes and products in the petrochemical field. The demand for Western technology in these areas is significant, not only for this reason, but also because these are areas in which the U.S.S.R. has sought to rapidly expand output. These two factors have combined to demand large expenditures of foreign currency for turnkey plants that will provide modern technology and rapidly expand the industry's productive capacity.

In the late 1950's and early 1960's, when the U.S.S.R. initiated its drive to import technology in chemicals, the Soviet chemical industry seemed likely to remain about 10 years behind the West in a number of areas. It was then taking 6 to 7 years to import and absorb technology that was already about 3 or 4 years old in the West. In recent years, however, the chemical industries of the West have experienced excess capacity and the rate at which new plants and equipment employing the latest technology have been coming onstream has slowed considerably. The chemical engineering companies that provide new technology and equipment have not slowed their innovations, and have been selling their latest technology to any customers in the market for new capacity. As a result, some of the new plants being built in the U.S.S.R. incorporate technology that is as advanced as that coming onstream in the West.

But despite significant contributions from Western plants, the Soviet chemical industry continues to lag considerably in the introduction of new products and technologies, and the output profile of the industry remains biased toward the production of chemicals based on older and simpler technologies.

In the case of plastics, for example, there has not been a single documented instance in which the U.S.S.R. first produced a major plastic material; in fact, the U.S.S.R. is usually the last industrialized economy to begin commercial production of each major group.[24] In synthetic fibers, total production in the U.S.S.R. between 1955 and 1973 expanded at a more rapid rate than in Western countries, but it took 11 years for synthetics to increase from 10 to 33 percent of all chemical fibers produced, while in the United States, Japan, Britain, and West Germany, this diffusion of new technology took only 5 to 8 years.[25] Even when synthetic fibers reached a significant share of total chemical fibers, output was dominated by those synthetics based on older technology.

These impressions were confirmed in the CIA's recent report on the sale of turnkey plants to the Soviet chemical industry and the share of Soviet chemical output accounted for by Western plants.[26] This study, based on a survey of more than 100 turnkey chemical plants purchased from the West between 1971 and 1977, concluded that the Soviet Union depends heavily on Western chemical technology. The U.S.S.R. placed orders for slightly more than $3.5 billion worth of turnkey chemical plants between 1971 and 1975, and ordered an additional $3 billion or more during the following 2 years. The study concluded that these imports did not lead to a noticeable advance in the level of overall plant technology in the U.S.S.R. Although since plants ordered as early as 1971 have only been in place for a few years, the effect of technological diffusion from them may only begin to show up over the next several years. The study also concluded that gains in overall efficiency and product quality have come more slowly and at greater cost than Soviet planners had anticipated.

The value of Western plants ordered between 1971 and 1975 equaled an estimated 20 to 25 percent of total Soviet investment in chemical industry equipment during that period, an amount that may have been higher than planners had in mind. When domestic and East European equipment suppliers were unable to meet commitments, the

U.S.S.R. was forced to increase orders from the West in order to meet planned output goals. The East European chemical industry has concentrated in the more traditional technology areas of basic chemicals and fertilizers. Soviet output based on plants from Eastern Europe is significant for several types of chemicals, with 20 percent of sulfuric acid output, 25 percent of ammonia output, and 40 percent of urea production in 1975.

In comparison, the CIA estimated that plants supplied by the West accounted in 1975 for 40 percent of the Soviet output of complex fertilizers, 60 percent of polyethylene production, and 75 to 85 percent of polyester fiber output. In addition, they were responsible for 72 percent of new ammonia production capacity to come onstream from 1971 to 1975, and 85 percent of the scheduled new ammonia capacity for 1976-80. Some plants supplied by Eastern Europe also incorporated some Western technology which was thereby transferred to the U.S.S.R. indirectly.

The largest share of chemical plants supplied to the U.S.S.R. from the West came from Italy (26.4 percent), followed by France (22 percent), West Germany (17.5 percent), the United States (14.3 percent), and Japan (14 percent). The prominence of Western European nations is largely explained by their willingness to accept product buy-back provisions in payment.

All Soviet orders for U.S. plants came while the U.S.S.R. had Export-Import Bank (Eximbank) credits available, but technology has also been supplied by American multinational firms with subsidiaries in countries that provide the U.S.S.R. with competitive financing. This means that although American chemical firms supply the technology, the United States does not receive the economic benefits of major equipment orders and is unlikely to do so until Eximbank financing is once again available to the U.S.S.R.

The CIA conjectures that the U.S.S.R. has had only limited success in attempts to

---

[24]Amman, Cooper, and Davies, op. cit., p. 275.
[25]Ibid., p. 53.
[26]Central Intelligence Agency, "Soviet Chemical Equipment Purchases From the West: Impact on Production and Foreign Trade," October 1978.

copy Western chemical technology, although it is still possible that the new ethylene plants being built by the U.S.S.R. might incorporate some features of larger ethylene plants that have been supplied by the West. The increasing complexity of modern equipment not only makes it more difficult for the technology in the equipment to be copied, but also makes it increasingly difficult to determine the origin of a given technology.

Technology transfer in the chemical sector has also been felt indirectly in other sectors, particularly agriculture. A recent study by Philip Hanson of the University of Birmingham attempted to measure the economic impact of Western technology in the Soviet mineral fertilizer industry by first estimating the increased fertilizer output that could be attributed to Western plants, and then estimating the increased agricultural output attributable to expanded supplies of these fertilizers. Hanson concluded that between 1970 and 1975 the Soviet Union achieved approximately 4 billion rubles of additional agricultural output by using fertilizer plants imported from the West and installed between 1960 and 1975, at a cost of approximately 2 billion rubles.[27]

All studies of the Soviet chemical industry conclude that the problems experienced by the U.S.S.R. are in developing technology and bringing it into industrial production in areas where technology is changing rapidly, where there must be close communication between research and production work, and where the number of unsuccessful experiments is high compared with the limited number of successful innovations. In the future, the Soviet chemical industry may need to choose between continued reliance on Western technology and turnkey plant capacity, or scaled-down targets for production growth, concentrating on increases based primarily on current technology.

Machine Tools

The Soviet Union has the largest stock of machine tools in the world; in the early 1970's, its inventory of metal-cutting machine tools was about one-third larger than that of the United States. Soviet metalforming equipment also outsizes comparable U.S. stock. When measured in terms of performance and capability, however, even Soviet specialists have admitted that American machine tools exceed their Soviet counterparts.[28]

Demand for machine tools still far exceeds supply in the U.S.S.R. This is due in part to the inefficient use of existing equipment. Soviet machine-tool output is dominated by relatively simple, general purpose machines, which are more easily built than the more complex equipment that machine-tool users increasingly demand. More than 60 percent of Soviet machine tools have been mass-produced with few design changes over many years.[29] In contrast, most machine tools in the United States are specialized models designed for a specific purpose and built in small quantities according to the needs of each customer.

The shortage of specialized machine tools, combined with the need for many plants to be self-sufficient, means that specialized machine tools built in the U.S.S.R. are often designed and produced by the users themselves. In these circumstances it is relatively unlikely that any machine-tool innovations will be diffused through the industry as rapidly as they would be if a regular machine-tool supplier had produced the innovation. The user-builder has no incentive to deploy new technology elsewhere.

Studies of the machine-tool industry concur that traditional Soviet machine tools—drills; lathes; boring, grinding, milling equipment; and transfer lines—do not differ sig-

[27]Philip Hanson, "The Impact of Western Technology: A Case-Study of the Soviet Mineral Fertilizer Industry," presented at the Conference on Integration in Eastern Europe and East-West Trade, Bloomington, Ind., October 1976.

[28]Amman, Cooper, and Davies, op. cit., p. 122.
[29]James Grant, "Soviet Machine Tools: Lagging Technology and Rising Imports," unpublished paper, p. 25.

nificantly from those used in the West, although Western models often perform better in terms of operating speed, tolerances, or durability. The greatest difference between Soviet and Western technology lies in the area of advanced machine tools, such as numerically controlled equipment. Due to technological lags in the Soviet electronics industry in the 1960's, Soviet numerically controlled machine tools began to fall increasingly behind the Western technology. In 1968 the Ministry of Machine Tools and the Ministry of the Aviation Industry—the latter an important user of numerically controlled equipment—decided to step up production. As a result, output of these machine tools in the U.S.S.R. jumped from 200 units in 1968—about 7 percent of U.S. production—to about 2,500 in 1971, exceeding the level of American output.[30]

This rate of growth was made possible by assistance from the West. Since 1968, the U.S.S.R. has signed agreements with firms in Japan, France, and West Germany. At the same time, Soviet cooperation with East European enterprises in this field has also increased. East Germany has been a leader in the development of computer numerical control having shown models at the Leipzig Fairs as early as 1972.[31] Computer numerical control, which first appeared in the United States in the late-1960's, allows a great deal of flexibility and precision. There are no indications that the Soviets have been able to improve on this technology, however.

Furthermore, even with these boosts, and despite lengthy effort, the U.S.S.R. has experienced a great difficulty in trying to copy Western gear-cutting and grinding technology. This may be an indication that the Soviet Union will continue to find it difficult to raise the productivity, reliability, and level of precision of conventional machine tools, and will have problems keeping abreast of technological development in sophisticated models.[32]

The latter have been restricted by Western export controls, but during the past few years, as export regulations on advanced machine tools have been liberalized, the share of advanced machine-tool imports has risen.

Motor Vehicles

The Ford Motor Company first helped provide technology and equipment for Soviet automobile plants at Gorky and Moscow in the 1920's; since then, the U.S.S.R. has continued to look to the West for assistance with motor vehicle production. The initial contract signed by Ford called for the company to transfer any new technology developed during the 9 contract years, yet the U.S.S.R. chose not to introduce the V-8 engine developed by Ford during this period, electing instead to stay with older and somewhat simpler technology. Soviet specialists reportedly recognized limits to their technological capabilities and the problems they might have in absorbing new technology.[33] These problems have persisted. Adequate R&D facilities have never been established in this field and the Soviets have difficulty keeping abreast of technological innovations.

Although the stock of trucks in Western economies is usually several times smaller than that of private automobiles, until a decade ago Soviet vehicle output was dominated by trucks. Owing to lack of production capacity, Soviet planners very early on restricted private ownership of automobiles. In the late-1960's, however, in response to a plan to increase worker incentives through major concessions to consumers, the decision was made to rapidly increase the production of cars.

In order to accomplish this, the Soviet Union's automobile industry received a massive infusion of Western technology. It

---

[30]Ibid., p. 20.
[31]Amman, Cooper, and Davies, op. cit., p. 190.
[32]Grant, op. cit., p. 38.

[33]John Hardt and George Holliday, "Technology Transfer and Change in the Soviet Economic System," in *Issues in East-West Commercial Relations*, Joint Economic Committee, January 1979, p. 71.

contracted with Fiat for a huge automobile plant at Tolgliatti . The Italians coordinated the selection and integration of technology from various sources. Some $550 million in Western equipment, primarily machine tools, were purchased from the West for the plant, with additional Soviet investment, including plant construction, coming to at least another $1 billion.[34]

Fiat was also asked to provide a large number of technicians and to train others in Italy. Ultimately 2,500 Western technicians assisted in equipment installation, training, and startup, and 2,500 Soviet technicians were trained in Italy.[35] This direct personal contact was instrumental in reducing the problems of absorbing the new technology.

One important test of the U.S.S.R.'s ability to absorb and diffuse the Fiat and other Western technology would measure improvements in the technology employed at the Tolgliatti plant and duplications of the technology at other plants. Significantly, the U.S.S.R. has twice chosen to renew the contract with Fiat, first in 1970 and again in 1975. It would appear from this that the technology employed at the plant has not been significantly improved upon by the Soviets and that further Western imports are needed. Moreover, in the Soviet motor vehicle industry, as in most other industries, output increases more from the expansion of existing plants than from the construction of new ones. The technical level of the expanded plant tends to be similar to that of the original plant, leading to growth, but little modernization.[36]

A desire for new production technology, and for a rapid expansion in capacity, led to a second major project involving the transfer of Western technology to the Soviet motor vehicle industry in the past decade. In building the huge Kama River truck plant with assistance from the West, the U.S.S.R. had hoped to entice a Western truck manufacturer to provide the same leadership that

Fiat had for the Tolgliatti plant. But no Western firm was willing to act as general contractor. This was probably due to a number of factors, including the size of the project and awareness of the difficulties experienced by Fiat in dealing with the Soviet system. At the time, the U.S. Secretary of Defense opposed having an American company act as contractor for a plant capable of producing vehicles that might eventually be used for military purposes. The U.S.S.R. therefore served as its own general contractor, selecting firms to supply the major components of the plant, who then chose subcontractors in turn.

Problems at Kama River appear not to have resulted from the choice of major suppliers, but from poor coordination and integration of technologies from different sources. This is a frequent problem and appears to be a major reason behind Soviet willingness to spend so much of its hard currency for Western turnkey plants. The U.S.S.R. is as much in need of expertise in integrating technologies and systems into efficient, highly automated plants as it is in need of new technology.

These industries—chemicals, machine tools, and motor vehicles—have been the most dependent on technology and production capacity of Western origin. Although Western technology has made crucial contributions in all three, it has neither eliminated Soviet lags with the West nor apparently much aided domestic abilities to absorb, diffuse, and improve on the technology. The Soviet computer industry has also been technologically dependent on the West, but Western export controls and corporate interests have limited the computer production capacity that could be imported by the U.S.S.R. The question of technology transfer has become vital in the Soviet oil and gas equipment industry, due to Soviet needs and the state of energy supplies worldwide. These two industries are reviewed in depth below, in discussions of the comparative level of technology in these industries, the extent of technology transfer from the West, the predominant forms that these transfers

---

[34]Ibid., p. 68.
[35]Ibid., p. 77.
[36]Ibid., p. 75.

have taken, and the overall impact of Western technology transfer on industry performance.

Computers[37]

To an increasing extent, the computer industry plays a key role in the overall planning, development, and capabilities of the Soviet economy. Because of the usefulness and interchangeability of computer systems in both civilian and military applications, the question of technology transfer is relevant to U.S. policy for both economic and security reasons.

The United States is presently the leading developer of computer technology, a position it has held since the early 1950's. For foreign producers, American dominance in the industry has meant not only extensive contact with American products and services, but also problems of competition from American firms in overseas markets. Restriction of American inroads by competing States would, in practical terms, have meant depriving themselves of the advantages that access to American technology could offer.

During the early years of the development of the U.S.-dominated international computer community, the U.S.S.R. remained at a distance. This choice reflected both Soviet desire to develop an indigenous capability and a narrow perception of the potential value of computers. In the late-1950's, however, the Soviet view of the computer began to change. Beyond its capabilities in the military sector, computer technology was now seen as crucial to low-level data processing and industrial process control.

The Soviets thus discovered that some contact with Western computer producers was necessary to develop a domestic computer industry suitable to the needs of their economy. While they possessed significant

Photo credits: Control Data Corporation

Soviet computer equipment displayed in a recent trade fair

domestic potential for hardware R&D, they chose to utilize the Western market mechanism to weed out those new processes and technologies that were not viable. This policy was particularly useful in the computer industry, with its rapid rate of technical innovation. Thus it is not surprising that the Soviets developed a close relationship with the Western world in this industry.

The Soviet-U.S. computer technology gap has grown over the years. In 1951, the first Soviet stored program electronic digital computer became operational, less than a year after its American counterpart. The machine was put into serial production only 2 years later, again less than a year behind the United States. These early successes suggested a substantial indigenous computer capability.

[37]See Seymour E. Goodman, "Soviet Computing and Technology Transfer: An Overview," in *World Politics*, vol. XXXI, No. 4, July 1979, pp. 539-570.

There was little transfer of technology during this period, despite a certain similarity between Western and Soviet hardware. Technical literature was the major vehicle for what little interaction took place.

Unlike the United States, the Soviet Union did not have a well-developed business equipment industry, nor an established organizational structure of user support. Close interaction between the designers and final users of equipment is typical with market-oriented firms like IBM; such relationships are virtually nonexistent in the U.S.S.R. This interaction is vital in the competitive and fast-changing business equipment market, and it provided a crucial advantage to U.S. firms in developing new and usable technology.

The Soviet military could have diverted sufficient resources into the computer industry to close the widening gap with the United States in the early 1960's; apparently it chose not to do so. While the Soviets followed the basic pattern of Western technical achievement, the pace of innovation in the U.S.S.R. fell far behind. It produced no major new practical contributions and it built functional equivalents of some Western products long after they had originally been introduced.

The Soviets began to change their attitude toward the computer in the early 1960's, when recordkeeping and data-processing tasks required the production of an upward-compatible series of computers. Such a system consists of a sequence of increasingly powerful computers that have been designed so that programs and data run on a smaller machine can also be run on the larger ones. The first Soviet attempt to produce such a series came in 1965. The U.S. functional counterpart to this machine had appeared in 1960. Here the U.S.S.R. initiated its policy of minimizing technological risk by using a proven U.S. system as a model for its own efforts.

In 1966-67, the Soviets began working on another upward-compatible series of computers, which attempted to copy the architecture of an IBM system that had appeared in 1965. The attempt was abandoned after the production of several machines.

In its next attempt, however, the U.S.S.R. organized a cooperative effort with five other CMEA countries—Bulgaria, East Germany, Hungary, Poland, and Czechoslovakia—all of which had computer industries. East Germany enjoyed access to IBM technology.

The fruits of this collaboration, the Ryad computers, began appearing in late-1972. They are not reverse-engineered from the IBM model; rather, they are functional duplications. The Soviet-led consortium required as long to design the Ryad computers, put them into production, and adopt them to IBM operating software as it took IBM to design, produce, and place the original family in operation. In spite of this, the Ryad system represents a significant achievement. It gives the U.S.S.R. and Eastern Europe a much improved indigenous capability for the production of computers, and has provided extensive experience in the design of computers based on foreign models. Ryad computers are still not produced at the rate at which IBM produced the original line, nor is the performance of the Ryad equipment strictly up to the standards of the IBM models. Nevertheless, U.S.S.R. and East European satisfaction with this program is indicated by the decision to move ahead with Ryad 2, also based on IBM models. By early 1977 most of these new models were well into the design stage, and the first prototypes of some models began appearing in 1978.[38]

The Ryad 2 program will concentrate on increased production of higher quality peripheral equipment, an area of significant technology lag and a source of past complaints by customers in the U.S.S.R. The core memory capacity for most Soviet computers is relatively small compared with the operating speed of the central processing

---

[38]N. C. Davis and S. E. Goodman, "The Soviet Bloc's Unified System of Computers," in *Computing Surveys,* June 1978, pp. 109-110.

unit, thus limiting system capabilities. It is hoped that core storage for Ryad computers in this new series should at least be doubled, if not quadrupled.[39]

The U.S.S.R. also cannot match the West in the quality and availability of magnetic tapes and disks. IBM introduced the first magnetic disk in the early 1960's, permitting the storage and readier availability of vast quantities of information compared to tape. The first Soviet computers to use disk storage may have appeared as early as 1970, but it was not until 1973 that such equipment regularly appeared with any models.[40]

Large memory capacities and input-output devices are important for a variety of data-processing applications. A larger proportion of Soviet input continues to be of the papertape and cardreader types, varieties that are being progressively phased out in the West. Soviet output devices, such as printers, plotters, and graphic displays, also leave much to be desired compared with Western systems. Much of the best Eastern-bloc input-output equipment is produced in Eastern Europe rather than in the U.S.S.R.; a number of models are produced under license from Western firms. Soviet software capabilities have been limited by each of the above-mentioned factors. By making use of the IBM operating system for the Ryad computers, the U.S.S.R. was able to gain access to software which required only minor modification for use on Ryad hardware, compared with what would have been involved in developing such software independently. Designing its own software is a major Soviet goal, yet it is questionable whether such copying aids this process.

Application programs tell the computer how to process data that is entered. Problems with hardware have held back application software development in the U.S.S.R., despite recent improvements. The lag between Soviet and Western software capabilities, as in other areas, has systemic origins.

The Soviets lack the Western motivation to look for more efficient and less expensive ways to accomplish given tasks.

Calculating the precise value of computers shipped from the West to the U.S.S.R. is difficult, due to the nature of Western and Soviet trade data. Neither provides breakdowns into categories for computers, and case-by-case information about sales is limited. One report, based on detailed trade data for each Western country that supplies computers to the U.S.S.R., has produced the figures shown in table 36. Several industry experts believe these figures to be misleadingly low, particularly as regards products transferred by American firms via their Western European subsidiaries. These sales are not completely accounted for in Department of Commerce statistics. Orders placed in 1977 and later indicate that the downward trend observed in 1977 has been reversed, and 1978 U.S. data show a near return to the record 1976 levels.

The commercial interests of Western computer manufacturers and export controls have together strictly limited the transfer of manufacturing technology to the U.S.S.R. Only one Soviet plant order—for purchase of a Japanese facility for production of minicomputer memory devices—has been recorded. In addition, Romania and Poland (and perhaps other East European nations) have purchased Western licenses for production of several computers and peripheral equipment.

Table 36.—Western and U.S. Computer Sales to the U.S.S.R., 1972-77 (in millions of dollars)

	Sales from United States	Total sales from the West
1972	$ 4.1	$ 16.1
1973	4.0	15.7
1974	3.7	20.1
1975	9.4	28.3
1976	17.2	41.6
1977	5.7	28.3
1972-77 total	$44.1	$150.1

SOURCE: IRD, Inc., *The Market for Computers in the PRC and the U.S.S.R.* (New Canaan, Conn.: January 1979).

---

[39]Amman, Davies, and Cooper, op. cit., p. 386.
[40]Davis and Goodman, op. cit., p. 98.

Although the Soviet Union has not acquired any licenses, some of the production technology obtained by Eastern Europe is likely to have been made available to it. The U.S.S.R. has thus derived its greatest benefits from importing computer systems to provide models of new technology to aid the Soviet computer R&D sector and provide capabilities otherwise unavailable. Export controls prevent the U.S.S.R. from importing the most advanced Western computers, although some sales have given them units with better reliability, software, and input-output capabilities than the best Soviet models.

Except for 1977, the 1970's have seen an upward trend in Soviet purchases of computers from the West (see table 36). Most computer sales are of systems costing several million dollars each. Users of large Western computer systems in recent years have included reservation systems for Intourist and Aeroflot, analysts of seismic data for geological prospecting, controllers of large industrial enterprises (particularly in the motor vehicle sector), and systems for inventory control and management. All these applications involve handling and managing large amounts of data. Soviet computers are less well-suited to such work in terms of memory and input/output capabilities, and the software required to perform such functions is frequently unavailable in the U.S.S.R. The purchase of these systems may sometimes be motivated as much by the desire to gain access to software as to hardware. The most important factor in a purchase decision, however, is generally the desire of the end-user management to obtain an entire system that, with a minimum of risk, will safely, effectively, and reliably address applications problems.

Scientific institutes and Government planners also buy Western computers to obtain good computer capability. These smaller sales receive much less press coverage than the headline-making orders for million-dollar computers. But as planners and scientific users have become more aware of the many uses to which computers can be put and given higher priority to the purchase of Western equipment, such sales have grown in importance.

Turnkey plants imported from the West also frequently include computers or sets of computers as part of the process control system. Almost without exception, the U.S.S.R. has insisted that plants imported from the West contain the latest process control and automation equipment. While this request may be partly motivated by the desire to obtain the embodied technology, it is also a reflection of the U.S.S.R.'s acute shortage of skilled operators for many industrial sectors; such automation is seen as an efficient means of reducing the labor requirements of new plants.

Because of its desire for maximum feasible self-sufficiency in such a strategic field, the U.S.S.R. cannot be expected to become a very large customer of Western computers. The Soviets will continue to rely on Western imports to meet certain needs. Such purchases may even reach a level several times higher than that of the past, but computer needs will compete with needs for other equipment and materials. Imports will tend to be restricted to those cases where the cost of doing the work without a computer is exceptionally high.

In addition to these constraints on the Soviet side, Western export license restrictions inhibit West-to-East computer sales. Often, the sales that are prohibited are the very ones which the U.S.S.R. desires most, i.e., they are sales of systems with those capabilities that the U.S.S.R. finds it most difficult to produce domestically. If export restrictions were eased, it is likely that the purchase of these systems would be of sufficiently high priority that hard currency would almost certainly be allocated for them. Under these conditions, the volume of such imports would probably rise sharply.

Computer sales to the U.S.S.R. tend to be won by those firms that are most aggressive in pursuing the Soviet market. Thus, the market share for American computers is much lower in the U.S.S.R. than in other markets around the world. The Japanese, like the Americans, have not yet pursued the Soviet computer market vigorously, but the West Europeans—particularly the British and French—have long sought involvement in the market.

The United States does enjoy a distinct advantage over competitors in the quality of its computer technology. This advantage is partially offset, however, by the strong disadvantage of uncertainty and delay due to export control. Only the United States fails to provide its companies with early indications that a license can or cannot be obtained. Only the United States will block a sale for political rather than strategic reasons. The United States takes longer than any other nation to approve a license, and regularly enforces stricter licensing regulations than those set by CoCom. As a result, American firms are sought as suppliers when they are able to provide products markedly better than those available from Japan or Western Europe, but not as suppliers of first choice when all else is equal. The difficulties experienced by American computer exporters lead to much of their business being handled out of Europe, since at least some of the problems are then avoided.

In many industries, the amount of time required for delivery is a factor in the selection of a supplier. American computer manufacturers should compete very well with suppliers from other Western countries in this regard, since U.S. firms often have more experience in putting together custom-designed systems. This potential advantage is frequently more than offset, however, by the regulatory delays a U.S. supplier may face. Even if the license is ultimately approved within a reasonably short time, the initial uncertainty of the outcome of the licensing procedure can chill the negotiations between an American computer supplier and the U.S.S.R. and can impose higher costs on the supplier, the Soviet Foreign Trade Organization negotiating the contract, and the Soviet user waiting for delivery of the equipment.

Financing is a factor only in those cases involving sales of computers for process control. The United States sells very few process control computers. The selection of turnkey plant suppliers is highly dependent on financing and on the willingness of the supplier to accept buy-back contracts for products produced at the plant. In both regards, the United States is at a disadvantage. Often, even though a plant is based on U.S. technology and incorporates an American license, it is financed and equipped by a Japanese or Western European firm. In such a case, the computer for process control, like all the other equipment for the plant, will come from the country that is supplying the credits for the plant. All U.S. turnkey plants that have been supplied to the Soviet chemical industry during the past few years resulted from orders that qualified for Eximbank credits, which have since been disallowed. No further chemical turnkey plants—and no process control computers for Soviet chemical plants—have been purchased from the United States since then.

It is difficult to assess the impact of Western computer sales on the economic performance of the U.S.S.R. The effect of any computer is difficult to measure in quantitative economic terms, but one can identify those areas of the Soviet economy that have benefited the most from Western computers. Western computers have had a strong impact on the motor vehicle manufacturing sector, as British and, more recently, American computers have been used to control production processes at a number of plants. Western computers have also become important for the analysis of seismic data, thus benefiting the identification of oil and gas reserves. Other sectors of the economy that have benefited include the chemical industry, from both the process control computers in imported turnkey plants, and the Ministry of the Chemical Industry's purchase of several computers to assist in the design of new chemical plants. Gosplan has received some Western computers, but has not used them with optimum efficiency. The greatest beneficiaries of Western imports have probably been scientific organizations, particularly those involved with nuclear physics.

In conclusion, virtually all major developments in Soviet computer technology have first taken place in the West. The U.S.S.R. has been a follower rather than an innovator in the computer technology field. Once a new technology has appeared in the West, the U.S.S.R. has usually succeeded in reproducing the technology domestically, although the timelag between Western and Soviet introduction of similar technologies has not diminished over time (see table 37).

As long as the U.S.S.R. continues in the role of follower, the technological lead of the West is assured. Even if the difficulties of moving swiftly through development stages into actual production of hardware are solved, the Soviets will still face difficulties in diffusing and effectively using the hardware they produce. Such problems do not lend themselves to ready solutions.

Oil and Gas

The U.S.S.R. is the world's leading producer of oil, and one of the largest suppliers of natural gas. Most of the equipment used for exploration, drilling, and extraction comes from within the U.S.S.R., with its relatively strong oil and gas equipment industry. The bulk of Soviet reserves of oil and gas is located in relatively shallow and very large fields, making it possible to reach high production levels without the most advanced technology.

But the Soviet concentration on these shallow deposits reflects the country's limited geological prospecting capabilities, which make the exploration of deeper reserves difficult. Recently, the U.S.S.R. has shown interest in acquiring more advanced Western prospecting equipment, such as sophisticated seismic mapping equipment and field units to assist in the recovery and analysis of seismic data. A number of computers have been sold to the U.S.S.R. to provide this analytical capability.

The turbodrill has long facilitated significant advances in the productivity of Soviet drilling. About 85 percent of Soviet drilling was done by turbodrills in the early 1960's; since 1970, the share has stabilized near 74 percent. Turbodrill technology was attractive because it permitted the industry to use pipe and tool joints which were readily available, while reducing breakdowns and increasing speed. Unfortunately, however, the drill loses effectiveness when deeper drilling is required. The high drill speed required for efficient use of the pumps that run the drill results in comparatively short drill-bit life, so the deeper the well, the more time lost in replacing bits. The power transfer to the bit also becomes less effective when used with jet bits. Finally, while good for drilling in hardrock formations, the drill is far less effective in soft formations. The Soviet oil and gas equipment industry has addressed these problems by providing improved designs for new turbodrills, rather than by increasing production of rotary drills, which are most common in the West, even though as early as 1960 some planners recommended development work on rotary drills.[41]

Table 37.—First Production of Comparable Soviet and American Computers[a]

American computer	Similar Soviet model	Date of appearance in the U.S.S.R.	Lag (in years)
IBM 650	Ural 1	1955	1
IBM 702	Ural 4	1962	7
IBM 1620	Nairi I	1964	4
IBM 7094	BESM-6	1966	4
IBM 360 series.	ES series	1972-3	6-8

Soviet lag in entering successive generations of computers

	First generation	Second generation	Third generation
First Soviet computer . .	1952	1961	1972
First American computer	1946	1957	1965
Lag (in years).	6	4	7

[a]Comparison of dates of first American commercial installation and first Soviet industrial production.

SOURCE: M. Cave, "Computer Technology," in *The Technological Level of Soviet Industry,* Amann, Cooper, and Davies. eds. (London, 1977).

---

[41]Robert Campbell, *Trends in the Soviet Oil and Gas Industry* (Baltimore, Md., 1976), pp. 20-22.

A 1977 CIA study of the Soviet oil industry pinpointed the inefficiency of Soviet drilling as a major reason for probable problems in meeting future production goals.[42] The CIA estimates that the U.S.S.R. will need 50 percent more drilling rigs by 1980 to meet its drilling targets. The U.S.S.R. hopes, however, to reach its increased drilling goals primarily through improved rig productivity.

The quality of Soviet drill bits has also been blamed for poor drilling performance. The U.S.S.R. recently agreed to purchase a turnkey drill-bit plant from U.S.-based Dresser Industries to help remedy this situation.

Soviet technology for wellhead equipment is reasonably good, although better wellhead equipment is reportedly needed when the oil or gas being extracted is particularly corrosive or under very high pressure. There has also been a lag in the U.S.S.R.'s development of multiple completion equipment. This equipment permits a number of producing wells to exist on the same structure.

Soviet oilfields are being depleted rapidly but with a relatively poor rate of recovery. The Soviet economic system, with its production quotas and demands for immediate results, is one reason why fields in the U.S.S.R. are exploited quickly. Soviets inject water into wells on about 80 percent of U.S.S.R. fields to increase immediate production rates. This practice, known as secondary recovery, increases field pressure and the flow rate of the well, and may increase the ultimate field recovery. According to the CIA, however, this method may also reduce the field's long-term production potential and result in a serious fluid-lifting problem. Centrifugal pumps must be installed to pump out the water and oil; while the Soviet Union produces such pumps, their capacity and service life do not match that of the equipment produced in the United States.

Alternatively, secondary recovery might involve the injection of detergents, polymers, steam, or carbon dioxide instead of

[42]See Central Intelligence Agency, *The Soviet Oil Industry,* April 1977; and *The Soviet Oil Industry: A Supplementary Analysis,* June 1977.

water. To learn more about these methods, the U.S.S.R. has increased its testing of such procedures and has imported equipment and material from the West.

Soviet experience and technology lag far behind that of the West in all phases of offshore work. The U.S.S.R.'s offshore drilling and production has been limited largely to activity on fixed platforms in shallow coastal waters of the Caspian and Baltic Seas, with only limited experience in jack-up drilling. The Soviet Union has avoided work further offshore because of technological difficulties and much higher production costs. The U.S.S.R. buys a larger share of its offshore equipment from the West than for any other phase of the oil and gas industry. U.S.S.R.-built equipment can only be used in limited water depths and for relatively shallow wells. The U.S.S.R. also lacks experience in subsea completion equipment, which is at the forefront of current Western technology, in underwater storage and transport, and in other advanced phases of offshore activity.

An offshore development project off Sakhalin Island, north of Japan, has produced the most active joint cooperation to date between the U.S.S.R. and the West. Japan is the U.S.S.R.'s principal partner in the project, although Gulf Oil plays a small part in it. In exchange for providing the technology and financing the exploration, the Western partners are assured a share of any resulting oil or gas production. The U.S.S.R.'s experience in this project will help it in further efforts to expand offshore drilling and production.

A similar arrangement will permit the joint development of gas onshore in Yakutia, in Eastern Siberia. For this project, Japanese and two American firms hold shares amounting to 50 percent of the project, with the U.S.S.R. retaining the other 50 percent. The progress on this project has been slow, largely because Eximbank financing for the American share of the cost is unavailable, and because sufficient gas reserves at the site to justify the project have yet to be proven. If successful, the project will entail the construction of a pipeline to the U.S.S.R.'s Pacific coast, a distance of some 3,100 kilometers. The U.S.S.R. claims one trillion cubic meters of reserves exist at Yakutia.

The Yakutia project will require sizable quantities of Western technology for the construction of the pipeline, and for drilling and extraction under extremely cold conditions. To exploit the field on its own, the Soviets would face much higher costs in both time and money, and time may be the critical factor. To meet increased production goals, the U.S.S.R. needs both increased supplies of equipment that is in short supply, and better technology. If the United States restricts the sale of certain types of equipment or technology, it is likely that the U.S.S.R. will seek it from other Western sources (see chapter IV).

The difficulties of measuring the amount of equipment and technology sold by the West to the Soviet oil and gas exploration and extraction sector is shown by the wide discrepancies between Soviet data and Western estimates, as shown in tables 38 and 39.

The Soviet-supplied data in table 38 excludes pumps, but this omission does not fully account for the discrepancies between it and CIA figures. The problem is further compounded by a third source, the New York-based consulting firm of Frost and Sullivan, whose recent study contained the following figures for U.S. sales of oil and gas exploration and extraction equipment to the

Table 39.—Breakdown of U.S. Oil and Gas Equipment Sales to the U.S.S.R. (1972-76) (in millions of dollars)

Category	Value
Pipelines	$304
Submersible oil pumps	148
Offshore and refining equipment	49
Other	49
Total	$550

SOURCE: Central Intelligence Agency, *The Soviet Oil Industry,* A Supplemental Analysis, June 1977.

U.S.S.R.: $3.7 million in 1973, $28.5 million in 1974, $10.9 million in 1975, and $34.0 million in 1976. The CIA data covered orders placed as sales, while the other two sources recorded actual deliveries. Subsequent investigation has shown that the CIA figure for submersible pumps was high by about $50 million, partly because of an order that was never filled.

It can be said with certainty that since 1976, the volume of Soviet orders for Western oil and gas equipment has risen significantly. There has also been a shift toward turnkey projects, either for plants to produce equipment or materials required by the industry, or for full-service contracts with firms to provide all equipment needs for an entire project. A recent order to a U.S. firm to supply gas equipment for wells in Western Siberia is an example of the latter.

The U.S.S.R. clearly realizes that it must import this equipment and technology to increase production of oil and natural gas at rates that meet domestic needs and allow it to sell surpluses to Eastern Europe and to the West, thereby earning hard currency. The sale of oil and gas has accounted for approximately half of all Soviet hard-currency earnings in recent years. These earnings are used for financing continued imports of Western grain, equipment, and technology.

Failure to meet oil and gas production goals would involve extreme costs in the loss of this earning power. But, if the U.S.S.R. were extremely concerned about its future oil and gas production, it would be logical for it to permit greater involvement of Western

Table 38.—Soviet Imports of Western Oil and Gas Exploration and Extraction Equipment (in millions of dollars)

	Purchases from United States	Total purchases from the West
1972	$ 4.6	$ 19.4
1973	4.3	23.5
1974	.5	9.0
1975	49.5	150.1
1976	40.6	226.5
1977	29.3	121.0

NOTE: These figures do not include turnkey manufacturing equipment.
SOURCE: Vneshnaya Torgovlaya (Soviet Trade Data), category 128.

firms in joint production projects to speed up development of reserves. Instead, the U.S.S.R. has chosen for the present to concentrate on acquiring equipment and technology beneficial to the long-run production capabilities of the country, with special emphasis on technology that requires only relatively short leadtimes to produce increases in output of oil or gas.

The selection of equipment and technology suppliers for the Soviet oil and gas industry is based on a number of factors, including financing and the kind and quality of technology. The oil and gas industry, as a major earner of hard currency, receives a very high priority when it comes to the allocation of foreign exchange for imports. When the technology offered by different suppliers is relatively the same, financing terms may determine the chosen supplier. In most cases, however, differences in technology will provide the basis of the choice. When a multinational firm can have equipment produced in a country that will provide better financing than the United States, the package becomes more attractive to the U.S.S.R.; American firms have done this a number of times.

The Carter administration decided in mid-1978 (during and presumably because of the Soviet dissidents' trials) to place all oilfield equipment on the Commodity Control List. This action may have affected Soviet perceptions of American firms as reliable suppliers. Although no sales of oil and gas equipment have been denied licenses since the order was given, in some cases the U.S.S.R. may have decided not to pursue negotiations with American firms to avoid the possibility that the license might be blocked for political reasons.

In other cases involving equipment such as seismic prospecting instruments or computers used to analyze seismic data, the stricter controls placed on American suppliers are more than offset by the superior American technology, which ensures that the American firm is the most likely choice as supplier.

In summary, the U.S.S.R.'s pattern of relying on Western technology to rapidly increase its capabilities in offshore operations and secondary recovery suggests that the primary interest of the U.S.S.R. in importing this equipment and technology is more to gain the productive capacity which the equipment represents than to obtain the opportunity to duplicate new technology. For the most part, oil and gas equipment imported from the West has not been integrated with Soviet equipment, partially because equipment purchases have primarily included complete units. This approach allows the U.S.S.R. to achieve the greatest possible productive capacity with the equipment it imports. The recent shift toward imports of turnkey plants will, however, increase the U.S.S.R.'s exposure to Western technology, and may speed the rate at which this equipment is absorbed by the Soviet industry.

It is still too early to tell how efficiently the U.S.S.R. will absorb most of the equipment and technology it has imported for oil and gas development. It may be expected that the rate of active oil and gas technology transfer between the West and the U.S.S.R. will increase in the future, particularly as Sakhalin Island, Yakutia, and other projects advance.

CONCLUSIONS

Western technology has made a marked impact on each of the Soviet industrial sectors considered here—chemicals, machine tools, automobiles, computers, and oil. Generalizations, however, either concerning aggregate economic effects of Western imports or motivations for importing Western technology are misleading. There are two basic rationales for importing foreign industrial technology and/or products: 1) such items could not, under any circumstances, be produced domestically and 2) it is economical to import rather than to produce domestically. But the role of imports in each particular industry is markedly different. Thus, a sophisticated and useful approach to sectoral impact of imports must recognize that between these two points lie a range of rationales for individual import decisions in any given sector. Motivations for foreign imports are closely associated with the capabilities of domestic industry. The range of categories of imports relative to domestic productive capacity is as follows:

1. technology and/or products that cannot be domestically developed or produced at any cost;
2. technology and/or products that can be developed or produced domestically at great cost in time and resources, and the lack of which create bottlenecks in other productive processes;
3. technology and/or products that can be developed at great expense in time and resources, but do not create bottlenecks;
4. more productive versions of technology or products similar to those already available in the U.S.S.R.; and
5. technologies that can lead to capacity increases in products equivalent to those available domestically or products providing marginal economic returns.

This range of choices may be regarded as a continuum, and the rationale for individual imports from the West may fall anywhere along it. Given the decision by Soviet planners to increase production in all the sectors under consideration, those imports that fall in the initial categories will be most beneficial in an economic sense. But products of technologies that the Soviets are incapable of producing at any cost are extremely rare. Most analysts have concluded that only time and commitment separate the Soviets from any given advance otherwise available to them through imports. At the opposite end

of the spectrum, it is highly unlikely that import decisions are often made for cases of marginal returns, both because a wide range of more productive processes are always available in the West and because of Soviet propensity to avoid expending hard currency on cases of doubtful return.

The relative role of Western imports in individual sectors may be determined by where in the range of import types purchases of Western products and processes cluster. In the chemical industry imports are generally used as a vehicle to acquire new equipment and processes that could be produced in the U.S.S.R., but at great R&D cost. Chemical output is also central in capacity increases in other crucial sectors—agriculture in particular. Imports in the chemical industry tend to occur in the higher range of import choices; equipment and processes acquired are consequently crucial to planned growth in the industry.

While imports in the automotive sector are made at all levels of the choice range, large imports tend to be made both for productivity and capacity increases. The Soviets are perfectly capable of producing automobiles with domestic technology, but Western imports increase the speed, efficiency, and overall capacity of their industry.

In the area of machine tools, productivity and capacity increases also appear to be the major factors behind imports. In this sector as well as in the oil industry, a relatively strong domestic industrial base exists. The Soviets have, however, planned large capacity increases in both. The fastest and most efficient way to accomplish this goal is through imports of Western capital, which transfer Western technological advance in addition to adding to capacity.

Soviet computer imports fall into the higher range of import types; R&D costs in this industry would be immense in the U.S.S.R. This is due both to the speed with which innovations are developed and the fact that they are often motivated by the needs of the user. A centrally planned economy is particularly unsuited to high levels of innovation in this industry. Soviet practice has been to wait for major innovations to be proven viable in Western markets before attempting to incorporate them into its own production.

In conclusion, the impact of Western imports differs significantly across sectors, both from a qualitative and quantitative point of view. There can be no doubt that economic benefits have accrued to all the industries under consideration as a result of imports from the West; the process by which this has been accomplished is complex and differs from industry to industry. It is clear that any policy aimed at affecting the economic impact of Western technology in the East must be tailored to achieve specific effects in specific industries.

CHAPTER XI

Western Technology in the People's Republic of China

CONTENTS

Western Technology in the People's Republic of China

THE HISTORY OF TECHNOLOGY IN CHINA

PREREVOLUTIONARY DEVELOPMENT

During the past 300 years, technological progress flourished in Western Europe and North America while China languished. That China was for so long bypassed by technological progress is one of history's great ironies, for few countries can match its long-term record of inventiveness and technological sophistication. There are many familiar examples of Chinese inventiveness including gunpowder, paper, and the compass. To these may be added many diverse artifacts and techniques, including oil refining, the chain-drive transmission, the segmental arch bridge, iron casting, the differential gear, deep drilling, the piston bellows, and the stirrup.[1] The sophistication and richness of early Chinese civilization staggered all who came into contact with it.

This outpouring of sophisticated techniques and devices resulted in significant changes in productive, military, and commercial activities. The introduction of new crop varieties and new strains of established crops had enormous repercussions, allowing the settlement of frontier areas to the South and permanently shifting the demographic center away from the former heartland of the Yellow River basin.[2] Use of the compass, along with fundamental innovations in shipbuilding and sailing techniques, led Chinese navigators perhaps beyond the Arabian Peninsula. Improved roadbuilding techniques and bridges of every description knitted the empire together and put every major city in regular communication with the imperial capital. In sum, traditional China went through many changes as new technologies elicited responses in productive efforts and administrative capabilities. This occurred despite substantial inertia created by established cultural and bureaucratic structures.

But while technological development provided a secure foundation for the governing elite and the civilization they created, new ways of doing things did not alter customary societal and political arrangements. In sharp contrast with the history of Europe, technological changes were not associated with fundamental changes in the social order or the rise of new classes that could threaten the existing order. Merchants remained politically and culturally subordinate, and Chinese cities remained firmly under the control of the existing Mandarin elite.

The traditional cultural and political system demonstrated a remarkable resilience in the face of extensive technological changes; the Confucian culture was deemed more valuable than the products of a new technology. Moreover, the cultural hegemony of Confu-

---

[1]See Joseph Needham, "Science and China's Influence on the World" in Raymond Dawson, ed., *The Legacy of China* (London: Oxford University Press, 1964).

[2]Ho Ping-ti, *Studies in the Population of China, 1368-1953* (Cambridge, Mass.: Harvard University Press, 1959), pp. 169-76; 183-92.

cianism helped to retard technological progress; the mental patterns fostered by the study of the Confucian classics were at considerable variance with those appropriate to the continuous development of new ways of doing things. Although there was not always outright hostility to technology, active involvement in technological matters fitted poorly with the dominant culture. It induced a reluctance to become involved with innovations of any sort, particularly those that centered on the improvement of menial existence.[3]

The literary and bookish nature of the elite culture in China stood in clear contrast to the empirical, practically oriented spirit necessary for the development of new technologies. The content of Confucian philosophy was not conducive to a spirit of conquering nature through the application of new devices and techniques. The primary concern of the scholar and the putative role of the official centered on learning the correct principles of human relationships, and using these principles for the maintenance of harmony, both between man and man, and man and nature. In contrast, technological change by its very nature disrupts existing relationships and dissolves harmony. Whatever the practical benefits of technological change, the disruption of the existing state of affairs could hardly be applauded by the traditional official who had been schooled in Confucian philosophy.

In sum, therefore, the main inhibiting cause (for the lack of development of science and technology) was the intellectual climate of Confucian orthodoxy, not at all favorable for any form of trial or experiment, for innovations of any kind, or for the free play of the mind. The bureaucracy was perfectly satisfied with traditional techniques. Since these satisfied its practical needs, there was nothing to stimulate any attempt to go beyond the concrete and the immediate.[4]

When stimulation to technological advance did come, it came in the form of a profound external threat. Technological stagnation was never absolute in China, but it was most sharply revealed when the Western powers attempted to extend their influence there. From the point of view of the West in the 18th and 19th centuries, China was poor, ignorant, and backward. It was China's misfortune to be in a period of dynastic decline during the time that an aggressive West (and later, Japan) bolstered by formidable technological prowess, was extending its influence across the seas. The technological stagnation that afflicted late traditional China might have continued were it not for the intervention of the West; indeed, it can be argued that the shock of Western and Japanese domination was essential if China was to rejuvenate. But China's inferiority to foreign powers left an enduring sense of powerlessness and debility; the development of technology became an indispensible part of efforts to regain a modicum of security and independence, especially relative to those nations that took part in its subjugation and humiliation.

The application of new technologies was essential to this endeavor. This in turn necessitated a greater receptiveness to ideas, processes, and materials that had been developed elsewhere. But nagging doubts attended the effort to transform China's technological order: could foreign products and techniques be adopted without their bringing a profound dislocation to Chinese culture? The answer was negative. Achieving a rapprochement between imported modernity and indigenous patterns of life proved difficult; technological progress could not be pursued in disregard of the cultural consequences. China's recent history can thus be seen as the search for "modernization with pride."[5]

---

[3]See Liu Ta-chung, "Economic Development of the Chinese Mainland, 1949-1965" in Ho Ping-ti and Tang Tsou, eds., *China's Heritage and the Communist Political System* (Chicago, Ill.: University of Chicago Press, 1968), pp. 134-5.
[4]Etienne Balazs, *Chinese Civilization and Bureaucracy* (New Haven and London: Yale University Press, 1964), p. 22.

---

[5]See Joseph Levenson, *Confucian China and Its Modern Fate: A Trilogy* (Berkeley and Los Angeles, Calif.: University of California Press, 1968).

The evident military superiority of the imperialist powers jarred the Chinese ruling elite into the realization that some adjustments would have to be made, and that the foreigners would have to be met by equivalent military strength. No serious ideological obstacles stood in the way of this; even the most conservative Confucian official recognized that military defense was a legitimate area of interest and concern.

Modern science and technology thus found their first home in China in the shipyards and armories established to counter the might of the West. A new generation of engineers, technicians, and scientists began to be trained in the Government's arsenal, where, after 1865, foreign instructors schooled their Chinese pupils in modern shipbuilding, metallurgy, and arms manufacture.[6] Despite earlier Chinese advances in nautical architecture, explosives, and ordnance, a new set of skills had to be developed as China attempted to preserve its national and cultural integrity.

The realization that improvements in defensive capabilities could not be pursued in isolation from other modernizing currents came slowly. At first, each step of military and technological modernization was justified in terms of its immediate importance for keeping out the foreigners. But the fact remained that the development of weapons and other items of military technology required at the least such supportive industries as railroads, steamships, and coal mines.[7] For "progressive" Chinese thinkers, the assimilation of Western science and technology took on an importance which transcended the strengthening of the military and its supporting infrastructure. Science and technology were to be the foundation of a new Chinese society erected on the remains of a crumbling civilization, and scientific inquiry and rational thought became "slogans

in an anticlerical war against superstition and authority."[8] Even after the collapse of China's last dynasty and the establishment of the Republic an aggressive belief in the intellectual and spiritual superiority of scientific ways of thought continued to challenge traditional religious and philosophical beliefs.[9]

There could be no easy incorporation of scientific thinking or technological application into the corpus of traditional Chinese elite thought and culture. Effecting the necessary changes in the basic elements of the traditional order posed a great threat to many Chinese, especially those whose self-identity and authority were bound up with the Confucian world view. The development of modern ways of doing things could not be initiated without thoroughgoing changes in Chinese culture and society, and the techniques necessary to preserve Chinese national integrity from the onslaughts of the West would paradoxically result in the complete conquest of the traditional way of life. The fact that the conquest would come from within would make it no less complete.

Some Chinese, however, endeavored to have it both ways, seeking to retain the essential features of Chinese civilization while at the same time acquiring and assimilating the foreign technologies necessary for China's resurgence. According to their prescription, new technologies and organizational structures could be incorporated for their utility, leaving unchanged the essence of Chinese civilization. The surface plausibility of this synthesis was quickly challenged by the more unregenerate members of the Chinese political and cultural elite. To them, a China that took the path of technological and organizational modernization would soon lose its way in its pursuit of foreign

---

[6]Charlotte Furth, *Ting Wen-chiang: Science and China's New Culture* (Cambridge, Mass.: Harvard University Press, 1970), p. 11.

[7]Albert Feuerwerker, The Chinese Economy, 1912-1949 (Ann Arbor, Mich.: University of Michigan Center for Chinese Studies, 1968), pp. 1-2.

[8]Furth, op. cit., p. 70.

[9]C.K. Yang, *Religion in Chinese Society* (Berkeley, Los Angeles, and London: University of California Press, 1970), pp. 363-7.

novelties: a commitment to acquire and assimilate modern technologies could only result in the dissolution of the established culture. These qualms were not enough to stem the tide of change, and there was little chance that the Confucian elite could maintain the world in which they had been so comfortable. By the beginning of the 20th century, China was ripe for fundamental change.

With its commitment to modernization and technological development, the People's Republic of China (PRC) could be viewed as the culmination in the transformation of Chinese culture, a process that began with the opium wars. This would be an oversimplification, for even today it is premature to conclude that modernization through the balancing of imported technologies with indigenous Chinese culture has been achieved. If anything, the problem is now even more acute, for cultural patterns in today's China are compounded of ancient tradition and modern socialist doctrine, and tensions between technological development and ideological patterns—both ancient and contemporary—persist.

Technology has not acted as an independent force which has autonomously transformed the traditional culture and shaped the postrevolutionary society. In traditional China, patterns of political and cultural domination in large measure determined the nature and extent of technological change. This is still the case, as the Communist government has manifestly committed itself to generating the policies that will guide the process of technological development. Yet it has never had a free hand in this matter; the possibilities of technological advance have been circumscribed by both the characteristics of Chinese society and the other goals of the leadership. The next section considers the evolution of technological policies in the PRC and the limitations on their implementation. Although the political actors and many of the fundamental goals are different from those of traditional China, technological change has remained firmly in a Chinese context.

POSTREVOLUTIONARY TECHNOLOGY POLICY

Although many changes have come to the PRC over the last three decades, there remains a certain continuity of goals and activities. Any analysis of technological policy must be cognizant of continuity as well as change. Most analyses of policy in China have given relatively little consideration to continuity; instead, most Western scholars have focused their attention on the apparently wide fluctuations in basic policies as the PRC has oscillated between periods of radical change and periods of retrenchment. From this perspective, "Maoist" methods of employing mass mobilization, popular initiative, and ideological incentives have been counteracted in succeeding phases by policies that were built around expertise, precise organization, and remunerative rewards. Although this characterization of social change in China has put policy changes in context, the policy cycle model can be taken as only a rough approximation of historical reality. Despite apparent shifts of considerable magnitude, there has remained a basic continuity of goals in China and a narrowing range of policies has been employed as a means of achieving these goals.[10] Moreover, in any given period, there are considerable divergencies between the specific policies employed in different economic and political realms (such as agriculture, education, and foreign relations). Even technological policy is itself too broad an analytic category to be easily fit into a broadly drawn policy cycle model; at any time, technological policy is really a cluster of separate policies which reflect the differential impact of general policies on specific sectors of the economy.[11] Finally, the chronology of policy changes in China is not an unending series of back and forth movements along a single axis; the history of the PRC has been a learning experience for the

[10]See Andrew Nathan, "Policy Oscillations in the People's Republic of China: A Critique," *China Quarterly* 68 (December 1976).

[11]See Alexander Eckstein, *China's Economic Revolution* (Cambridge, Mass.: Cambridge University Press, 1977), p. 85.

Chinese people and the leadership alike. Neither a "Maoist" nor a "pragmatic" model of development is likely to be taken as an inclusive blueprint for social change and economic progress, and the years to come will likely see continued efforts to produce a workable resolution of the conflicts between egalitarian participation and managerial direction, mass motivation and professional expertise, and ideological incentives and monetary rewards.

When the Communists gained control over the Chinese mainland in 1949, these dilemmas were faint concerns. After years of foreign invasion and civil war, China was in desperate need of economic reconstruction. Providing a modicum of economic organization and meeting the minimal subsistence needs of a war-ravaged populace were in themselves staggering tasks. In addition, the Communists had set a more ambitious task—the transformation of a backward and easily exploited country into a strong and self-sufficient Socialist nation.

Technological development was to be an integral part of this transformation. In addition to restructuring economic institutions and relationships through land reform, the nationalization of key industries, and the construction of a central-planning apparatus, the new Government began to take the first steps toward the formidable task of modernization of production. Despite some promising starts in industrial development during the 1930's,[12] China had scarcely risen above the agriculture-and-handicrafts economy which had endured for centuries. To build a foundation for China's economic modernization, 156 large industrial projects were designed and built with the direct assistance of the Soviet Union. China's isolation from the capitalist world made Soviet assistance crucial, and the U.S.S.R. responded with one of the largest programs of economic aid and technology transfer in history.[13]

It can be argued that the pattern of economic development that emerged under Soviet sponsorship was in fact ill-suited to China's developmental needs. The Chinese embarked on an essentially Stalinist strategy which emphasized the rapid development of heavy industry through the construction of the large industrial projects under Soviet patronage. At this time, the Government plowed 20 to 25 percent of the country's total output into investment, a percentage similar to the Soviet Union and other Communist countries during the first years of their existence.[14] This investment was confined to a narrow part of the economy; in 1952, industry received nearly 40 percent of investment funds, and heavy industry was allocated 76 percent of this.[15]

The distribution of investment funds was reflected in the choice of technologies. Few efforts were made to develop and apply technologies that could make use of China's abundant labor supply or be tied into the agricultural economy. The Chinese received state-of-the-art technologies from the Soviets, particularly in the area of steelmaking,[16] but modernity in the emerging industrial sector underscored the continued backwardness of farm technologies. In the agricultural sector investment rates remained astonishingly low. Less than 8 percent of the State budget was invested in agriculture during the first 5-year plan (1953-57), and of this, a sizable percentage went toward the construction of hydraulic projects only margin-

---

[12]See John K. Chang, *Industrial Development in Pre-Communist China: A Quantitative Analysis* (Chicago, Ill.: Aldine, 1969).

[13]Robert Dernberger, "Economic Development and Modernization in China," in Frederic Fleron, ed., *Technology and Communist Culture: The Socio-Cultural Impact of Technology Under Socialism* (New York and London: Praeger, 1977), p. 250.

[14]K. C. Yeh, "Soviet and Communist Chinese Industrialization Strategies," in Donald W. Treadgold, ed., *Soviet and Chinese Communism: Similarities and Differences* (Seattle and London: University of Washington Press, 1967), p. 339.

[15]Jan S. Prybyla, *The Political Economy of Communist China* (Scranton, Pa: International Textbook Company, 1970), p. 56.

[16]M. Gardner Clark, *The Development of China's Steel Industry and Soviet Technical Aid* (Ithaca, N.Y.: Cornell University Press, 1973), p. 6.

ally connected to the needs of farm production.[17]

Technological change would have to come to the countryside if the perennial problem of feeding China's people was ever to be solved. But this realization was slow in coming, and stipulated technological priorities bore little relevance to agricultural needs. In September 1956, a 12-year plan for the development of science and technology was formulated. Although the details of the plan were never released, 12 areas earmarked for future technological advance were published:

1. peaceful uses of atomic energy,
2. radio and electronics,
3. jet propulsion,
4. automation and remote control,
5. petroleum and scarce mineral exploration,
6. metallurgy,
7. fuel technology,
8. power equipment and heavy machinery,
9. the harnessing of the Yellow and Yangtze Rivers,
10. chemical ferilizer and the mechanization of agriculture,
11. prevention and eradication of diseases, and
12. problems of basic theory in natural science.[18]

As the list indicates, the main thrust of scientific and technological research and application was to be directed at the development of technologies essential to the operation of a sophisticated modern economy. With the exception of items 9 and 10, none of the areas selected for intensive development had any immediate relationship to the needs of the agricultural economy, nor were they congruent with the development of labor-intensive modes of production. In sum, ambitions seemed to have exceeded economic and political realities; after decades of submission

and inferiority, China was determined to catch up with the scientific and technological accomplishments of the West.

The limitations of the Chinese economy soon overwhelmed these ambitions. Although industrial growth during the 5-year plan period was impressive, the agricultural sector continued to lag. Collectivization and other administrative rearrangements had not resulted in an increase of peasant income[19] and by 1957 Chinese agriculture was in the hold of diminishing returns as the farm sector began to exhaust its traditional sources of growth. Although industrial development was impressive, it was apparent that future economic progress in that sector was tied to economic progress in agriculture.

Improved agricultural productivity was of foremost importance in the economic development of the countryside, but at first the Government sought to generate it in an oblique fashion, using political mobilization as a means of coaxing out the "productive forces" which had hitherto lain dormant. The Government accurately gauged that a technological revolution in crop growing was ultimately dependent on the widespread development of water-control projects. In order to effect the necessary changes in irrigation, the Communist leadership began to make massive efforts to tap China's seemingly limitless supply of rural labor for the building of dams, canals, wells, and reservoirs. During the slack farming season in the winter of 1957-58, 100 million people worked an average of 130 days each on hydraulic projects of this kind.[20] It was hoped that the construction of these projects would make a

---

[17]Li Choh-ming, *Economic Development of Communist China: An Appraisal of Five Years of Industrialization* (Berkeley, Calif.: University of California Press, 1959), p. 53.

[18]Richard P. Suttmeier, *Research and Revolution: Science Policy and Societal Change in China* (Lexington, Mass.: Lexington Books, 1974), pp. 60-1.

[19]Thomas P. Berstein, "Leadership and Mass Mobilization in the Soviet and Chinese Collectivization Campaigns of 1929-30 and 1955-56: A Comparison," *China Quarterly* 31 (July-September 1967), p. 35.

[20]Karl A. Wittfogel, "U.S.S.R. and Mainland China: Agrarian Systems," in W.A. Douglas Jackson, ed., *Agrarian Policies and Problems in Communist and Non-Communist Countries* (Seattle and London: University of Washington Press, 1971), p. 46. Alva Lewis Erisman presents a much lower figure of nine million man days: cf. "China: Agricultural Development, 1949-71," in *People's Republic of China: An Economic Assessment* (Washington, D.C.: U.S. Government Printing Office, 1972), p. 126. In any event, statistical ambiguity was a hallmark of the Leap.

major contribution to the Great Leap Forward, which sought to transform China's economy in the space of a few years.

So massive a mobilization of rural labor had to be complemented by fundamental changes in the organization of rural society. At first, special water conservancy organizations were formed. These and other forms of cooperatives were used to deploy peasant labor for projects that exceeded the boundaries of the village. Eventually most of these units were amalgamated into people's communes. In the summer of 1958, the communes began to proliferate, and by the end of the year 99.1 percent of the peasant households were reportedly incorporated into them.

The establishment of the communes was paralleled by a major effort to decentralize political authority and administration. Initiative in economic and technological matters was passed to lower administrative levels. Industry, agriculture, commerce, education, and defense became the concerns of commune administrators, thus vastly expanding the responsibilities of local cadres.

This expanded role for local-level leadership had important ramifications for technological policies and their implementation. Party cadres began to assume many of the technical and managerial tasks previously held by Government officials.[21] This did not simply mean the imposition of political control over the activities of technical and managerial personnel; political cadres, it was hoped, would become experts in their own right, the cadres' personal involvement with productive labor would result in the synthesis of "red" and "expert."[22] The transfer of political cadres to basic-level posts, where they could participate in labor while overseeing day-to-day operations, was a key policy of the Great Leap Forward.

First-hand involvement with local economic affairs was an absolute necessity; with the radical decentralization measures of the Great Leap Forward, the formal national planning apparatus and the statistical work that supported it were dismantled. The detached and rationalized style appropriate to administration by Government officials and specialized experts was forsaken in favor of a "combat style" through which leaders were to form an intimate association with the people, share in their struggles, and make policy on a largely *ad hoc* basis.[23] The Great Leap Forward was a time of politically mandated uncertainty and disjointed or nonexistent planning.

The technological consequences of the Great Leap Forward were soon evident. The 12-year plan for the development of science and technology was all but forgotten as uncoordinated local initiatives became the basis of technological change. Each locality began to push its own programs of economic and technological development through the communes. Although great emphasis continued to be placed on the rapid development of the heavy industrial sector in the cities, the countryside now became the scene of myriad labor-intensive projects. In the cities, heavy industry continued to be the primary recipient of State-supplied capital and technical assistance; in the countryside productive enterprises were to be developed through the mobilization of local labor and resources. The urban enterprises established during the 5-year plan period would continue to grow through the conscious fostering of economic dualism, while new small-scale, labor-intensive ones took root in the countryside. There were few connections between the two sectors.[24]

The most dramatic example of local initiative in technological and economic develop-

[21]Franz Schurmann, *Ideology and Organization in Communist China* (Berkeley and Los Angeles, Calif.: University of California Press, 1968), p. 171.

[22]New China News Agency, "CCP CC Directive on Physical Labor" in *Survey of China Mainland Press*, No. 1532.

[23]Such a mode of leadership is often found among circumstances of danger and uncertainty. See Alvin W. Gouldner, *Patterns of Industrial Bureaucracy* (New York: Free Press, 1957), pp. 105-16.

[24]Eckstein, op. cit., p. 57.

ment were the "backyard blast furnaces" which were erected for the local production of steel. Coinciding with the commune movement in the summer of 1958, thousands of furnaces sprung up each day in the countryside; by September, 350,000 were in operation.[25] This effort was a sad failure: the steel produced was of such low quality that it had few applications. Other efforts at decentralizing production were more successful, however, and the foundation of local industries for the production of cement, agricultural chemicals, and energy was laid during this period. Meanwhile, the urban industrial sector continued to emphasize capital-intensive higher technology methods of production.

In the critical agricultural sector, however, the policies embodied in the Great Leap failed. A key assumption—that mobilized labor could be effectively tapped without complementary inputs—soon proved fallacious. Agriculture continued to languish in the absence of modern inputs and techniques, and China found itself in the grips of a severe economic crisis. This was exacerbated in the middle of 1960 when the Sino-Soviet rift widened and Soviet economic and technical support missions were withdrawn.

The Chinese leadership now had little choice but to retrench, and mass mobilization under the leadership of technically unsophisticated political cadres was abandoned as the most appropriate road to economic progress. A recentralized economic system with strong inputs from technical experts once again took precedence over efforts to create an autarkic rural economy through collective labor and political leadership.

Although the policies of the Great Leap Forward had been specifically constructed to provide an improved productive base in the countryside, the rural economy remained critically deficient in both the kind of technologies employed and the capital investment necessary for their application.

In 1962, the critical importance of the farm sector was officially recognized and a

slogan declaring agriculture the "foundation" of the economy was adopted. Industrial production was more closely tied to agricultural needs, and the central Government began importing complete plans from Japan and Western Europe for fertilizer manufacture. At the same time, the rural sector was better enabled to make effective use of these inputs. Agricultural taxes were eased and the price relationships between industrial and agricultural goods altered in favor of the latter, thereby increasing the incentives for increasing crop production.[26]

Industrial policy now shifted, and a more rationalized and coordinated approach to increasing production was taken, stressing technological innovation over untrammelled mobilization of labor. The responsibilities and authority of managers and technicians were expanded. The political cadres who had supervised local production and technological innovation were replaced by engineers and managers with the specialized skills necessary for modernizing China's economy.

Yet engineers still did not have a free hand to design projects according to the standards of technical competence alone. Efforts to synthesize mass participation in technological development with the more regularized procedures of engineers, scientists, and technical specialists continued. During the design reform campaign which began in late 1964, engineers became more closely involved with actual production work. They were expected to break away from accepted engineering traditions, established during the period of Soviet influence, to engage in on-the-spot design work in close collaboration with the workers and managers in the plant. This was expected to result in a combination of technical experts, managers, and shop-floor workers who could pool their talents in order to solve technical problems.[27]

---

[25]Clark, op. cit., p. 69.

[26]Eckstein, op. cit., p. 60.
[27]See Genevieve Dean, "A Note on the Sources of Technological Innovation in the People's Republic of China," *Journal of Development Studies* 9, 1, (October 1972), pp. 190-193.

Despite this effort to unite all segments of the Chinese work force, tensions remained, and Mao and his followers feared that the extension of expertise and administrative power would result in the formation of new social classes and new sources of oppression. In late 1965, Mao launched the Great Proletarian Cultural Revolution as a means of redressing the inequalities reemerging in Chinese society. Although the cultural revolution did not take up the issue of technological policy directly, it did have important repercussions on the people and institutions directly responsible for promoting technological development. The Chinese Academy of Sciences came under heavy criticism from "Mao Zedong Thought Propaganda Teams," which had been organized by the army under the guidance of Lin Biao. The State Scientific and Technological Commission fared even worse. Its Chairman Nie Rong-Zhen was sharply attacked, and for a while the Commission was threatened with dissolution; an ad hoc Science and Education Group under the State Council seemed likely to assume its functions.[28]

For all the strident attacks on the entrenched centers of technological expertise, however, political control over technological change was not a prime issue during the course of the cultural revolution. The chief struggles of the period were directed at political (rather than technical) cadres who, it was claimed, had become entrenched in their offices, separated themselves from the masses, and forgot their revolutionary roots.

The convening of the Ninth Party Congress in April 1969, signaled an official end to the militant phase of the cultural revolution. The ensuing period was marked by conflicting political signals and uncertainty in technological policy. Imports of machinery and whole plants began to increase dramatically by 1973,[29] hardly a hallmark of the success of the radicals' efforts to insulate China from foreign influences. The fundamental changes that the cultural revolution wrought in education and research activities generated considerable discontent, including complaints about the excessively practical orientation of research efforts and the absence of adequate basic research. In 1972, the Chinese press featured articles calling for the improvement of science curricula and for more attention to the conduct of theoretical research.[30]

On the other hand, the cultural revolution left an inhibiting policital legacy. Although technological policy was never publicly debated, the postcultural revolution climate was not conducive to bold new thrusts in the technological realm. Ideological rectitude seemed to count more than economic expansion, and with Mao's succession very much in doubt, few leaders were willing to challenge basic Maoist tenets about the primacy of political and ideological mobilization.

This impasse was seemingly broken in January 1975, when Chou En-lai delivered an important speech at the Fourth National People's Congress. In it, he asserted that China's prime task lay in the "four modernizations" (in agriculture, industry, defense, and science and technology). These would pave the way for China's sustained development as a "powerful socialist nation" by the beginning of the next century.[31]

Chou's assessment of the pressing need for technological modernization was seconded by Deng Xiaoping, who had reemerged from his cultural revolution disgrace. Deng used his newly regained influence to aggressively push for technological modernization, even at the expense of class struggle and party domination over technical work. But with Mao's health steadily weakening and China's political course in doubt, Deng was made the target of continual attacks in the Chinese press. He was

---

[28]Suttmeier, op. cit., pp. 104-5.

[29]See Hans Heymann, Jr., *China's Approach to Technology Acquisition: Part III, Summary Observations* (Santa Monica, Calif.: The RAND Corporation, 1975), p. 7.

[30]See Thomas Fingar and Genevieve Dean, *Developments in PRC Science and Technology, October-December, 1976* (Stanford, Calif.: U.S.-China Relations Program, 1977), p 5.

[31]Chou En-lai, *Cheng-fu Kung-tso Pao- Kao* (Report on the Work of the Government) (Peking: Shen-wu Yin-shu Kuan, 1975), p. 16.

again purged in April 1976, and his report, which called for an expanded and more autonomous role for the Academy of Sciences, was labeled a "poisonous weed."[32] Deng's alleged enthusiasm for importing advanced

technologies from abroad in return for raw material exports,[33] was also criticized. This policy was in direct opposition to the autarkic economy advocated by his Maoist opponents.

---

[32]Thomas Fingar and Genevieve Dean, *Developments in PRC Science and Technology, January-March 1977* (Stanford, Calif.: U.S.-China Relations Program, 1977), pp. 5-7.

[33]See Hung Yuan, "Ultra-Right Essence of Teng Hsi p'ing's Revisionist Line," *Hung Ch'i* 10, (1976), in *Foreign Broadcast Information Service*, Oct. 8, 1976, p. E1.

PRESENT TECHNOLOGY ACQUISITION POLICY

Mao Zedong died in 1976; by the beginning of 1977, it was clear that a return to the principles of the four modernizations could be anticipated. Indeed, throughout 1977 there were abundant signs of departure from many of the practices and policies of the previous 10 years. In February 1978, a statement of the goals of the four modernizations program was made by Premier Hua Guofeng in his "Report on the Work of the Government" delivered to the first session of the Fifth National People's Congress. Hua's report contained the outlines of an ambitious 10-year economic development plan designed to produce an independent and comprehensive national industrial and economic system.

Among other things, this plan provided for enormous increases in agricultural output; agroscientific research; increases in the value of industrial output of 10 percent per year to 1985; the development of transport, communications, postal, and telecommunications networks big enough to meet growing industrial and agricultural needs; and 120 large-scale industrial projects. It was followed by plans for investment in science and technology including theoretical research and the establishment of nuclear power stations, the development of satellites, laser research, genetic engineering, and integrated circuit and computer applications.

To foreign observers, this program was exceedingly ambitious and it seemed to lack a sense of priorities. Doubts within China

Photo credit: Ben Lanwu, Xinhua News Agency
Unloading soybean on a thrashing ground

about its viability began to deepen, and evidence that the objectives of early 1978 were to be scaled down began appearing later that year and in early 1979. Ultimately, the Chinese decided to postpone, defer indefinitely, or cancel a number of agreements with foreign companies for plants and equipment. The most striking case was the postponement of a final agreement with Japan on the $2 billion Baoshan steel complex.

Northeast China farm introduces agricultural machines from America

A U.S.-made sprayer at work

Photo credits: Ben Lanwu, Xinhua News Agency

Combine harvesters working in a soybean field of the Friendship Farm

In redefining the objectives of the four modernizations, Beijing has begun to stress the concepts of "balance" and "proportionate development," and the thinking associated with the economic administration characteristic of the 1950's is now being praised. "Three major balances," have been identified as the foundation for economic policy: the balance in the expenditures and revenues of the State budget; the balance in the issuance and withdrawal of bank credits; and the balance in the supply and demand of materials, including the balance of receipts and payments of foreign exchange. It is claimed that during the last 30 years, when these balances were observed, the economy progressed nicely; when they were not observed, development slowed. According to this line of reasoning, serious imbalances have now occurred between development in agriculture and industry, particularly heavy industry; within industry itself (for instance, between the fuel, power, and raw materials industries and the processing industries; among farming, forestry, and animal husbandry; and between food crops and industrial crops); between investments and capital construction and the availability of manpower, materials and financial resources; between accumulation and consumption; and within accumulation itself (e.g., between accumulation for productive purposes and accumulation for nonproductive purposes).[34]

It is safe to assume that fears of new "imbalances" are behind the current efforts to scale down the four modernizations, and that balance and proportion in development are the objectives now being sought. This conclusion is borne out by the main points of the revised program which is intended to check the spurt of investments in heavy industry, and return to the principle of priority for agriculture and light industry. There have been reports that the proportion of investment in iron and steel will be reduced: past investments have not yielded expected returns because of serious shortages of electricity, transport, and other necessary supporting infrastructure. Major investments in harbor construction are to be postponed. Agriculture is to receive the most attention, although the grand objectives for agricultural mechanization announced in early 1978 have been abandoned. Instead, it is likely that investment will foster greater differentiation of agricultural production based on the comparative characteristics of different regions, and that more attention will be given to cash crops and to industries supporting agriculture. The performance of Chinese agriculture is of utmost importance to the four modernizations; successful performance in this sector would release foreign exchange for other purposes.

The serious problems of Chinese infrastructure development will be addressed by additional investments in coal, electric power, oil, transport, and building materials. Some of the 120 key industrial projects identified in the original four modernizations program will be postponed, and only a limited number will be in operation by 1985. Housing construction is to be accelerated. New attention also is to be given to light industry, both to satisfy domestic aspirations, and to make light industry the chief export and foreign exchange earner over the short run. Although foreign borrowing has not been completely ruled out, the Chinese are stressing that future investments should come mainly from domestic savings.

In spite of these cutbacks, it is clear that foreign trade and the acquisition of foreign technology will play an important role in Chinese modernization programs, although not as great a role as the euphoric estimates of 1978 might have suggested. China's foreign trade has fluctuated somewhat over time, but it has tended to rise consistently through the 1970's. Figure 15 shows the Chinese balance of trade from 1950 to 1976, including the shares of trade that non-Communist countries have enjoyed since the early 1960's.

In addition to agreements with individual foreign companies, China has entered into trade agreements during the last 2 years

[34]*Beijing Review,* May 11, 1979, pp. 17-18.

Figure 15.—China: Balance of Trade, 1950-76

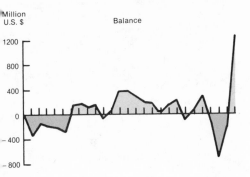

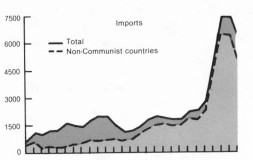

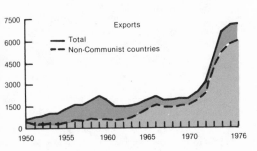

SOURCE: Richard E. Batsavage and John L. Davie, "China International Trade and Finance, in *Chinese Economy Post Mao*, Joint Economic Committee, 1978, p. 709.

with a number of countries. The most dramatic example is a $20 billion agreement with Japan calling for Chinese imports of technology in exchange for oil and coal. These agreements indicate that at least through 1985 China will continue to have an active interest in foreign trade and particularly in foreign technology. A recent estimate of China's budget for imports of foreign technology through 1985 shows that $11.4 billion was contracted for in 1978, and that negotiations to acquire additional technology could amount to as much as $59 billion by 1985. China's foreign technology budget is summarized in table 40.

While revisions in the four modernization programs may lead to a scaling down of Chinese foreign technology purchases, there will undoubtedly be major acquisitions in the next 5 to 8 years. According to a recent Central Intelligence Agency (CIA) report, the areas that China has designated for closest attention are as follows:

1. **Iron and Steel.**—It is expected that this sector will require the greatest expenditure of foreign exchange. The plans for doubling China's steel pro-

Table 40.—China's Foreign Technology Budget, 1978-85
(in billions of dollars)

Category	Contracted for in 1978[a]	Under negotiation[b] (rounded estimates)
Iron and steel . . . .$	0.1	$19.0
Coal	4.0	1.6
Other mining and processing . .	2.4	4.4
Ports	2.1	1.1
Petrochemical plants and equipment.	0.5	2.3
Hotels	0.5	3.0
Shipping	0.5	—
Petroleum	0.4	3-5
Electronics	0.3	0.1
Agriculture-related plants and equipment.	0.1	0.6
Power development	0.2	3.0
Fisheries	—	2.0
Aircraft	—	1.0
Other transportation	—	10-15
Construction plants and equipment.	0.2	0.5
Textile plants and equipment . .	—	0.1
Miscellaneous machinery and machine tools.	0.1	0.2
Total.	$11.4	$52-59

Data as of Apr. 15, 1979.
[a]Excludes approximately $1.5 billion in projects contracted after mid-December 1978, which were postponed until financing is arranged.
[b]Some projects under negotiation will be completed after 1985.

SOURCE: *The China Business Review*, March-April 1979, p. 57.

duction to 60 million tons by 1985 can only be achieved by importing major new steel complexes. Contracts under discussion have included a $15 billion, 10-million ton complex near Tianjin in Hubei Province, and a $2 billion major facility at Baoshan near Shanghai. The latter was to be a complete plant purchased from Japan, a deal which has now been delayed but not canceled.

2. **Coal and Electric Power.**—China's original plan called for the opening of 8 new major coal mines and 30 power stations by 1985. A coal industry development project worth $4 billion has been under discussion with West Germany. Discussions have been held with France for the construction of two nuclear powerplants, but no contracts have resulted. Priority in electric power development will probably go to conventional sources of power, including major new hydroelectric stations.

3. **Transportation.**—This has been a hitherto neglected area, and Beijing has been moving aggressively toward reversing this situation. Reportedly, since 1976 more than $1 billion has been set aside for foreign equipment in road, rail, water, and pipeline transportation, and negotiations valued at $4 billion have been conducted for imports of truck, automobile, railroad car, and locomotive plants and for shipping facilities.

4. **Petrochemicals and Synthetic Fibers.**—Approximately $3 billion in contracts for technology in petrochemical and synthetics fields have been signed.

5. **Communications and Electronics.**—This is another sector that has been neglected in the past but which now has a high priority. China evidently wishes to leapfrog toward the most modern communications systems and has entered into discussions with the United States for a communications satellite. It has also concluded an agreement with Japan for manufacturing facilities for color television sets. Part of this deal included the purchase of an integrated circuit plant.

6. **Nonferrous Metals.**—Signed agreements exist in this area with Japan and the United States for a copper mining concentration complex and a copper smelter. Of the 120 key projections in Hua's original formulation of the 4 modernizations, 9 dealt with nonferrous metals complexes.

7. **Construction.**—China's construction industry has also been somewhat neglected in the past and has become a major bottleneck for the expansion of other sectors that require massive construction programs. As a result, China has been interested in cement, asbestos plate, insulation, and construction equipment. These have a projected value of about three-fourths of a billion dollars.

8. **Petroleum and Gas.**—Both because of domestic energy shortages and because of the potential of this sector as a foreign exchange earner, the petroleum and gas industry has received high priority. Since 1976, $500 million has been spent on oil and gas exploration and exploitation equipment.

9. **Machine Building.**—China is interested in foreign forges, foundaries, press lines, and cutting tools which could lead to whole plant contracts totaling more than $70 million.

10. **Instruments and Controls.**—China's economic leaders have, in the last few years, shown a renewed interest in quality control and standardization of products. As a result there has been considerable interest in measuring instruments and analytic devices, and control mechanisms for quality control.

11. **Agriculture.**—Thus far, a relatively small amount of agricultural machinery has been purchased, primarily for

use in the large-scale State farms in the north. With the reduction of mechanization goals in the revised version of the four modernizations program, machinery imports in this sector could be less than originally anticipated. On the other hand, China will probably continue to be interested in turnkey projects for fertilizers and insecticides.[35]

---

[35]Central Intelligence Agency, *China: Post-Mao Search for Civilian Industrial Technology*, National Foreign Assessment Center, 1979, pp. 4-6.

In addition, China has shown considerable interest in engineering and service contracts intended to supply design-engineering know-how and managerial skills. Such contracts provide for foreign firms to design, act as prime contractor, and supervise the development of various projects. The transportation and mining sectors have received the most attention: contracts have been signed with West Germany and the United Kingdom for coal mine development, and a major service contract for railroad development has been under discussion with Japan. Negotiations have also been conducted with Denmark on port construction.

THE NATURE OF THE CHINESE ECONOMY

When the PRC was established, Chinese leaders looked to the Soviet Union for guidance and to the Soviet economy as a model to emulate. Although this close relationship lasted only until 1956, the Chinese and Soviet economies still resemble each other. Both are characterized by central planning, and both tend to focus on physical output as a criterion for evaluating performance. This has resulted in less attention to the value of production efficiency.

While there are similarities in the two economies there are also significant differences, The Chinese economy is both more decentralized and far more ideologically driven than the Soviet. As a consequence, China has had a looser approach to economic planning than has the U.S.S.R., and at times ideologically induced decisions have superceded planner preferences. Decentralization stems from 1956 to 1957, when dissatisfaction with the Soviet model began to surface. As previously noted, the first significant decentralization occurred as part of the Great Leap Forward in 1958, when a wide range of industries and financial resources were placed under provincial control and Communist Party committees at the province and subprovincial levels were assigned a more direct role in running economic activities.

This had many negative results and was modified during the early 1960's. It is the second model of decentralization which largely persists today. The significant differences from the Great Leap model are first, that professional managers and engineers share responsibility with Party committees in the leadership of enterprises; and second, that responsibility for finance and commerce is held closely by the central Government in contrast to the Great Leap years when this responsibility was relinquished. It should be noted that China's search for a balance between centralized and decentralized forms of economic management is closely related to the search for a balance between economic efficiency (which has tended to lead to decentralization) and the distribution of wealth and economic equity (which has tended to support the continuation of an active central role in interprovincial redistributive revenue-sharing).[36]

There have also been attempts to accommodate the power interests of influential provincial leaders. It is possible to think of the Chinese industrial economy as a three-

---

[36]Nicholas R. Lardy, *Economic Growth and Distribution in China* (Cambridge, Mass.: Cambridge University Press, 1978).

tiered phenomenon, the three tiers being "center," "local," and "collective." Although the collective sector is subject to planning, it experiences a measure of autonomy not enjoyed by the other two tiers because it is entitled to use aftertax profits in a locally discretionary fashion. Establishing and upgrading rural industry have been important investment options for the collective sector.

The "local" sector includes enterprises under provincial, municipal, and county control. Unlike those in the collective sector, these enterprises are all under State ownership, and, despite the nomenclature, are subject to a significant amount of central influence. Local sector output forms a substantial part of the total value of Chinese industrial output.

Despite considerable decentralization, the central Government continues to be important. It maintains control over much heavy and strategic industry; it acts as the allocator of such priority goods as energy, key raw materials, capital goods, military equipment, important exports, staple food products, and cotton; and it controls interprovincial trade.[37] The center maintains control over the economy through such mechanisms as material supply planning; the fiscal and banking systems; and control over prices, wages, and foreign trade.[38]

According to a recent statement by the State Planning Commission (SPC), the center's responsibilities include:

1. the guidelines and policies of the national economy; 2. output targets for major industrial and agricultural products; 3. basic construction investment and major construction projects; 4. allocation of important materials; 5. the purchase and allocation of key commodities; 6. the state budget and the issuance of currency; 7. the number of workers and employees to be added and the total wage bill; 8. the prices of major industrial and agricultural products.[39]

SPC plays a key role in the planning system of the Chinese economy. It is generally believed that the preparation of economic plans involves both vertical or functional planning, and horizontal or territorial planning.[40] This dual approach is consistent with the concept of dual control over the economy in which enterprises are subject both to central ministries and a unit of local government. At the central level there are a large number of economic ministries that oversee the operation of the economy. These are organized along functional lines. Enterprises throughout the country that have the same functional identification come under their jurisdiction. One stream of planning activity apparently occurs in this vertical system.

On the other hand, there is a second stream of planning activity that is territorial, centered around the province. Planning is a process of aggregation and disaggregation of data and tasks. Whereas in the vertical stream the aggregation and disaggregation is performed by ministries and subordinate enterprises, in the territorial stream this is done by provinces and by the enterprises within provincial boundaries. It is the responsibility of SPC to aggregate on a national basis the information produced in these two streams of activity.

The planning process is thought to involve three main stages which the Chinese refer to as "sent down twice, reported up once."[41] The process begins when SPC issues "control numbers" for the coming year. These include value of output, amount of investment, number of workers, and total wage bill and are determined on functional and territorial bases. These flow down through the system, territorially and functionally, generating more specific planning information which is

[37]Gordon Bennett, *China's Finance and Trade* (M. E. Sharpe, 1978), p. 74.
[38]Lardy, op. cit., pp. 15-16.
[39]Quoted in Nai-Ruenn Chen, "Economic Modernization in Post-Mao China; Policies, Problems, and Prospects," in Joint Economic Committee, *Chinese Economy Post-Mao* (Washington, D.C.: U.S. Government Printing Office, 1978).
[40]Lardy, op. cit., pp. 15-16.
[41]Ibid.

"reported up" and eventually gets back to SPC. This upward and downward flow of information is by no means a purely technical operation; it is highly political, characterized by bargaining and tradeoffs, and probably resembles the U.S. budget process. After the "reporting up once" stage, SPC aggregates the information and produces a final plan for the economy.

Provisions for innovation, presumably also including the incorporation of foreign technology, are folded into the plans during this process. But as with the Soviet economy, significant structural constraints on innovation are related to the system of management. China has seen a variety of approaches to enterprise management since the 1950's; it has been the subject of political dispute and ideological debate, centering around the problem of whether "reds" or "experts" should run factories. Additional issues have concerned whether production norms should predominate in management, and the use of material incentives. Managerial systems have varied from the Soviet style "one man management" to management by Party committee. Since 1978, the role and authority of professional managers and expert engineers has been emphasized, although the Party still has final responsibility and many, if not most, managers are members of Party committees.

Regardless of managerial form, it is important to note the criteria used by higher authorities to evaluate management, and the performance of firms in fulfilling the objectives of the plans. Performance criteria have changed over 30 years, but there have always been multiple, sometimes contradictory, criteria. For example, a 1972 report lists performance criteria as quantity, quality, variety, cost relationship, labor productivity, profits, and funding; plus numbers of workers and total wage bill.[42] The targets were output, variety, quality, consumption of raw materials, fuel and power, labor pro-

ductivity, costs, profits, and working capital ratio.[43]

While enterprises are expected to meet all of these targets, from time to time there has been a tendency for enterprises to focus only on physical output in the belief that the planning system is biased in this direction. Attention to this target has led to losses in efficiency, product quality, and innovation:

> Innovation is another frequent casualty of the drive to raise output. Research and experimentation require the attention of engineers and skilled workers whose services are also needed to maintain high levels of output. Complaints that "many enterprise leaders who are very concerned about plan fulfillment pay little attention to new product work" fall on deaf ears as long as quantity rules supreme. Conflict between research and production persists. In 1971, at Shanghai's Hung Ch'i shipyard, "Under the condition of the urgency of the task of production, there were some people including some members of the basic level revolutionary committee who said, 'the task of production is already so heavy, where is there time for carrying out innovation?' There are even some who said, 'The task of production is a hard target, but the task of scientific research is a soft target.' "[44]

Although there are signs that in some respects, particularly in terms of intersectoral communication, the Chinese economy is more innovative than the Soviet economy, it nevertheless imposes technically conservative norms on enterprise managers. In the Soviet case, this has inhibited not only indigenous innovation, but also the effective absorption of foreign technology.[45] (See chapter X.)

As discussed above, one of the central Government's reactions to the chaos of the Great Leap Forward was to reassert tight financial control over the economy. This con-

---

[42]See Thomas G. Rawski, "China's Industrial System," in Joint Economic Committee, *China; A Reassessment of the Economy,* 1975.

[43]Chen, op. cit.
[44]Rawski, op. cit., p. 182.
[45]North Atlantic Treaty Organization, *East-West Technological Cooperation* (Brussels, 1976), and North Atlantic Treaty Organization, *The U.S.S.R. in the 1980's. Economic Growth and the Role of Foreign Trade* (Brussels, 1978).

trol has continued, and it is exercised primarily through the State budget and the banking system. State enterprises are expected to remit most of their aftertax profits to the center, and these remittances make up a substantial share of the State budget. Therefore, for capital construction and other forms of investment, enterprises must rely on investment resources returned to them via the State budget. This system, known as unified revenues and unified expenditures, has resulted in a lack of a direct connection between investment funds and economic performance, and has also been a source of inefficiency. Recent policies have attempted to overcome this problem by shifting from budgeted capital construction funding to loans supervised by the People's Bank.

Bank loans have, therefore, been used for overcoming production difficulties, and will be used more for working capital than for investment in capital goods. Reportedly, however, enterprises do try to get around banking regulations and use bank loans for capital stock. Because of the importance attached to physical output norms, enterprise managers seek investment funds. When these funds are available through the State budget, only depreciation, not interest, is charged. To meet production quotas, firms therefore have a natural tendency to pursue a relatively costless strategy of seeking vertical integration, especially the acquisition of well-stocked machine shops which can repair equipment received from the outside, or retool existing machinery.[46]

Such patterns of management have produced defects in the Chinese economy, including weak quality control, neglect of innovation, excessive vertical integration, and stockpiling—problems endemic to Soviet-style economies.[47] Not surprisingly, the Chinese have sought remedies for these kinds of problems. During the early 1960's, a number of reforms were implemented, and some of the current discussion seems inspired by the events of this period.

The reforms of the early 1960's were soundly denounced as revisionist during the cultural revolution. In fact, a series of these measures came to be called "the five soft daggers." They included the use of profits as the chief success indicator for enterprises; the subordination of "politics"—especially political campaigns and ideological remolding sessions—to production; the widespread use of material incentives; reliance on experts in factory management; and movement towards greater enterprise autonomy.[48]

Current economic thinking shows many similarities to the "five soft daggers." The leadership is concerned with achieving greater economic efficiencies and higher quality goods. It therefore feels that production goals should be expressed in value terms (such as profits) rather than mere physical output. The idea of "production first" also is now widely discussed; Deng Xiaoping has suggested that in the current historical period, putting production first is the key political task. Material incentives in the form of bonuses and wage increases have been reintroduced in the past 2 years (although not without problems), and expertise in factory management is being stressed with a consequent downplaying of the role of Party committees. Finally, new approaches are being taken to the principle of enterprise autonomy. In addition, it appears that at least one leading economist of the 1960's, Sun Yefang, has been rehabilitated following his cultural revolution disgrace, and has been asked to take the lead in designing six regional economic systems. The establishment of regional economic units above the province level is a step towards the assertion of greater central control over the local economy and the simplification of planning to achieve better economic results.

In an attempt to free up economic activity from restrictive bureaucratic control exercised by central ministries, China has reestablished industrial companies (reminiscent of the "trusts" of the early 1960's that were

[46]Rawski, op. cit., p. 183.
[47]Ibid.

[48]Bennett, op. cit., pp. 31 ff.

denounced during the cultural revolution), at least 15 of which have been established during the last 2 years. Companies are designed to rationalize internal commerce and contracts and to cut across vertical bureaucratic lines. They have responsibility for planning, production, and coordination in their respective fields. In addition, there are corporations which are "economic accounting units" and operate on the basis of profit and loss.[49] These serve as consultants for their respective parent ministries and foreign trade corporations on matters relating to the import of foreign technology. In any particular field there will be one corporation and several companies. As of late 1978, a partial list of corporations included: The China Agricultural Machinery Corporation, The China Cereal and Oils Corporation, The China Chemical Construction Corporation, The China Chemical Fiber Corporation, The China Coal Industry Technique and Equipment Corporation, The China Cotton Spinning and Weaving Corporation, The China Feedstuffs Corporation, The China Geological Exploration Corporation, The China Oil and Gas Exploration and Development Corporation, The China Petroleum Corporation, The China Radio Equipment Corporation, The China Railway and Technical Equipment Corporation, The China Seed Company, The China Shipbuilding Corporation, and The China Waste Materials Reclamation Corporation.

Recently, these have entered into direct contracts with foreign vendors. For example, the Chemical Construction Corporation was responsible for importing hydrocracking know-how from Lummus and purchasing eight synthetic ammonia plants from Kellogg in 1974. The China National Chemical Fibers Corporation evaluates petrochemical fiber technology from Europe, the United States, and Japan, and recommends which technologies China should buy.[50]

The reintroduction of the corporation form of organization has also been accompanied by calls for greater enterprise specialization designed to break down excessive patterns of vertical integration. Some Chinese economists are beginning to discuss a version of market socialism in which enterprise specialization would be matched by a much greater reliance on market mechanisms in the industrial economy. Currently they are at least officially forbidden. The approach to economic organization would also make use of contracts to enforce interenterprise agreements, and would rely more heavily on the banking system both for supplying investment capital and for enforcing economic discipline. These proposed reforms are at various stages of discussion and implementation. While some of them do in fact appear to be steps in the direction of great efficiency, quality consciousness, and innovativeness, they are unquestionably subject to opposition in a number of quarters, including provincial authorities and corporation officials.[51]

In sum, China's planning and management practices, and indeed the structure of the economy itself, are now in a state of flux, and it is difficult to predict how far or how fast economic reform might go. Those members of the leadership, including many of China's economists, favoring an aggressive pursuit of the four modernizations, undoubtedly favor significant economic reform. But there are others in the leadership who, while wishing to promote the four modernizations, are, for reasons of self interest and ideology, hesitant to encourage significant economic reform. This is particularly true of reforms that will emphasize profit, market exchanges, and more rationalized management practices. The ongoing debate bears watching. It will have a major impact at least on the pace, if not the character, of the four modernizations program, and will affect the extent of China's reliance on and capacity to absorb foreign technology.

---

[49]*China Business Review*, September/October 1978, pp. 21 ff.

[50]Ibid., p. 22.

[51]Ibid., March/April 1979, p. 56.

THE STRUCTURE OF ECONOMIC DECISIONMAKING

At present, the highest levels of economic policymaking take place in the State Council on the Government side, and in the Politburo of the Central Committee of the Communist Party on the Party side. The pinnacle of power in China is the standing committee of the Politburo of the Central Committee. While the body is clearly the locus of decision for issues of strategic national importance, studies of policymaking in China indicate that participation in policymaking is considerably broader. Of particular interest are various kinds of sectorial work conferences (that is, national conferences in a given functional area involving Central Committee members, and central Government and provincial government officials) and various kinds of meetings of the Central Committee itself. The State Council is China's cabinet. It interprets policy coming from the Party and oversees the work of the Government. Most high Government officials are also members of the Politburo, while most ministers are members of the Central Committee. Under the State Council are a series of specialized agencies, ministries, and commissions. Of particular importance in economic policy are the State commissions dealing with economic and technical affairs. These are SPC and the State Science and Technology Commission (STC).

SPC is the center of economic policymaking after the Central Committee and the State Council. The other planning bodies follow the SPC's policy; it approves the budgets of the other commissions which then make detailed plans within these budgetary guidelines. Reportedly, SPC has bureaus for production, foreign trade, communications, agriculture, and construction, as well as other bureaus corresponding to each ministry. It also has field offices at the provincial and municipal levels.[52]

STC, abolished during the cultural revolution, was reestablished in October 1977. A

measure of the importance the regime attaches to science and technology in the four modernizations can be seen in that fact that a member of the Politburo was made minister in charge. STC is known to have a bureau for foreign affairs, a planning bureau, a policy research office, a bureau for energy and petrochemicals, a bureau for computers and machinery, and five or six other bureaus of unknown functions. These functions probably reflect the eight priority areas of the national science plan which was announced in early 1978—agriculture, energy, materials, computers, lasers, space, high energy physics, and genetic engineering. Prior to the cultural revolution, STC had an important role in coordinating national R&D activities, in science and technology planning, and in setting national standards.

The Chinese planning process involves both vertical plan development through the functional ministries and horizontal plan development through the provinces. While STC develops its own plans, it presumably also acts to provide specialized advice and information to SPC in the formulation of national plans. It is SPC that has the key role in turning national economic policy as specified by the Party and the State Council into a set of priorities, although little is known about how and by what criteria this is done. Nor is it entirely clear how technological requirements are set, although decision on incremental technological improvements and evaluation of the utility of technologies result from the downward and upward flow of planning information. The process of setting major new nonincremental technological requirements is less clear. On issues of major national importance, the State Council and the Central Committee undoubtedly play important roles, but information available to, and the strategic position of, SPC should ensure it a lead role in identifying at the policy level significant areas of technological need. In evaluating that need, particularly from the technical side, SPC has at

[52] Ibid., p. 14.

its disposal the competencies of the State Capital Construction Commission (described below) and STC. The latter can draw on the resources of the Academy of Science, ministerial research institutes, and professional societies.

DECISIONMAKING ON FOREIGN TECHNOLOGY

Predictably, given the multiple bureaucratic actors involved in making Chinese economic policy and managing the economy, there is no single or simple pattern of decisionmaking concerning the importation of foreign technology. Moreover, little detailed knowledge of this process is available in the West. Nevertheless with the growth in Chinese foreign trade and in the number of contacts relating to foreign technology, understanding of the decisionmaking process is increasing. Figure 16 represents a recent attempt by the CIA to summarize the decisionmaking procedure.

According to the CIA, the procedure begins with an initial request for foreign technology from provincial governments, individual plants, or industrial ministries. In addition, China has recently resurrected various types of industrial corporations that typically would be the end users of technology, and which undoubtedly generate their own requirements for foreign technology. Requests then go to SPC which approves the initiation of an investigation into obtaining technology. The SPC behavior presumably is guided by national economic policy directed from the State Council and the Central Committee, and by its own preliminary estimates and initial control figures for imports and exports based on overall economic goals and statistical information it has received from various sources.[53] SPC works with the Ministry of Foreign Trade (MFT) setting priorities for foreign technology. Reportedly, MFT has responsibility for mapping out general import and export plans that accord both with foreign policy and with existing contractual commitments to foreign trading partners.

Also taken into account are the nature of import and export commodities, world market conditions, domestic demand and export capability, and available foreign currency and external credits.[54]

According to the CIA, the State Economic Commission (SEC) also may initiate requests for foreign technology. SEC is responsible for the implementation of production plans approved by SPC on an annual and quarterly basis. It coordinates the supply and demand of industrial raw materials, energy, and other inputs. If domestic sources are unavailable it can turn to imports to overcome shortages. The SEC's activities in foreign trade are thus limited more to raw materials and other industrial inputs, than to plant and equipment orders.[55] But according to information received during a recent SEC mission to the United States, the Commission also has a bureau of technical affairs, and may, therefore, have a more active role in assessing and evaluating technology than had been thought.

An important actor which is not included in the CIA scheme is the State Capital Construction Commission (SCCC), which probably coordinates with SPC in the early stages of the decision process. SCCC has the lead role in overseeing the administration of projects exceeding a certain size. SCCC coordinates the work of various ministries—for example, the Ministry of Metallurgical Industry with regard to steel investments. Once the SCCC's budget receives SPC approval, it coordinates investment plans of ministries, and in turn approves their investment budgets. SCCC has liaison bureaus for

[53]Gene T, Hsiao, *The Foreign Trade of China: Policy, Law, and Practice* (Berkeley, Calif.: The University of California Press, 1977), p. 74.

[54]Ibid.
[55]*China Business Review*, March/April 1979, p. 14.

Figure 16.—China: Technology Import Decisionmaking Procedure

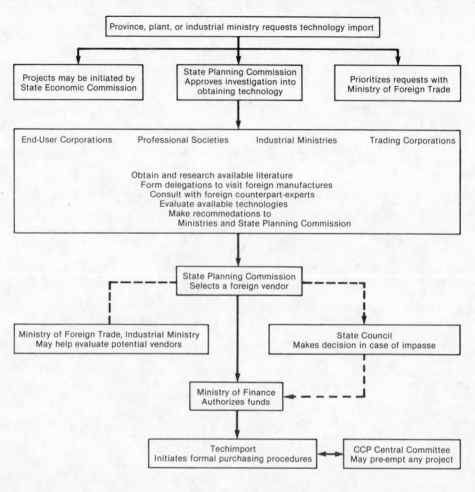

SOURCE: Central Intelligence Agency, National Foreign Assessment Center, China: Post-Mao Search for Civilian Industrial Technology, 1979, p. 8.

each ministry except agriculture. These include bureaus of coal, petrochemicals, machine building, petroleum, railroad, metallurgy, and water conservation and power. SCCC maintains branches at the local level.[56]

Unlike SEC, SCCC can initiate requests for plant equipment orders from abroad. It is likely that SCCC has become considerably more influential since the beginning of 1978. Some have speculated that it was the motive force behind the reconsideration of the four modernizations and the move toward more "balanced" and proportionate development as it began to aggregate the enormous capital construction requirements coming from various ministries.

[56]Ibid.

Once authorization for beginning an investigation for obtaining technology has been approved, the action shifts to other institutions. Until recently a lead role in this stage was played by the foreign trading corporations. These fall administratively under MFT which transmits to them guidelines for specific import and export plans. The foreign trade corporations add detail to the guidelines, and return them to MFT. Once approved, they become part of the general trade plans and, in turn, part of the national economic plan. Finally, after approval of the national plan by the State Council, MFT assumes ultimate responsibility for supervising the national foreign trade corporations in executing their specific plans.[57] However, at the search stage, industrial ministries, professional societies, and end-user corporations are also important. As the CIA has pointed out, these groups obtain and research available literature, form foreign delegations to visit foreign manufacturers, consult with foreign counterpart experts, evaluate foreign technology, and finally make recommendations to the ministries and SPC.

The Chinese have shown increasing sophistication in this search procedure, and a great deal of technology and technical information is transferred during this search stage. The Chinese have been making a coordinated national effort to accumulate as much written material as possible. The China National Publication Import Corporation procures technical titles from Western booksellers, and the Beijing Document Service was recently established to serve as a focus for a national technical information system linking the libraries of the Chinese Academy of Sciences, research institutes, ministries, and other organizations. The Chinese are interested in using the latest information-handling technology to provide a high-speed technical information network on a national scale.

The last 2½ years have seen an enormous growth in Chinese technical delegations traveling abroad to Western Europe, Japan, and the United States. Some 2,000 Chinese technicians and officials have traveled in this capacity in 1977 and 1978. Members of these delegations appear to their Western contacts to be knowledgeable and well-informed. It has been suggested that these delegations are logically arranged in a serial fashion beginning first with "survey groups" which visit potential vendors to discuss technology and prices. They are the window shoppers who then report back their findings. Follow-on "study groups" then visit only a few selected firms chosen on the basis of the reports. These study groups engage in a more detailed investigation oriented toward the specific applicability of foreign technology to specific Chinese needs, and may be composed of individuals with greater technical expertise. The reports of the study groups become the basis for selecting a vendor.[58]

In addition, the Chinese have been inviting many Western firms and Western experts to present technical seminars in China. Reportedly, more than 250 such seminars were held during the first 6 months of 1978.[59]

The results of these technical survey activities are widely discussed in China within the economic ministries. Additional technical advice is supplied by professional societies such as the China Chemical Society or the China Civil Engineering Society. Such societies occupy a strategic position in networks of information since they have contacts with foreign counterparts on one hand and they are linked to domestic industrial ministries and research institutes on the other. China's R&D efforts are vertically organized by production ministries, the Academy of Sciences, and the higher educational system. The professional societies on the other hand, draw their members from the vertically organized R&D sectors, and from production units. They thus appear to offer a significant horizontal linkage that may overcome some of the worst forms of "bureaucra-

---

[57]Hsiao, op. cit., pp. 74-5.

[58]CIA, op. cit., pp. 3-4.
[59]Ibid., pp. 2-3.

tism" induced by strong vertical organizations.

Once the search and survey are completed, a recommendation is made to the ministries and the Planning Commission and the purchase is then factored into national economic plans. As the CIA scheme shows, MFT is again involved at this stage. SCCC also probably plays an important role in approving the selection of the foreign vendor. Depending on the scale of the project, the State Council may also be asked to intervene in the final stage of the decisionmaking process. Once approval of the project has been given, the Ministry of Finance, working through the Bank of China, must authorize funds.

Although it had been assumed that foreign exchange was closely controlled by the center, in the last few years individual enterprises may have had greater discretion than has been thought. Institutes of the Academy of Sciences have their own foreign exchange budgets, and reportedly allocations of foreign exchange to local production units for the purchase of foreign equipment have recently doubled. However, China has apparently been dissatisfied with authorization and control procedures for the use of foreign exchange which make it possible for buyers and users of technology to minimize their consultations with the Bank of China.[60] As a result, it was announced in April 1979, that a new General Administration of Exchange Control has been established directly under the leadership of the State Council.[61]

The final phase of the technology acquisition process is the initiation of formal purchasing procedures by a foreign trade corporation. The corporation most directly involved with acquiring foreign technology and responsible for importing whole plants and high-technology equipment has been the China National Technical Import Corporation (TECHIMPORT). While the use of TECHIMPORT and other corporations may facilitate central control, particularly control over foreign exchange, in the past these corporations have prevented direct contacts between vendors and end users. The Chinese seem now to be moving away from the use of middlemen by encouraging users to enter into direct negotiations with vendors. The foreign trade corporations may then take responsibility for administrative details. The first such contract signed with a U.S. firm was between Atlantic Richfield and the China Petroleum Corporation. The contract was signed on March 17, 1979, and provides for seismic survey work in offshore waters.

Another organization with a role in China's technology acquisition procedures is the China Committee for the Promotion of International Trade. Officially a nongovernmental body which nevertheless maintains close ties with MFT, it has helped establish and maintain contacts with foreign trading firms and agencies, and has also had a role in sponsoring foreign travel and exhibits in China. Its U.S. counterpart, particularly in the period prior to normalization and direct government-to-government contacts, is the National Council for U.S.-China Trade.

A final, and possibly significant, omission from the CIA scheme is STC. It is likely that, at least for some types of technology, this commission would play a role both in the initial stages and in the search and evaluation stages.

It is difficult to assess the effectiveness of China's decisionmaking processes for importing foreign technology. It is possible that the initial surge of interest in foreign technology in 1977 and 1978 exceeded China's decisionmaking capacity, and that one of the reasons for retrenchment evident in early 1979 was the realization that the multiple demands for foreign technology coming from the various sectors of the economy would exceed the PRC's ability to pay and to absorb. It is curious, however, that discussions with foreign firms should have gone as far as they did without apparent central control. One explanation for these events is that the new policies were intended to shake up the economic structure, and/or were part of strategy to encourage Chinese

[60]*China Business Review*, March/April 1979, p. 3.
[61]*Beijing Review*, Apr. 20, 1979, p. 2.

economic managers to survey the technology available on world markets. An alternative explanation is that things did get out of control, and that only when SPC and SCCC began to aggregate the demands did China's top leaders begin to realize that the country's capacity to pay for and absorb technology was being taxed. The charge of inadequate management of the technology acquisition program is supported by the establishment of the General Administration of Exchange Control in April of this year.

The Chinese are clearly seeking a balance between central control and flexibility. The sanctioning of contracting authority for end-user corporations is a step in the direction of greater flexibility. It is also possible that with the greater liberalization of the economy described above, additional flexibility in foreign trade and technology acquisition will follow. For instance there have been reports of industries under local government control entering into product payback schemes with foreign firms on their own authority. It is likely, therefore, that a greater measure of decentralization in foreign trade can be expected. There are two caveats to this generalization, however. First, the central Government is likely to retain careful control over foreign exchange. Second, acquisition of whole plants and high-technology items is also likely to remain in centralized hands.

Apart from exchange control functions, the two main centers of authority for foreign trade appear to be SPC and SCCC, which reportedly "direct the work of the Ministry of Foreign Trade."[62] Undoubtedly the tasks confronting SPC and SCCC are numerous and complex since they have major roles in directing the domestic economy as well. Reportedly the personnel at SPC are of the highest quality. Unfortunately, it is impossible to determine the views of the SPC staff on the series of important matters of economic policy confronting China: the mix of heavy industry, light industry, and agriculture investments; the degree of technological

self-reliance; the relative share of effort going to military work; etc. Officially, SPC carries out national economic policy handed down from above, but because of its strategic position in economic management it is presumably also in a position to influence the agenda of the top policymakers, there affecting policy itself.

Some sense of the operation of the technology acquisition procedures can be extracted from the following examples, which, however, are not intended to be typical examples of China's negotiating strategies.

Decisionmaking for acquiring foreign technology needed in Chinese scientific research is nicely illustrated in a recent report on the efforts of the Institute of Oceanography to procure an advanced research vessel. This project is reportedly one of the top 25 priority items in the new science development plan. The decisionmaking process involved an initiative from the Institute, approval from the Academy's central offices, and authorization from the State Council. The first phase took less than 1 year and was completed in December 1977. With this high-level authorization, the Institute began detailed studies of its objectives and its domestic capabilities for reaching those objectives, and concluded that the ship should be procured abroad. An interdisciplinary and interorganizational team was then formed to investigate the state-of-the-art in Japan and the United States. The delegation was abroad for nearly 6 weeks. Its recommendations will require the approval of the Academy of Sciences, which in turn will pass it on to the Ministry of Foreign Trade, and finally to the China National Machinery Import and Export Corporation, which will contact potential suppliers.[63]

A second example concerns a $5.2 million contract between TECHIMPORT and the High Voltage Engineering Corporation (HVEC) of Burlington, Mass., for the purchase of an HI-13 tandem accelerator to be

---

[62]*China Business Review*, March/April 1979, p. 14.

[63]Ibid., May/June 1978, pp. 9 ff.

used by the Atomic Energy Institute of the Chinese Academy of Sciences. In 1975, the Chinese initiated correspondence with the American firm. The corporation attracted Chinese attention because one of its founders had been Robert Van de Graff, a well-known figure in modern physics and the inventor of the Van de Graff accelerator. Nothing came of these initial contacts until 1978, presumably because of the confusion in Chinese scientific circles occasioned by the succession struggle involving the Gang of Four. In 1978, however, TECHIMPORT invited a representative from HVEC to Peking for preliminary talks. In July 1978, TECHIMPORT sent its own 10-man delegation to the United States to visit the firm and also one of its chief competitors. In August, HVEC was invited to Peking to discuss the possible installation of the HI-13 machine, and it sent its top people in engineering and management.

Photo credit: U.S.-China Trade Council

13-million volt HI-13 Tandem accelerator, sold by High Voltage Engineering Corporation to the PRC

What was expected to be a 2-week visit lasted 32 days. The U.S. team met with its Chinese counterparts 35 times during this period, the Chinese inevitably outnumbering the Americans. Representatives from the Academy of Sciences as well as from TECHIMPORT participated. The scientist members of the Chinese team produced a barrage of questions and kept seeking an improved version of the HI-13. According to the American side, the Chinese objectives were to extract the maximum technical information and to wear down the Americans with highly technical questions in order to enable TECHIMPORT to get the best possible contract terms. The patience of the Americans wore thin, and on the second occasion that they threatened to leave without concluding the contract, the Chinese moved quickly to an agreement and the contract was promptly prepared. In the final 2 days while the contract was being typed, the Chinese asked the Americans whether they would be prepared to meet with a second group of end users who were also interested in their equipment. HVEC agreed, and this led to preliminary discussions on a second purchase.

The endurance of HVEC paid off. Not only was it able to incorporate favorable terms into the contract, but there is now potential for further business. Once assurance has been given that the equipment is not subject to U.S. or CoCom export controls, the Chinese will put down 15 percent of the price and will make installments at 6-month intervals with the last two 5-percent payments coming after demonstration and a 12-month performance evaluation.[64]

A third example involves an engineering service contract between TECHIMPORT and Kaiser Engineers of Oakland, Calif. This contact was initiated when Kaiser sent a representative to China to give a paper on coal mine maintenance. At this time, the firm was not actively soliciting business in China. In March 1978, however, the Chinese contacted the firm via the PRC liaison office in Washington, to discuss the possibility of

developing iron ore mines. Kaiser first ascertained that the Chinese were aware that they were purely an engineering service concern and then supplied the Chinese with information about their firm. After this material was thoroughly reviewed, TECHIMPORT provided Kaiser with sufficient information about its requirements to enable it to organize an appropriate team to go to China the following month.

As with the previous case, the Chinese subjected the Kaiser representatives to exhaustive questioning. Among the questioners was the Minister of the Metallurgical Industry. The Chinese were apparently satisfied with Kaiser's answers, for 1 month after the visit it obtained the contract from TECHIMPORT to develop two mines. One was an old mine that needed upgrading; the other involved opening mining operations at a new iron ore deposit. The contract is for 1 year and provides for one lumpsum payment. According to the contract, Kaiser is to do the general engineering, although there are also possibilities for detailed engineering developments. Kaiser has begun sending teams to China and also has been receiving Chinese engineers for training in California. Although Kaiser dealt primarily with the mining and metallurgical technology specialists at TECHIMPORT, they also had contact with the Chinese Society of Metals, confirming again the importance of professional societies in technology acquisition decisions.

Although there is undoubtedly considerable variation in China's approach to negotiations, on the basis of these and other cases, a few generalizations about the style of acquiring technology can be tentatively advanced. First, when technical requirements enter the decisionmaking process, efforts will be made, as in the oceanographic vessel case, to assess the possibility and desirability of meeting the need with indigenous technology. Second, these and other cases indicate that China prepares for its searches and negotiations with great care. Third, efforts are made to extract the maximum amount of information from the negotiations, and increasingly to use contractual arrangements that will also yield as much information and experience to the Chinese as possible. For instance, it has been reported that China wants to change from fixed-fee to cost-plus contracts, which offer more intimate contacts with vendors. Fourth, the Chinese are hard bargainers seeking to get as much for their money as possible. As HVEC and other cases show, they sometimes use delaying tactics and gruelling negotiating sessions to wear down the seller in order to get better terms. Finally, China's search for technology and negotiating tactics may have political and diplomatic, as well as economic and technical objectives.

In conclusion, mention should be made of the technology transfers and knowledge flows that will result from the exchanges of students and scholars. While this area has generally been beyond the scope of export control legislation, it is important to recognize that such exchanges are mechanisms of transfer. The Chinese have estimated that as many as 10,000 students and scholars could be sent abroad for training by 1985. The exchange agreement with the United States provides for a target of 500 to 700 Chinese to come to this country to study during the first year of the agreement. Most of those who have come thus far have been midcareer research scientists. It should be noted that in keeping with the highly decentralized and pluralistic system of higher education in the United States, student and scholar exchanges are at present largely beyond most kinds of central U.S. policy controls.

THE ROLE OF WESTERN TECHNOLOGY IN THE CHINESE ECONOMY

Western technology has become economically significant in the PRC only in the past decade. For this reason, and because hard economic data for the PRC is extremely scarce (the population of the country was unknown until 1979), attempts at macroeconomic analyses of the economy would be meaningless exercises. The role of foreign trade and the technology component in that trade can only be treated anecdotally. In this area the most meaningful generalizations that can be made concern the potential, rather than past, impacts of trade with the West.

Chinese foreign trade presently involves the exchange of crude oil, coal, ores, foodstuffs, simple machine tools, textiles, bicycles and other manufactures for equipment and technologies for oil exploration, coal mining, steelmaking, chemical fertilizers, power generation, petrochemicals, and a small number of consumer goods. In addition, after 1961, bad harvests forced the PRC to import several million tons of grain from non-Communist countries, and it has continued this practice as an economical way of feeding its large northern cities.

The PRC reaps substantial gains from its comparative advantage in foreign trade. It exports items that have a high labor and natural resource content, and imports products that it could produce, if at all, only with great expenditure of high-technology resources. Its exports—specialty foods, silk, textiles, and high-grade handicrafts—command a high price abroad but are accorded low-priority domestically; with these earnings, it imports wheat, steel mill products, and electronics—items that command a comparatively low price in world markets, but are of great usefulness in running and expanding the economic machinery at home.

Foreign technology will provide the cutting edge of the general program for economic modernization discussed above. This technology will be most useful in the urban industries where large-scale plants are engaged in basic industry and military production. These plants typically are under central control, mass producing standardized products of tried and proven design. They tend not to be highly innovative. The quality of their products and production efficiency stand to benefit greatly from the import of modern process equipment and complete plants. In contrast, the medium- to small-scale enterprises that are typically under provincial or municipal control have been described as innovative and dynamic. Some of these are also important potential and actual end users of such foreign technology as production equipment and prototypes for adoption and copying.

In the 1950's embodied technology in the form of imported industrial goods from the U.S.S.R., particularly in complete plant purchases, was crucial in equipping basic industries such as iron and steel, transport, and mining. Imports were also responsible for the creation of virtually new industries, including machine building, electrical power generation, chemicals, and crude oil production. Soviet withdrawal, halfway through the agreed program, virtually halted China's industrial progress in some sectors.

The expansion of industrial imports in the 1970's is important, but not in the same way as that of the 1950's. The scale of total imports is now relatively small; for instance, the average value of imports of machinery and transport between 1970 and 1973 was below the average for 1952 to 1960. Further, since domestic output of these commodities has increased several times between the two phases, the imported share of the total value of deliveries of machinery and transport equipment has fallen dramatically.

But despite the low overall volume of industrial imports, their importance to the economy cannot be overemphasized. Not only have Chinese imports of complete plants risen from the 1950's, but these imports are crucial in the context of particular industries. In fact, nearly all of the $2.6 billion spent between 1973 and 1976 has been to support two industries: chemicals and steel. In chemicals, imports are adding over a third to the capacity of the chemical fertilizer industry, and in the case of manmade fibers and petrochemicals, imports are practically creating new industries.

The situation with respect to steel is different. This is a well-established industry that, due to planning errors, is technically inefficient and unable to satisfy domestic demand. Imports of finished steel products now account for over 90 percent of total Chinese steel purchases, paid for at the expense of machinery. The effect of steel plant imports should be to increase steel finishing capacity by a third as compared to the early 1970's.

The industrial imports are more important than their quantitative level suggests precisely because of their component of embodied technology. They have the capacity to provide a cumulative, quantitative improvement in Chinese industry. The Chinese, like the Soviets, acquire foreign technology in many ways. They read literature; they send specialists to study abroad; they encourage foreign firms to give expositions in China. They also purchase prototypes that they try to copy in substantial quantities. All of this is useful, although prototype copying has been more difficult than expected. For example, in 1963, the Chinese purchased a Dutch urea plant that they planned to replicate in a twin plant. They were unable to do so, and in general prototype copying has not so far proven feasible as a solution to technology acquisition. The Chinese have, therefore, had to purchase substantial quantities of equipment and to obtain the technical assistance necessary to adapt and integrate specialized equipment into their industrial systems. The remainder of this section deals with two industries in which this process has been facilitated by U.S. industrial exports; computers and oil drilling and exploration equipment.

COMPUTERS

The computer industry in the PRC has received relatively little attention in the West. Western computers did not begin to appear in China until several years ago, and relatively little information on this industry is available. Western sources are primarily limited to a series of trip reports which record the observations of delegations that have traveled in China in the past 5 years,[65] and a 1973 CIA report on computers in the PRC. The latter provided details on production facilities and on performance characteristics of many domestically produced computers.[66] The CIA material reveals the extent to which the Chinese computer industry before 1960 relied on Soviet assistance; the trip reports often provide excellent information on the present technical level of these computers. Nevertheless, discussion of this industry must remain incomplete and inconclusive.

[65]These are available through the National Council for U.S.-China Trade.
[66]Central Intelligence Agency, *The Computer Industry in the People's Republic of China*, National Foreign Assessment Center, 1973.

China produced its first computer in 1958 from designs provided by the U.S.S.R. Subsequent models were also based on Soviet designs or prototypes. The Sino-Soviet split occurred before the development of Soviet computers based on transistors, but the PRC was able to continue its domestic development. It produced a second-generation computer in 1965, only about 3 years after the introduction of transistorized computers in the U.S.S.R. A computer based on integrated circuits appeared within about 2 years of such models in the U.S.S.R. In 1974, the PRC announced the production of its first computer capable of 1 million operations per second. (In comparison, American computers were then operating with speeds of about 12 million operations per second.) The Chinese are presently capable of building computers with speeds higher than any reported Soviet computer in production.

Photo credit: Hsinhua News Agency

Testing calculators produced domestically in the PRC

This might suggest that the Chinese have been able to surpass the U.S.S.R. in computer technology, but the impression is misleading. The U.S.S.R. is capable of producing relatively large quantities under conditions of serial production, but virtually all Chinese computers are prototypes or small batch models. The stock of computers in the U.S.S.R. is at least 10, if not 20, times that of the PRC. Only a few factories in China produce more than a few computers each year, and only one of these is equipped with modern automated equipment for assembly and testing of computer components and final products.

As was the case in the U.S.S.R., the quality of Chinese computer peripherals and software lags significantly behind the performance of the central processing unit, thus limiting the effectiveness of the computer. This is an area that is currently receiving high priority, as is the development of better production capabilities to meet the demand for larger quantities of higher quality integrated circuits. At present, China's integrated circuits are about 7 years behind the state-of-the-art in the West, and the lag in production technology and production capacity is even greater. Although China has been able to close some of the gap in its production of advanced integrated circuits, it has done so only in small-scale laboratory production. The methods used to achieve these results are outdated, the production process inefficient, and the number of devices that can be manufactured is limited. In order to meet this shortcoming, the PRC attempted to purchase a Japanese turnkey plant for the production of integrated circuits (ICs). This plant was to be part of a deal that included facilities for the production of color television sets. The IC portion of the project was blocked by CoCom restrictions. Nonetheless, the Chinese have begun to close the gap in the production of ICs. As their capability in this area increases, computer production will also begin to expand more rapidly. But, while the availability of better ICs is important to the production of new computers with very high speeds, the most important

advantage of their increased availability will be to reduce the size and production costs of those computers already available in limited quantities in China.

Essential to the effective use of those computers produced in the PRC are improvements in the quality and range of peripheral devices. Available primary memory, for instance, lags far behind both Western and Soviet computers. Although access time is reasonably good by Western standards, the limited core memory is a great problem, especially in light of an even greater lag behind the West in other online storage capabilities.

Magnetic drums and magnetic tapes continue to be the predominant form of other online storage for Chinese computers. Magnetic disk use did not appear on a prototype computer until late in 1977. Thus far, there are believed to be only two models capable of making use of disks for storage, and those disks that are available have a capacity one-third that of the best disks now being produced in the U.S.S.R. and Eastern Europe.

In addition to problems with insufficient memory capacity, Chinese computers are limited in the rate at which information can be fed into the computer and put out by the system. The principal forms of input continue to be keyboard and papertape readers. There has been no evidence of the use of cardreaders, which would represent a significant improvement over the current state of input technology. This is a particular liability for problems that involve large data-handling requirements, or frequent updating of data banks. Such tasks are also difficult for Chinese computers since they place a high demand on the availability of a range of high-quality output devices.

The Chinese produce standard line printers in quantities that seem to meet their demand but, at 600 to 800 lines per second, the unit is slow by Western standards. A newer model, which employs electrostatic printing technology and has a speed of 1,800 lines per minute, has appeared in the most advanced computer prototypes, but is unlikely to be widely available soon. The availability of other terminals and output devices is similarly limited, and those that are produced employ early technology.

The small number of standardized computer models has delayed the development of software, a situation similar to that in the U.S.S.R. prior to the introduction of Ryad. Efforts are now focused on extending the repertoire of the languages most commonly used in the West and on training programers in adapting programs for use in different models.

The architecture of the first series of Chinese minicomputers seems to have been based on an American prototype, but the second series was almost certainly designed by the Chinese themselves—after extensive examination of the architecture of both IBM and Control Data computers. The goal for this second series is to achieve software compatability and program interchangeability, but there are indications that different factories producing the same model are using different hardware designs.

Since the cessation of extensive transfers of technology from the U.S.S.R., the Chinese computer industry has received very limited assistance from abroad, in either manufacturing technology or computers themselves. In fact, no country appears to have transferred computer manufacturing technology to China since 1960. This partly explains the comparatively slow progress of the Chinese in serial production relative to their progress in the design of prototypes. Limited imports of Western computers took place during the mid-1960's, but no complete or detailed survey of the sales from this period, or for the 1970's, is available.

France and Britain both sold several computers to the PRC between 1964 and 1967, before imports of virtually all Western technology were blocked by the policies of the cultural revolution. These systems went to several different end users for seismic data

analysis, process control, plant automation, and medical research. The two largest are believed to be in Beijing's Central Statistical Office. The primary source of demand throughout the 1970's has continued to be plant automation and data handling or analysis.

Accurate figures for computer imports during the 1970's are unavailable, but it is unlikely that major orders were placed before 1973, and these have remained quite limited since. Orders during 1973 and 1974 probably amounted to about $6 million to $7 million per year, rising about $25 million for each of the next several years. During 1978, these figures rose dramatically, primarily as a result of a single $69 million contract with an American firm, but imports much in excess of $100 million per year are considered unlikely.

China desires to be as self-sufficient as possible, but difficulties in serial production of computers are likely to persist. Imports of individual Western models will contribute much by providing computers with large data-handling capabilities for networks that involve data transmission and the use of remote terminals, and for various business and advanced scientific uses. The acquisition of software and the increased exposure of Chinese programers and users to Western technology will also aid the Chinese in the advancement of their own software capabilities. Imported computers will be placed in high-priority areas, but less-than-critical users will have to await improvement in domestic manufacturing capabilities. Thus, China has little interest in purchasing many small- or medium-size computers. These will be produced domestically; until then, end users will do without.

A shift in the primary motivation behind the selection of computer imports appears to have occurred in recent years. The demand for computers with specific applications, particularly in the all-important petroleum industry, indicates a preoccupation with the specific services rendered by individual computers rather than more general software acquisition. Most other computers imported by the Chinese have been for process control at imported turnkey plants, or for end uses for which domestic computers are not particularly well-suited—analyses of weather data, computers and terminals for the Bank of China, and air traffic control.

On the other hand, some imports are motivated primarily by the desire to obtain prototypes for domestic production. Minicomputers probably fall into this category. The secondary benefits of all imports, moreover, lie in the exposure they provide to terminals, peripherals, software, etc. Any Western computer, therefore, may ultimately help in the design of better Chinese products.

Western export controls place a severe constraint on the purchase of computer manufacturing technology. They also limit the sale of systems with certain performance characteristics. Relaxation of these regulations would almost certainly result in the allocation of greater amounts of foreign currency for the purchase of large, advanced Western computers in high-priority sectors. Access to these systems would provide technology surpassing China's present technical frontier and specific capabilities to benefit the entire economy.

Little is known about the criteria employed in selecting Western suppliers. As in the U.S.S.R., the most important factor seems to be the desire to obtain the best available technology. In many cases this would point to American suppliers, despite the difficulty of obtaining U.S. export licenses. For example, a Japanese firm recently contracted to supply computers for the Chinese Meteorological Center, but only after an American computer firm withdrew from negotiations for export control reasons. When the contract was signed by the Japanese, the United States tried to block the sale through CoCom. Safeguards were added and the sale was ultimately approved. At the moment, the Chinese have no difficulties obtaining financing, although this may change as their hard-currency debt grows (see chapter III). Thus, the only other factor influencing choice of supplier may be the special relationship between the PRC and Japan. The Japanese are willing to provide the Chinese with large amounts of sophisticated technology. In the future, this may include manufacturing technology for the production of computers or peripheral equipment. The Japanese are more likely to agree to supply such technology than any other Western country (see chapter IX) and further enjoy the advantage of their experience with the character alphabet. It is likely, then, that the two most important sources of computer technology for the Chinese will be firms from the United States and Japan.

It is still too early to determine the impact of Western computer technology on the Chinese economy. Although the imports of the 1960's clearly aided the design of later domestically produced computers, no Chinese models seen since closely resemble Western prototypes. The major effect of Western technology so far has been as a source of information and starting point for R&D within the domestic computer industry. The lack of direct transfers of manufacturing technology is the most important factor affecting future impacts.

It will take time for the Chinese to make full use of imported technology. Some Western computers remain advanced beyond the understanding of many of the people working with them. The adaptation and diffusion of the technology embodied in accompanying software is likely to be more rapid than that of the hardware. Again, however, there will be significant lags before the availability of this software has an appreciable effect on the ability of the Chinese to generate their own software capabilities.

OIL AND GAS

The oil industry, like most other sectors of the economy, was extremely underdeveloped in 1949, when the PRC came into existence. Levels of production were very low and China had relied heavily on imports to equip its small domestic industry. Potential reserves were believed to be inconsequential, and much of the country had not been fully explored.

During the 1950's, due primarily to the efforts of the U.S.S.R., this picture changed considerably. One of the most important forms of assistance provided by the Soviet Union was in training. Between 1950 and 1958, some 8,000 skilled workers, 6,500 students, and about 1,000 industry experts from China were trained in the Soviet Union. The data does not show how many of these were involved in the oil industry, but it is known that about 450 Soviet petroleum experts were sent to the PRC to provide technical assistance.

In addition, the U.S.S.R. provided large amounts of equipment to the Chinese oil industry. Imports from the U.S.S.R. and Eastern Europe accounted for nearly 65 percent of all equipment supplied to this sector during the first 5-year plan (1953-58).

By 1954, China had begun expanding its production of basic tools and parts (e.g., oil-extraction drilling tools, pumps, and small compressors), but it was not until the U.S.S.R. helped it to establish several large plants for the production of oilfield equipment that China's output began to expand significantly. By 1958, using Soviet models, the Chinese had increased production of oil industry equipment.

All Soviet technical help was withdrawn and equipment sales virtually ceased in 1960, although Romania remained an important source of oil and gas technology. The loss of Soviet aid was sorely felt, however, particularly in the area of geological prospecting. The U.S.S.R. had provided extensive help in carrying out a number of serial magnetic, gravity, and seismic surveys. These had led to the identification of the fields that have since become the center of China's industry.

The most important oil reserves in China now, and for the past two decades, have been the Taching fields in the northeast part of the country. The U.S.S.R. began prospecting in this area in 1955, and drilling began in 1958. Fortunately, most of the wells drilled during the first years were relatively shallow and did not tax the capacities of domestically produced rigs.

After 1960, research programs aimed at improving technological levels in the oil industry were begun and many of these centered on the analyses of the available foreign equipment. These efforts paid off. From 1962 to 1963, China's output of oil equipment increased by more than 60 percent. By the following year, 1963 production was more than doubled.

But the equipment now being produced, primarily for shallow drilling, was patterned primarily on Soviet and Romanian design which lagged behind the technology employed in the West. The Chinese made only very limited use of equipment imports from the West before the 1970's. Although small purchases were made from Japan and France there were no direct sales by the United States until after 1972, although there is evidence that some American equipment designs were used by the Chinese to aid in the development of domestic designs. This technology was acquired either through equipment sales from third parties, or through foreign equipment incorporating U.S. technology.

Detailed knowledge of equipment and oil production in China is limited in the West, where, again, trip reports are among the few sources of information. Furthermore, because most Western oil technology was not acquired until after 1972, its full impact on production and the level of domestic equipment development has not yet become apparent.

According to most Western oil industry experts, the technology being employed by the Chinese for geological prospecting, drilling, and production is about at the level of U.S. technology circa 1950. The Chinese themselves estimate that their technical capabilities in various phases of the oil indus-

Photo credit: U.S.-China Trade Council

Drilling platform in the Pohai Gulf

try lag behind Western state-of-the-art by 15 to 20 years. Often, Chinese oil equipment technology has been copied or adapted from the dated technology of the U.S.S.R. or Eastern Europe. China, therefore, experiences many of the same problems in this sector as the U.S.S.R., although often on a larger scale. Geological prospecting technology is unsophisticated, and this limits the usefulness of other activities. Further, the Chinese seem to lag the Soviet Union in deep-drilling capabilities, the level of most well-completion equipment and automation of production facilities could be significantly improved and technical assistance is sorely needed in offshore operations.

Much of China's oil reserves still lie undiscovered, and domestic prospecting equipment is inadequate to this task. The geology of China's three largest fields, Taching, Shengli, and Takang, involves complicated structures, the result of unpredictable fracturing which has left oil dispersed in a number of small pools. Efficient exploration requires sophisticated prospecting equipment, including such equipment for the collection and analysis of seismic data as computers with specially designed software. The Chinese have already imported several of these, in order both to obtain the capabilities of the equipment itself and to gain access to the software. It will be some time before domestic production is adequate to meet the demand for large-scale high-quality prospecting equipment.

Most of the geological prospecting equipment imported by China thus far has come from the United States, which sells the best technology available in the field. The National Council for U.S.-China Trade has estimated that such sales totaled nearly $40 million between 1973 and mid-1977.

At the end of 1977, it was reported that workers at the Shengli oilfields had both greatly increased drilling speeds and reduced costs by 50 percent by using domestically produced synthetic diamond drill bits and high-pressure jet drilling techniques. Western drilling and well-completion technologies that might lead to further improvements are mud technology and cementing. Drilling-mud lubricates the cutting bit, aids in removing waste material from the well, and seals the borehole. In the West, a variety of chemical additives for the mud are available, but a more primitive practice is to mix soil with the fluid being pumped into the well. The Chinese primarily use the latter technique.

There are also indications that significant improvements could be made in the cementing operations necessary for well completion. Present cementing methods are predominantly either domestic or imported from Romania or France.

There is no detailed information on quality of Chinese wellhead equipment or the number of operating wells. Whatever their number, several production problems are being encountered. Most Chinese oil has a high paraffin content. This means that extractive equipment must be heated during the winter to maintain oil flow. Other production problems arise from the high water content of the oil and from complex geological structures.

By early 1979, 85 percent of the oil output of at least one field was measured and recorded by computer, but there have been no reports of automatic metering at other fields, and very little evidence of automation of any other aspects of field operations.

The Chinese claim to be relatively advanced in their understanding and use of water-flooding as a means of increasing well production, but there is evidence, for example, that injection control and the monitoring of injection performance remains limited. In the West, water injection is used heavily in mature or declining fields. In both the U.S.S.R. and the PRC, however, water injection is often used at new fields and with recently completed wells. This may increase the initial rate of production, but it also reduces the ultimate recovery rate from the reserve. Because the reserves at China's major oilfields are fragmented into small pools, it is difficult to know whether low pressures at a particular well is symptomatic of the state of the field as a whole. Low wellhead pressure would indicate that rapid growth rates from the field cannot be achieved by modern technology alone.

Initially, China depended heavily on the U.S.S.R. for assistance with drilling and extraction of onshore reserves; the Soviets provided both equipment and assistance in the manufacture of equipment. Eastern Europe's technological contribution to the Chinese industry was relatively minor at this time. The role of the West was virtually nonexistent.

During the 1960's, after the U.S.S.R.'s importance to China as a source of drilling and extraction equipment drastically declined, the Chinese continued to rely on the Soviet equipment already in place, using it as models for domestic production. In the 1970's, China turned increasingly to the West for both equipment and technology to improve and expand domestic production, but as yet there have been no sales of turnkey plants for production of oilfield equipment. Earlier, Chinese desire for self-sufficiency led them to accept delivery of equipment without participating in training on operation and maintenance. More recent sales have included such training, however.

Most sales of drilling and extraction equipment from the West to the PRC have come from American firms, but Western sales of onshore drilling and extraction equipment through mid-1976 totaled only about $13.4 million. The pace of these purchases has since been stepped up. The Chinese oil equipment market is now highly competitive, and many Western companies competing in it are reluctant to discuss the details, particularly the value, of recent sales. Accounts of oil equipment sales can, therefore, be assumed to understate the true volume of equipment and technology transferred.

An upswing in Chinese equipment purchases for drilling and extraction of onshore reserves has occurred over the past few years. In 1978, the Chinese bought a wide variety of drilling and extraction equipment. One French company has signed a $22.9 million contract for drilling equipment, but all other major sales are believed to have been made by U.S. companies. Sales have included $10 million worth of drill bits, drilling tools, submersible pumps, down-hole instruments, and wellhead equipment. In some cases, China has chosen to pay a higher price in order to gain access to the latest technology. For example, it has purchased drill bits with tungsten-carbide cutting structures and a new bearing design. These are said to last three to four times as long as previous designs and are particularly important to help China with deep drilling.

China also gained access to a variety of other equipment to help improve drilling performance and extraction capabilities. It has purchased $100,000 worth of petroleum-handling tools, $7 million worth of workover rigs and blowout preventors, as well as unknown values of down-hole instruments, wellhead and well-completion equipment, well-testing instruments, and other equipment. Much of this has been bought in limited quantities,

suggesting that China is seeking primarily to import prototypes for duplication.

The total value of 1978 contracts for which values are known is about $50 million. These figures indicate that China is likely to continue looking to the West as a primary source of drilling and extraction technology, although Romania may continue to be an important source of technology and equipment for drilling rigs and cementing equipment. While purchases from the West are made primarily for the embodied technology, purchases from Romania are used to obtain supplies of currently needed equipment.

Serious difficiencies exist in Chinese prospecting, drilling, and extraction methods and equipment, yet China has successfully developed several major oilfields, and reached a production level of about 200 million tons per year. This is about 25 percent of U.S. output. Extensive reserves of oil and possibly natural gas, still largely unexplored, are believed to exist in fields in the interior of China and in its offshore waters. To exploit the interior fields, China must use deep-drilling equipment. Heretofore, little deep drilling has been carried out; the development of these fields has been delayed, primarily by the tremendous cost involved in building long-distance pipelines to transport the oil.

Delays in offshore production result more from technological problems. The Chinese (as the Soviet) oil industry lags farthest behind the West in offshore technology. China's first offshore drilling activity began only in the late 1960's, when limited drilling was undertaken from fixed platforms in shallow water in the Pohai Gulf. The Chinese have relied extensively on foreign technology for offshore geological prospecting capabilities, purchasing entire vessels as well as smaller pieces of equipment. Two of three prospecting vessels in use in offshore waters in 1976 were outfitted by copying Western equipment.

A large program to import Western offshore equipment was begun in late 1972, starting with the purchase of a Japanese jack-up drilling rig and accompanying workship. The value of orders, all placed with the West within little more than a year, came to more than $175 million, a figure greatly exceeding the total value of all Chinese equipment then in use for offshore exploration and development. Most purchases were for equipment that embodied high technology, but in some cases—dredges, for example—the technology is not sophisticated. Some of the PRC's decisions to import rather than build such equipment can be attributed to a desire to begin using the equipment quickly. This approach typifies the Chinese use of Western imports to fill production gaps as much as to fill technology gaps.

The orders placed in 1972 and 1973 for oil industry equipment ultimately resulted in a sharp drop in foreign exchange reserves, and the drive for offshore equipment contracts from the West greatly abated thereafter. The decision to cut back imports was partially the result of traditional Chinese conservative fiscal policy, but the political uncertainties as rival political factions battled for position before and after the death of Mao Zedong were also factors. This political conflict made planners more cautious, particularly since reliance on foreign products and technology was a central point over which the two sides disagreed.

Despite this retrenchment, equipment imports for offshore needs totaled at least another $125 million through mid-1977, reflecting the priority accorded them. They continue to dominate oil industry equipment purchases. Over the past 2 years, Chinese orders for Western offshore equipment have again risen sharply. The PRC decided to make much greater use of Western technology in high-priority industries as Chinese foreign exchange reserves recovered and the industry had more time to assimilate foreign technology that had already been imported. Preliminary estimates for the 2-year period from mid-1977 to mid-1979 indicate Chinese orders for more than $275 million in offshore equipment, about 80 percent in drilling equipment.

Geological prospecting has generated a great deal of interest as the first area in which Western firms have considered going beyond equipment sales and training. A number of major U.S. oil companies, along with representatives from Japan and several European countries, have discussed the possibility of assisting the Chinese directly with exploration, and even production, of offshore reserves. The most important roles in such cooperative ventures are likely to be played by Japan and several American oil firms.

By early in 1974, three U.S. oil companies—Exxon, Phillips, and Union Oil of California—had signed a "group shoot" exploration contract, in partnership with the foreign-owned firms of Shell, Elf-Aquitaine, and British Petroleum (project leader); and Atlantic Richfield became the first American firm to sign a contract for sole exploration of Chinese prospects. Other companies are also expected to enter the market, since it appears that the Chinese are willing to permit extensive Western involvement in offshore development. Reportedly, the Chinese have not imported any turnkey plants for the production of offshore or onshore oil equipment, but moves in this direction may be underway.

The Chinese have clearly made a serious commitment to the development of their offshore oilfields. They realize that the active acquisition of Western equipment, technology and experience will significantly speed the development process. It may be that Chinese experience in this stage of offshore development will help to shape the size and direction of Western involvement in China's onshore oil program.

Until 1972, the amount of Western technology transferred to the oil industry was limited by conscious policy decision. Yet even then the Chinese recognized their need for such technology. They had relied heavily on the U.S.S.R. during the 1950's, and later on Romania, particularly for deep-drilling equipment. There is ample evidence that individual pieces of equipment have been copied by the Chinese from Western, and particularly American, models. Several oil industry experts believe that the Chinese could make very effective use of greater transfers of technology from the West in several areas of the onshore oil industry. The Chinese have reportedly made detailed studies of the state of Western technology, and are likely to make a greater effort in the future to increase its acquisition. Shortages of seismic equipment and computerized field units are posing serious problems for the geological prospecting sector in locating deeper reserves. Improvements in drilling instruments could produce important changes in performance by raising the effectiveness of the limited number of available rigs. The

Chinese have limited experience with secondary recovery methods other than water injection. Greater experience with alternative techniques could produce results without a major need for additional equipment.

It is still too soon to assess Chinese capabilities to absorb the technology embodied in purchased equipment. Some experts point to the low level of experience of the average oil industry workers and express doubts as to the level at which this equipment can be operated and maintained. Imported equipment used in technical schools and colleges for training may ameliorate this problem.

CONCLUSIONS

Transfers of foreign technology have influenced both the level and direction of economic progress in the PRC. In the immediate postrevolutionary period, Soviet equipment and expertise contributed heavily to Chinese industrialization. With the intensification of the Sino-Soviet rift in 1960 this source of technology was eliminated. Imports of technology did not again play a significant role in Chinese growth until the mid-1970's—this time, however, through transfers of plant, equipment, and associated technology from the West. Japan has been the major beneficiary of the process; thus far, the United States has succeeded in garnering only a small (7 to 8 percent) share of PRC imports.

China's imports from the West have been crucial to the development of key industrial

sectors such as steel and petrochemicals. China's modernization drive, although significantly less ambitious now than as originally announced in February 1978, depends on imports of plant and associated technology to play an important role in strengthening the industrial infrastructure, raising productivity in the agricultural sector, and in the exploration and development of energy resources. While Japan will undoubtedly benefit most from this drive in terms of increased export receipts, the United States can significantly increase its share of Chinese purchases through a normalization of trade relations as well as extension of official export credit facilities.

Appendix

Export Administration Act of 1979

PUBLIC LAW 96-72—SEPT. 29, 1979   93 STAT. 503

Public Law 96-72
96th Congress

An Act

To provide authority to regulate exports, to improve the efficiency of export regulation, and to minimize interference with the ability to engage in commerce.

Be it enacted by the Senate and House of Representatives of the United States of America in Congress assembled,

SHORT TITLE

SECTION 1. This Act may be cited as the "Export Administration Act of 1979".

FINDINGS

SEC. 2. The Congress makes the following findings:

(1) The ability of United States citizens to engage in international commerce is a fundamental concern of United States policy.

(2) Exports contribute significantly to the economic well-being of the United States and the stability of the world economy by increasing employment and production in the United States, and by strengthening the trade balance and the value of the United States dollar, thereby reducing inflation. The restriction of exports from the United States can have serious adverse effects on the balance of payments and on domestic employment, particularly when restrictions applied by the United States are more extensive than those imposed by other countries.

(3) It is important for the national interest of the United States that both the private sector and the Federal Government place a high priority on exports, which would strengthen the Nation's economy.

(4) The availability of certain materials at home and abroad varies so that the quantity and composition of United States exports and their distribution among importing countries may affect the welfare of the domestic economy and may have an important bearing upon fulfillment of the foreign policy of the United States.

(5) Exports of goods or technology without regard to whether they make a significant contribution to the military potential of individual countries or combinations of countries may adversely affect the national security of the United States.

(6) Uncertainty of export control policy can curtail the efforts of American business to the detriment of the overall attempt to improve the trade balance of the United States.

(7) Unreasonable restrictions on access to world supplies can cause worldwide political and economic instability, interfere with free international trade, and retard the growth and development of nations.

(8) It is important that the administration of export controls imposed for national security purposes give special emphasis to the need to control exports of technology (and goods which

[Margin notes: Sept. 29, 1979 [S. 737]; Export Administration Act of 1979.; 50 USC app. 2401 note.; 50 USC app. 2401.]

93 STAT. 504

contribute significantly to the transfer of such technology) which could make a significant contribution to the military potential of any country or combination of countries which would be detrimental to the national security of the United States.

(9) Minimization of restrictions on exports of agricultural commodities and products is of critical importance to the maintenance of a sound agricultural sector, to achievement of a positive balance of payments, to reducing the level of Federal expenditures for agricultural support programs, and to United States cooperation in efforts to eliminate malnutrition and world hunger.

DECLARATION OF POLICY

SEC. 3. The Congress makes the following declarations:

(1) It is the policy of the United States to minimize uncertainties in export control policy and to encourage trade with all countries with which the United States has diplomatic or trading relations, except those countries with which such trade has been determined by the President to be against the national interest.

(2) It is the policy of the United States to use export controls only after full consideration of the impact on the economy of the United States and only to the extent necessary—

(A) to restrict the export of goods and technology which would make a significant contribution to the military potential of any other country or combination of countries which would prove detrimental to the national security of the United States;

(B) to restrict the export of goods and technology where necessary to further significantly the foreign policy of the United States or to fulfill its declared international obligations; and

(C) to restrict the export of goods where necessary to protect the domestic economy from the excessive drain of scarce materials and to reduce the serious inflationary impact of foreign demand.

(3) It is the policy of the United States (A) to apply any necessary controls to the maximum extent possible in cooperation with all nations, and (B) to encourage observance of a uniform export control policy by all nations with which the United States has defense treaty commitments.

(4) It is the policy of the United States to use its economic resources and trade potential to further the sound growth and stability of its economy as well as to further its national security and foreign policy objectives.

(5) It is the policy of the United States—

(A) to oppose restrictive trade practices or boycotts fostered or imposed by foreign countries against other countries friendly to the United States or against any United States person;

(B) to encourage and, in specified cases, require United States persons engaged in the export of goods or technology or other information to refuse to take actions, including furnishing information or entering into or implementing agreements, which have the effect of furthering or supporting the restrictive trade practices or boycotts fostered or imposed by any foreign country against a country friendly to the United States or against any United States person; and

[Margin note: 50 USC app. 2402.]

(C) to foster international cooperation and the development of international rules and institutions to assure reasonable access to world supplies.

(6) It is the policy of the United States that the desirability of subjecting, or continuing to subject, particular goods or technology or other information to United States export controls should be subjected to review by and consultation with representatives of appropriate United States Government agencies and private industry.

(7) It is the policy of the United States to use export controls, including license fees, to secure the removal by foreign countries of restrictions on access to supplies where such restrictions have or may have a serious domestic inflationary impact, have caused or may cause a serious domestic shortage, or have been imposed for purposes of influencing the foreign policy of the United States. In effecting this policy, the President shall make every reasonable effort to secure the removal or reduction of such restrictions, policies, or actions through international cooperation and agreement before resorting to the imposition of controls on exports from the United States. No action taken in fulfillment of the policy set forth in this paragraph shall apply to the export of medicine or medical supplies.

(8) It is the policy of the United States to use export controls to encourage other countries to take immediate steps to prevent the use of their territories or resources to aid, encourage, or give sanctuary to those persons involved in directing, supporting, or participating in acts of international terrorism. To achieve this objective, the President shall make every reasonable effort to secure the removal or reduction of such assistance to international terrorists through international cooperation and agreement before resorting to the imposition of export controls.

(9) It is the policy of the United States to cooperate with other countries with which the United States has defense treaty commitments in restricting the export of goods and technology which would make a significant contribution to the military potential of any country or combination of countries which would prove detrimental to the security of the United States and of those countries with which the United States has defense treaty commitments.

(10) It is the policy of the United States that export trade by United States citizens be given a high priority and not be controlled except when such controls (A) are necessary to further fundamental national security, foreign policy, or short supply objectives, (B) will clearly further such objectives, and (C) are administered consistent with basic standards of due process.

(11) It is the policy of the United States to minimize restrictions on the export of agricultural commodities and products.

GENERAL PROVISIONS

SEC. 4. (a) TYPES OF LICENSES.—Under such conditions as may be imposed by the Secretary which are consistent with the provisions of this Act, the Secretary may require any of the following types of export licenses:

50 USC app. 2403.

(1) A validated license, authorizing a specific export, issued pursuant to an application by the exporter.

(2) A qualified general license, authorizing multiple exports, issued pursuant to an application by the exporter.

(3) A general license, authorizing exports, without application by the exporter.

(4) Such other licenses as may assist in the effective and efficient implementation of this Act.

(b) COMMODITY CONTROL LIST.—The Secretary shall establish and maintain a list (hereinafter in this Act referred to as the "commodity control list") consisting of any goods or technology subject to export controls under this Act.

(c) FOREIGN AVAILABILITY.—In accordance with the provisions of this Act, the President shall not impose export controls for foreign policy or national security purposes on the export from the United States of goods or technology which he determines are available without restriction from sources outside the United States in significant quantities and comparable in quality to those produced in the United States, unless the President determines that adequate evidence has been presented to him demonstrating that the absence of such controls would prove detrimental to the foreign policy or national security of the United States.

(d) RIGHT OF EXPORT.—No authority or permission to export may be required under this Act, or under regulations issued under this Act, except to carry out the policies set forth in section 3 of this Act.

(e) DELEGATION OF AUTHORITY.—The President may delegate the power, authority, and discretion conferred upon him by this Act to such departments, agencies, or officials of the Government as he may consider appropriate, except that no authority under this Act may be delegated to, or exercised by, any official of any department or agency the head of which is not appointed by the President, by and with the advice and consent of the Senate. The President may not delegate or transfer his power, authority, and discretion to overrule or modify any recommendation or decision made by the Secretary, the Secretary of Defense, or the Secretary of State pursuant to the provisions of this Act.

(f) NOTIFICATION OF THE PUBLIC; CONSULTATION WITH BUSINESS.—The Secretary shall keep the public fully apprised of changes in export control policy and procedures instituted in conformity with this Act with a view to encouraging trade. The Secretary shall meet regularly with representatives of the business sector in order to obtain their views on export control policy and the foreign availability of goods and technology.

NATIONAL SECURITY CONTROLS

50 USC app. 2404.

SEC. 5. (a) AUTHORITY.—(1) In order to carry out the policy set forth in section 3(2)(A) of this Act, the President may, in accordance with the provisions of this section, prohibit or curtail the export of any goods or technology subject to the jurisdiction of the United States or exported by any person subject to the jurisdiction of the United States. The authority contained in this subsection shall be exercised by the Secretary, in consultation with the Secretary of Defense, and such other departments and agencies as the Secretary considers appropriate, and shall be implemented by means of export licenses described in section 4(a) of this Act.

Publication in Federal Register.

(2)(A) Whenever the Secretary makes any revision with respect to any goods or technology, or with respect to the countries or destinations, affected by export controls imposed under this section, the Secretary shall publish in the Federal Register a notice of such revision and shall specify in such notice that the revision relates to controls imposed under the authority contained in this section.

(B) Whenever the Secretary denies any export license under this section, the Secretary shall specify in the notice to the applicant of the denial of such license that the license was denied under the authority contained in this section. The Secretary shall also include in such notice what, if any, modifications in or restrictions on the goods or technology for which the license was sought would allow such export to be compatible with controls imposed under this section, or the Secretary shall indicate in such notice to which officers and employees of the Department of Commerce who are familiar with the application will be made reasonably available to the applicant for consultation with regard to such modifications or restriction, if appropriate.

[margin: Export license denial, notice.]

(3) In issuing regulations to carry out this section, particular attention shall be given to the difficulty of devising effective safeguards to prevent a country that poses a threat to the security of the United States from diverting critical technologies to military use, the need to take effective measures to prevent the reexport of critical technologies from other countries to countries that pose a threat to the security of the United States. Such regulations shall not be based upon the assumption that such effective safeguards can be devised.

[margin: Regulatory safeguards for U.S. security.]

(b) POLICY TOWARD INDIVIDUAL COUNTRIES.—In administering export controls for national security purposes under this section, United States policy toward individual countries shall not be determined exclusively on the basis of a country's Communist or non-Communist status but shall take into account such factors as the country's present and potential relationship to the United States, its present and potential relationship to countries friendly or hostile to the United States, its ability and willingness to control retransfers of United States exports in accordance with United States policy, and such other factors as the President considers appropriate. The President shall review not less frequently than every three years in the case of controls maintained cooperatively with other nations, and annually in the case of all other controls, United States policy toward individual countries to determine whether such policy is appropriate in light of the factors specified in the preceding sentence.

(c) CONTROL LIST.—(1) The Secretary shall establish and maintain, as part of the commodity control list, a list of all goods and technology subject to export controls under this section. Such goods and technology shall be clearly identified as being subject to controls under this section.

(2) The Secretary of Defense and other appropriate departments and agencies shall identify goods and technology for inclusion on the list referred to in paragraph (1). Those items which the Secretary and the Secretary of Defense concur shall be subject to export controls under this section shall comprise such list. If the Secretary and the Secretary of Defense are unable to concur on such items, the matter shall be referred to the President for resolution.

[margin: Regulations.]

(3) The Secretary shall issue regulations providing for review of the list established pursuant to this subsection not less frequently than every 3 years in the case of controls maintained cooperatively with other countries, and annually in the case of all other controls, in order to carry out the policy set forth in section 3(2)(A) and the provisions of this section, and for the prompt issuance of such revisions of the list as may be necessary. Such regulations shall provide interested Government agencies and other affected or potentially affected parties with an opportunity, during such review, to submit written data, views, or arguments, with or without oral presentation. Such regulations shall further provide that, as part of

[margin: Submittal of written data, views, or arguments.]

such review, an assessment be made of the availability from sources outside the United States, or any of its territories or possessions, of goods and technology comparable to those controlled under this section. The Secretary and any agency rendering advise with respect to export controls shall keep adequate records of all decisions made with respect to revision of the list of controlled goods and technology, including the factual and analytical basis for the decision, and, in the case of the Secretary, any dissenting recommendations received from any agency.

[margin: Review.]

(d) MILITARILY CRITICAL TECHNOLOGIES.—(1) The Secretary, in consultation with the Secretary of Defense, shall review and revise the list established pursuant to subsection (c), as prescribed in paragraph (3) of such subsection, for the purpose of insuring that export controls imposed under this section cover and (to the maximum extent consistent with the purposes of this Act) are limited to militarily critical goods and technologies and the mechanisms through which such goods and technologies may be effectively transferred.

(2) The Secretary of Defense shall bear primary responsibility for developing a list of militarily critical technologies. In developing such list, primary emphasis shall be given to—

(A) arrays of design and manufacturing know-how,

(B) keystone manufacturing, inspection, and test equipment, and

(C) goods accompanied by sophisticated operation, application, or maintenance know-how,

which are not possessed by countries to which exports are controlled under this section and which, if exported, would permit a significant advance in a military system of any such country.

(3) The list referred to in paragraph (2) shall be sufficiently specific to guide the determinations of any official exercising export licensing responsibilities under this Act.

(4) The initial version of the list referred to in paragraph (2) shall be completed and published in an appropriate form in the Federal Register not later than October 1, 1980.

[margin: Publication in Federal Register.]

(5) The list of militarily critical technologies developed primarily by the Secretary of Defense pursuant to paragraph (2) shall become a part of the commodity control list, subject to the provisions of subsection (c) of this section.

(6) The Secretary of Defense shall report annually to the Congress on actions taken to carry out this subsection.

[margin: Report to Congress.]

(e) EXPORT LICENSES.—(1) The Congress finds that the effectiveness and efficiency of the process of making export licensing determinations under this section is severely hampered by the large volume of validated export license applications required to be submitted under this Act. Accordingly, it is the intent of Congress in this subsection to encourage the use of a qualified general license in lieu of a validated license.

(2) To the maximum extent practicable, consistent with the national security of the United States, the Secretary shall require a validated license under this section for the export of goods or technology only if—

[margin: Validated licenses, requirement.]

(A) the export of such goods or technology is restricted pursuant to a multilateral agreement, formal or informal, to which the United States is a party and, under the terms of such multilateral agreement, such export requires the specific approval of the parties to such multilateral agreement;

(B) with respect to such goods or technology, other nations do not possess capabilities comparable to those possessed by the United States; or

(C) the United States is seeking the agreement of other suppliers to apply comparable controls to such goods or technology and, in the judgment of the Secretary, United States export controls on such goods or technology, by means of such license, are necessary pending the conclusion of such agreement.

(3) To the maximum extent practicable, consistent with the national security of the United States, the Secretary shall require a qualified general license, in lieu of a validated license, under this section for the export of goods or technology if the export of such goods or technology is restricted pursuant to a multilateral agreement, formal or informal, to which the United States is a party, but such export does not require the specific approval of the parties to such multilateral agreement.

(4) Not later than July 1, 1980, the Secretary shall establish procedures for the approval of goods and technology that may be exported pursuant to a qualified general license.

(f) FOREIGN AVAILABILITY.—(1) The Secretary, in consultation with appropriate Government agencies and with appropriate technical advisory committees established pursuant to subsection (h) of this section, shall review, on a continuing basis, the availability, to countries to which exports are controlled under this section, from sources outside the United States, including countries which participate with the United States in multilateral export controls, of any goods or technology the export of which requires a validated license under this section. In any case in which the Secretary determines, in accordance with procedures and criteria which the Secretary shall by regulation establish, that any such goods or technology are available in fact to such destinations from such sources in sufficient quantity and of sufficient quality so that the requirement of a validated license for the export of such goods or technology is or would be ineffective in achieving the purpose set forth in subsection (a) of this section, the Secretary may not, after the determination is made, require a validated license for the export of such goods or technology during the period of such foreign availability, unless the President determines that the absence of export controls under this section would prove detrimental to the national security of the United States. In any case in which the President determines that export controls under this section must be maintained notwithstanding foreign availability, the Secretary shall publish that determination together with a concise statement of its basis, and the estimated economic impact of the decision.

(2) The Secretary shall approve any application for a validated license which is required under this section for the export of any goods or technology to a particular country and which meets all other requirements for such an application, if, the Secretary determines that such goods or technology will, if the license is denied, be available in fact to such country from sources outside the United States, including countries which participate with the United States in multilateral export controls, in sufficient quantity and of sufficient quality so that denial of the license would be ineffective in achieving the purpose set forth in subsection (a) of this section, subject to the exception set forth in paragraph (1) of this subsection. In any case in which the Secretary makes a determination of foreign availability under this paragraph with respect to any goods or technology, the Secretary shall determine whether a determination of foreign availability under paragraph (1) with respect to such goods or technology is warranted.

(3) With respect to export controls imposed under this section, any determination of foreign availability which is the basis of a decision

[margin notes: Qualified general license. Review. Export controls maintenance. Validated license approval.]

to grant a license for, or to remove a control on, the export of a good or technology, shall be made in writing and shall be supported by reliable evidence, including scientific or physical examination, expert opinion based upon adequate factual information, or intelligence information. In assessing foreign availability with respect to license applications, uncorroborated representations by applicants shall not be deemed sufficient evidence of foreign availability.

(4) In any case in which, in accordance with this subsection, export controls are imposed under this section notwithstanding foreign availability, the President shall take steps to initiate negotiations with the governments of the appropriate foreign countries for the purpose of eliminating such availability. Whenever the President has reason to believe goods or technology subject to export control for national security purposes by the United States may become available from other countries to countries to which exports are controlled under this section and that such availability can be prevented or eliminated by means of negotiations with such other countries, the President shall promptly initiate negotiations with the governments of such other countries to prevent such foreign availability.

(5) In order to further carry out the policies set forth in this Act, the Secretary shall establish, within the Office of Export Administration of the Department of Commerce, a capability to monitor and gather information with respect to the foreign availability of any goods or technology subject to export controls under this Act.

(6) Each department or agency of the United States with responsibilities with respect to export controls, including intelligence agencies, shall, consistent with the protection of intelligence sources and methods, furnish information to the Office of Export Administration concerning foreign availability of goods and technology subject to export controls under this Act, and such Office, upon request or where appropriate, shall furnish to such departments and agencies the information it gathers and receives concerning foreign availability.

(g) INDEXING.—In order to ensure that requirements for validated licenses and qualified general licenses are periodically removed as goods or technology subject to such requirements become obsolete with respect to the national security of the United States, regulations issued by the Secretary may, where appropriate, provide for annual increases in the performance levels of goods or technology subject to any such licensing requirement. Any such goods or technology which no longer meet the performance levels established by the latest such increase shall be removed from the list established pursuant to subsection (c) of this section unless, under such exceptions and under such procedures as the Secretary shall prescribe, any other department or agency of the United States objects to such removal and the Secretary determines, on the basis of such objection, that the goods or technology shall not be removed from the list. The Secretary shall also consider, where appropriate, removing site visitation requirements for goods and technology which are removed from the list unless objections described in this subsection are raised.

(h) TECHNICAL ADVISORY COMMITTEES.—(1) Upon written request by representatives of a substantial segment of any industry which produces any goods or technology subject to export controls under this section or being considered for such controls because of their significance to the national security of the United States, the Secretary shall appoint a technical advisory committee for any such goods or technology which the Secretary determines are difficult to evaluate because of questions concerning technical matters, worldwide availability, and actual utilization of production and technology, or

[margin notes: Information gathering. Regulations. Site visitation requirements, removal.]

licensing procedures. Each such committee shall consist of representatives of United States industry and Government, including the Departments of Commerce, Defense, and State and, in the discretion of the Secretary, other Government departments and agencies. No person serving on any such committee who is a representative of industry shall serve on such committee for more than four consecutive years.

Membership.

(2) Technical advisory committees established under paragraph (1) shall advise and assist the Secretary, the Secretary of Defense, and any other department, agency, or official of the Government of the United States to which the President delegates authority under this Act, with respect to actions designed to carry out the policy set forth in section 3(2)(A) of this Act. Such committees, where they have expertise in such matters, shall be consulted with respect to questions involving (A) technical matters, (B) worldwide availability and actual utilization of production technology, (C) licensing procedures which affect the level of export controls applicable to any goods or technology, and (D) exports subject to multilateral controls in which the United States participates, including proposed revisions of any such multilateral controls. Nothing in this subsection shall prevent the Secretary or the Secretary of Defense from consulting, at any time, with any person representing industry or the general public, regardless of whether such person is a member of a technical advisory committee. Members of the public shall be given a reasonable opportunity, pursuant to regulations prescribed by the Secretary, to present evidence to such committees.

Term of office.

(3) Upon request of any member of any such committee, the Secretary may, if the Secretary determines it appropriate, reimburse such member for travel, subsistence, and other necessary expenses incurred by such member in connection with the duties of such member.

Travel and other expenses, reimbursement.

(4) Each such committee shall elect a chairman, and shall meet at least every three months at the call of the chairman, unless the chairman determines, in consultation with the other members of the committee, that such a meeting is not necessary to achieve the purposes of this subsection. Each such committee shall be terminated after a period of 2 years, unless extended by the Secretary for additional periods of 2 years. The Secretary shall consult each such committee with respect to such termination or extension of that committee.

Termination.

(5) To facilitate the work of the technical advisory committees, the Secretary, in conjunction with other departments and agencies participating in the administration of this Act, shall disclose to each such committee adequate information, consistent with national security, pertaining to the reasons for the export controls which are in effect or contemplated for the goods or technology with respect to which that committee furnishes advice.

Export controls maintenance.

(6) Whenever a technical advisory committee certifies to the Secretary that goods or technology with respect to which such committee was appointed has become available in fact, to countries to which exports are controlled under this section, from sources outside the United States, including countries which participate with the United States in multilateral export controls, in sufficient quantity and of sufficient quality so that requiring a validated license for the export of such goods or technology would be ineffective in achieving the purpose set forth in subsection (a) of this section, and provides adequate documentation for such certification, in accordance with the procedures established pursuant to subsection (f)(1) of this section, the Secretary shall investigate such availability, and if

such availability is verified, the Secretary shall remove the requirement of a validated license for the export of the goods or technology, unless the President determines that the absence of export controls under this section would prove detrimental to the national security of the United States. In any case in which the President determines that export controls under this section must be maintained notwithstanding foreign availability, the Secretary shall publish that determination together with a concise statement of its basis and the estimated economic impact of the decision.

Determination, publication.

(i) MULTILATERAL EXPORT CONTROLS.—The President shall enter into negotiations with the governments participating in the group known as the Coordinating Committee (hereinafter in this subsection referred to as the "Committee") with a view toward accomplishing the following objectives:

Coordinating Committee, functions.

(1) Agreement to publish the list of items controlled for export by agreement of the Committee, together with all notes, understandings, and other aspects of such agreement of the Committee, and all changes thereto.

(2) Agreement to hold periodic meetings with high-level representatives of such governments, for the purpose of discussing export control policy issues and issuing policy guidance to the Committee.

(3) Agreement to reduce the scope of the export controls imposed by agreement of the Committee to a level acceptable to and enforceable by all governments participating in the Committee.

(4) Agreement on more effective procedures for enforcing the export controls agreed to pursuant to paragraph (3).

(j) COMMERCIAL AGREEMENTS WITH CERTAIN COUNTRIES.—(1) Any United States firm, enterprise, or other nongovernmental entity which, for commercial purposes, enters into any agreement with any agency of the government of a country to which exports are restricted for national security purposes, which agreement cites as intergovernmental agreement (to which the United States and such country are parties) calling for the encouragement of technical cooperation and is intended to result in the export from the United States to the other party of unpublished technical data of United States origin, shall report the agreement with such agency to the Secretary.

(2) The provisions of paragraph (1) shall not apply to colleges, universities, or other educational institutions.

(k) NEGOTIATIONS WITH OTHER COUNTRIES.—The Secretary of State, in consultation with the Secretary of Defense, the Secretary of Commerce, and the heads of other appropriate departments and agencies, shall be responsible for conducting negotiations with other countries regarding their cooperation in restricting the export of goods and technology in order to carry out the policy set forth in section 3(9) of this Act, as authorized by subsection (a) of this section, including negotiations with respect to which goods and technology should be subject to multilaterally agreed export restrictions and what conditions should apply for exceptions from those restrictions.

(l) DIVERSION TO MILITARY USE OF CONTROLLED GOODS OR TECHNOLOGY.—(1) Whenever there is reliable evidence that goods or technology, which were exported subject to national security controls under this section to a country to which exports are controlled for national security purposes, have been diverted to significant military use in violation of the conditions of an export license, the Secretary for as long as that diversion to significant military use continues—

(A) shall deny all further exports to the party responsible for that diversion of any goods or technology subject to national security controls under this section which contribute to that particular military use, regardless of whether such goods or technology are available to that country from sources outside the United States; and

(B) may take such additional steps under this Act with respect to the party referred to in subparagraph (A) as are feasible to deter the further military use of the previously exported goods or technology.

(2) As used in this subsection, the terms "diversion to significant military use" and "significant military use" means the use of United States goods or technology to design or produce any item on the United States Munitions List.

"Diversion to significant military use" and "significant military use".

FOREIGN POLICY CONTROLS

Sec. 6. (a) Authority.—(1) In order to carry out the policy set forth in paragraph (2)(B), (7), or (8) of section 3 of this Act, the President may prohibit or curtail the exportation of any goods, technology, or other information subject to the jurisdiction of the United States or exported by any person subject to the jurisdiction of the United States, to the extent necessary to further significantly the foreign policy of the United States or to fulfill its declared international obligations. The authority granted by this subsection shall be exercised by the Secretary, in consultation with the Secretary of State and such other departments and agencies as the Secretary considers appropriate, and shall be implemented by means of export licenses issued by the Secretary.

50 USC app. 2405.

(2) Export controls maintained for foreign policy purposes shall expire on December 31, 1979, or one year after imposition, whichever is later, unless extended by the President in accordance with subsections (b) and (e). Any such extension and any subsequent extension shall not be for a period of more than one year.

Expiration date.

(3) Whenever the Secretary denies any export license under this subsection, the Secretary shall specify in the notice to the applicant of the denial of such license that the license was denied under the authority contained in this subsection, and the reasons for such denial, with reference to the criteria set forth in subsection (b) of this section. The Secretary shall also include in such notice what, if any, modifications in or restrictions on the goods or technology for which the license was sought would allow such export to be compatible with controls implemented under this section, or the Secretary shall indicate in such notice which officers and employees of the Department of Commerce who are familiar with the application will be made reasonably available to the applicant for consultation with regard to such modifications or restrictions, if appropriate.

Export license denial.

(4) In accordance with the provisions of section 10 of this Act, the Secretary of State shall have the right to review any export license application under this section which the Secretary of State requests to review.

Export license application, review.

(b) Criteria.—When imposing, expanding, or extending export controls under this section, the President shall consider—

(1) the probability that such controls will achieve the intended foreign policy purpose, in light of other factors, including the availability from other countries of the goods or technology proposed for such controls;

(2) the compatibility of the proposed controls with the foreign policy objectives of the United States, including the effort to counter international terrorism, and with overall United States

policy toward the country which is the proposed target of the controls;

(3) the reaction of other countries to the imposition or expansion of such export controls by the United States;

(4) the likely effects of the proposed controls on the export performance of the United States, on the competitive position of the United States in the international economy, on the international reputation of the United States as a supplier of goods and technology, and on individual United States companies and their employees and communities, including the effects of the controls on existing contracts;

(5) the ability of the United States to enforce the proposed controls effectively; and

(6) the foreign policy consequences of not imposing controls.

(c) Consultation With Industry.—The Secretary, before imposing export controls under this section, shall consult with such affected United States industries as the Secretary considers appropriate, with respect to the criteria set forth in paragraphs (1) and (4) of subsection (b) and such other matters as the Secretary considers appropriate.

(d) Alternative Means.—Before resorting to the imposition of export controls under this section, the President shall determine that reasonable efforts have been made to achieve the purposes of the controls through negotiations or other alternative means.

(e) Notification To Congress.—The President in every possible instance shall consult with the Congress before imposing any export control under this section. Except as provided in section 7(g)(3) of this Act, whenever the President imposes, expands, or extends export controls under this section, the President shall immediately notify the Congress of such action and shall submit with such notification a report specifying—

Contents of report.

(1) the conclusions of the President with respect to each of the criteria set forth in subsection (b); and

(2) the nature and results of any alternative means attempted under subsection (d), or the reasons for imposing, extending, or expanding the control without attempting any such alternative means.

Such report shall also indicate how such controls will further significantly the foreign policy of the United States or will further its declared international obligations. To the extent necessary to further the effectiveness of such export control, portions of such report may be submitted on a classified basis, and shall be subject to the provisions of section 12(c) of this Act.

(f) Exclusion for Medicine and Medical Supplies.—This section does not authorize export controls on medicine or medical supplies. It is the intent of Congress that the President not impose export controls under this section on any goods or technology if he determines that the principal effect of the export of such goods or technology would be to help meet basic human needs. This subsection shall not be construed to prohibit the President from imposing restrictions on the export of medicine or medical supplies, under the International Emergency Economic Powers Act. This subsection shall not apply to any export controls under this section under this subsection which is in effect on the effective date of this Act.

50 USC 1701 note.

(g) Foreign Availability.—In applying export controls under this section, the President shall take all feasible steps to initiate and conclude negotiations with appropriate foreign governments for the purpose of securing the cooperation of such foreign governments in controlling the export to countries and consignees to which the

PUBLIC LAW 96-72—SEPT. 29, 1979   93 STAT. 515

United States export controls apply of any goods or technology comparable to goods or technology controlled under this section.

(h) INTERNATIONAL OBLIGATIONS.—The provisions of subsections (b), (c), (d), (f), and (g) shall not apply in any case in which the President exercises the authority contained in this section to impose export controls, or to approve or deny export license applications, in order to fulfill obligations of the United States pursuant to treaties to which the United States is a party or pursuant to other international agreements.

(i) COUNTRIES SUPPORTING INTERNATIONAL TERRORISM.—The Secretary and the Secretary of State shall notify the Committee on Foreign Affairs of the House of Representatives and the Committee on Banking, Housing, and Urban Affairs of the Senate before any license is approved for the export of goods or technology valued at more than $7,000,000 to any country concerning which the Secretary of State has made the following determinations:

Notification to congressional committees.

(1) Such country has repeatedly provided support for acts of international terrorism.

(2) Such exports would make a significant contribution to the military potential of such country, including its military logistics capability, or would enhance the ability of such country to support acts of international terrorism.

(j) CRIME CONTROL INSTRUMENTS.—(1) Crime control and detection instruments and equipment shall be approved for export by the Secretary only pursuant to a validated export license.

(2) The provisions of this subsection shall not apply with respect to exports to countries which are members of the North Atlantic Treaty Organization or to Japan, Australia, or New Zealand, or to such other countries as the President shall designate consistent with the purposes of this subsection and section 502B of the Foreign Assistance Act of 1961.

22 USC 2304.

(k) CONTROL LIST.—The Secretary shall establish and maintain, as part of the commodity control list, a list of any goods or technology subject to export controls under this section, and the countries to which such controls apply. Such goods or technology shall be clearly identified as subject to controls under this section. Such list shall consist of goods and technology identified by the Secretary of State, with the concurrence of the Secretary. If the Secretary and the Secretary of State are unable to agree on the list, the matter shall be referred to the President. Such list shall be reviewed not less frequently than every three years in the case of controls maintained cooperatively with other countries, and annually in the case of all other controls, for the purpose of making such revisions as are necessary in order to carry out this section. During the course of such review, an assessment shall be made periodically of the availability from sources outside the United States, or any of its territories or possessions, of goods and technology comparable to those controlled for export from the United States under this section.

Periodical review.

SHORT SUPPLY CONTROLS

SEC. 7. (a) AUTHORITY.—(1) In order to carry out the policy set forth in section 3(2)(C) of this Act, the President may prohibit or curtail the export of any goods subject to the jurisdiction of the United States or exported by any person subject to the jurisdiction of the United States. In curtailing exports to carry out the policy set forth in section 3(2)(C) of this Act, the President shall allocate a portion of export licenses on the basis of factors other than a prior history of exportation. Such factors shall include the extent to which a country engages

50 USC app. 2406.

Export licenses, allocation.

PUBLIC LAW 96-72—SEPT. 29, 1979   93 STAT. 516

in equitable trade practices with respect to United States goods and treats the United States equitably in times of short supply.

(2) Upon imposing quantitative restrictions on exports of any goods to carry out the policy set forth in section 3(2)(C) of this Act, the Secretary shall include in a notice published in the Federal Register with respect to such restrictions an invitation to all interested parties to submit written comments within 15 days from the date of publication on the impact of such restrictions and the method of licensing used to implement them.

Publication in Federal Register.

(3) In imposing export controls under this section, the President's authority shall include, but not be limited to, the imposition of export license fees.

Fees.

(b) MONITORING.—(1) In order to carry out the policy set forth in section 3(2)(C) of this Act, the Secretary shall monitor exports, and contracts for exports, of any good (other than a commodity which is subject to the reporting requirements of section 812 of the Agricultural Act of 1970) when the volume of such exports in relation to domestic supply contributes, or may contribute, to an increase in domestic prices or a domestic shortage, and such price increase or shortage has, or may have, a serious adverse impact on the economy or any sector thereof. Any such monitoring shall commence at a time adequate to assure that the monitoring will result in a data base sufficient to enable policies to be developed, in accordance with section 3(2)(C) of this Act, to mitigate a short supply situation or serious inflationary price rise or, if export controls are needed, to permit imposition of such controls in a timely manner. Information which the Secretary requires to be furnished in effecting such monitoring shall be confidential, except as provided in paragraph (2) of this subsection.

7 USC 612c-3.

(2) The results of such monitoring shall, to the extent practicable, be aggregated and included in weekly reports setting forth, with respect to each item monitored, actual and anticipated exports, the destination by country, and the domestic and worldwide price, supply, and demand. Such reports may be made monthly if the Secretary determines that there is insufficient information to justify weekly reports.

Weekly or monthly reports.

(3) The Secretary shall consult with the Secretary of Energy to determine whether monitoring or export controls under this section are warranted with respect to exports of facilities, machinery, or equipment normally and principally used, or intended to be used, in the production, conversion, or transportation of fuels and energy (except nuclear energy), including, but not limited to, drilling rigs, platforms, and equipment; petroleum refineries, natural gas processing, liquefaction, and gasification plants; facilities for production of synthetic natural gas or synthetic crude oil; oil and gas pipelines, pumping stations, and associated equipment; and vessels for transporting oil, gas, coal, and other fuels.

(c) PETITIONS FOR MONITORING OR CONTROLS.—(1)(A) Any entity, including a trade association, firm, or certified or recognized union or group of workers, which is representative of an industry or a substantial segment of an industry which processes metallic materials capable of being recycled with respect to which an increase in domestic prices or a domestic shortage, either of which results from increased exports, has or may have a significant adverse effect on the national economy or any sector thereof, may transmit a written petition to the Secretary requesting the monitoring of exports, or the imposition of export controls, or both, with history to such material, in order to carry out the policy set forth in section 3(2)(C) of this Act.

(B) Each petition shall be in such form as the Secretary shall prescribe and shall contain information in support of the action requested. The petition shall include any information reasonably available to the petitioner indicating (i) that there has been a significant increase, in relation to a specific period of time, in exports of such material in relation to domestic supply, and (ii) that there has been a significant increase in the price of such material or a domestic shortage of such material under circumstances indicating the price increase or domestic shortage may be related to exports.

(2) Within 15 days after receipt of any petition described in paragraph (1), the Secretary shall publish a notice in the Federal Register. The notice shall (A) include the name of the material which is the subject of the petition, (B) include the Schedule B number of the material as set forth in the Statistical Classification of Domestic and Foreign Commodities Exported from the United States, (C) indicate whether the petitioner is requesting that controls or monitoring, or both, be imposed with respect to the exportation of such material, and (D) provide that interested persons shall have a period of 30 days commencing with the date of publication of such notice to submit to the Secretary written data, views, or arguments, with or without opportunity for oral presentation, with respect to the matter involved. At the request of the petitioner or any other entity representative of producers or exporters of such material, the Secretary shall conduct public hearings with respect to the subject of the petition, in which case the 30-day period may be extended to 45 days.

(3) Within 45 days after the end of the 30- or 45-day period described in paragraph (2), as the case may be, the Secretary shall—

(A) determine whether to impose monitoring or controls, or both, on the export of such material, in order to carry out the policy set forth in section 3(2)(C) of this Act; and

(B) publish in the Federal Register a detailed statement of the reasons for such determination.

(4) Within 15 days after making a determination under paragraph (3) to impose monitoring or controls on the export of a material, the Secretary shall publish in the Federal Register proposed regulations with respect to such monitoring or controls. Within 30 days following the publication of such proposed regulations, and after considering any public comments thereon, the Secretary shall publish and implement final regulations with respect to such monitoring or controls.

(5) For purposes of publishing notices in the Federal Register and scheduling public hearings pursuant to this subsection, the Secretary may consolidate petitions, and responses thereto, which involve the same or related materials.

(6) If a petition with respect to a particular material or group of materials has been considered in accordance with all the procedures prescribed in this subsection, the Secretary may determine, in the absence of significantly changed circumstances, that any other petition with respect to the same material or group of materials which is filed within 6 months after consideration of the prior petition has been completed does not merit complete consideration under this subsection.

(7) The procedures and time limits set forth in this subsection with respect to a petition filed under this subsection shall take precedence over any review undertaken at the initiative of the Secretary with respect to the same subject as that of the petition.

[Margin notes: Publication in Federal Register. Hearings. Publication in Federal Register. Publication in Federal Register. Petitions, consolidation.]

(8) The Secretary may impose monitoring or controls on a temporary basis after a petition is filed under paragraph (1)(A) but before the Secretary makes a determination under paragraph (3) if the Secretary considers such action to be necessary to carry out the policy set forth in section 3(2)(C) of this Act.

(9) The authority under this subsection shall not be construed to affect the authority of the Secretary under any other provision of this Act.

(10) Nothing contained in this subsection shall be construed to preclude submission on a confidential basis to the Secretary of information relevant to a decision to impose or remove monitoring or controls under the authority of this Act, or to preclude consideration of such information by the Secretary in reaching decisions required under this subsection. The provisions of this paragraph shall not be construed to affect the applicability of section 552(b) of title 5, United State Code.

(d) DOMESTICALLY PRODUCED CRUDE OIL.—(1) Notwithstanding any other provision of this Act and notwithstanding subsection (u) of section 28 of the Mineral Leasing Act of 1920 (30 U.S.C. 185), no domestically produced crude oil transported by pipeline over right-of-way granted pursuant to section 203 of the Trans-Alaska Pipeline Authorization Act (43 U.S.C. 1652) (except any such crude oil which (A) is exported to an adjacent foreign country to be refined and consumed therein in exchange for the same quantity of crude oil being exported from that country to the United States; such exchange must result through convenience or increased efficiency of transportation in lower prices for consumers of petroleum products in the United States as described in paragraph (2)(A)(ii) of this subsection, or (B) is temporarily exported for convenience or increased efficiency of transportation across parts of an adjacent foreign country and reenters the United States) may be exported from the United States, or any of its territories and possessions, unless the requirements of paragraph (2) of this subsection are met.

(2) Crude oil subject to the prohibition contained in paragraph (1) may be exported only if—

(A) the President makes and publishes findings that exports of such crude oil, including exchanges—

(i) will not diminish the total quantity or quality of petroleum refined within, stored within, or legally committed to be transported to and sold within the United States;

(ii) will, within 3 months following the initiation of such exports or exchanges, result in (I) acquisition costs to the refiners which purchase the imported crude oil being lower than the acquisition costs such refiners would have to pay for the domestically produced oil in the absence of such an export or exchange, and (II) not less than 75 percent of such savings in costs being reflected in wholesale and retail prices of products refined from such imported crude oil;

(iii) will be made only pursuant to contracts which may be terminated if the crude oil supplies of the United States are interrupted, threatened, or diminished;

(iv) are clearly necessary to protect the national interest; and

(v) are in accordance with the provisions of this Act; and

(B) the President reports such findings to the Congress and the Congress, within 60 days thereafter, agrees to a concurrent resolution approving such exports on the basis of the findings.

(3) Notwithstanding any other provision of this section or any other provision of law, including subsection (u) of section 28 of the Mineral Leasing Act of 1920, the President may export oil to any country

[Margin notes: Temporary monitoring or controls. Exportation, conditions. Report to Congress. 30 USC 185.]

PUBLIC LAW 96-72—SEPT. 29, 1979   93 STAT. 519

pursuant to a bilateral international oil supply agreement entered into by the United States with such nation before June 25, 1979, or to any country pursuant to the International Emergency Oil Sharing Plan of the International Energy Agency.

(e) REFINED PETROLEUM PRODUCTS.—(1) No refined petroleum product may be exported except pursuant to an export license specifically authorizing such export. Not later than 5 days after an application for a license to export any refined petroleum product or residual fuel oil is received, the Secretary shall notify the Congress of such application, together with the name of the exporter, the destination of the proposed export, and the amount and price of the proposed export. Such notification shall be made to the chairman of the Committee on Foreign Affairs of the House of Representatives and the chairman of the Committee on Banking, Housing, and Urban Affairs of the Senate. [margin: Export license applications, notification of congressional committees.]

(2) The Secretary may not grant such license during the 30-day period beginning on the date on which notification to the Congress under paragraph (1) is received, unless the President certifies in writing to the Speaker of the House of Representatives and the President pro tempore of the Senate that the proposed export is vital to the national interest and that a delay in issuing the license would adversely affect that interest.

(3) This subsection shall not apply to (A) any export license application for exports to a country with respect to which historical export quotas established by the Secretary on the basis of past trading relationships apply, or (B) any license application for exports to a country if exports under the license would not result in more than 250,000 barrels of refined petroleum products being exported from the United States to such country in any fiscal year.

(4) For purposes of this subsection, "refined petroleum product" means gasoline, kerosene, distillates, propane or butane gas, diesel fuel, and residual fuel oil refined within the United States or entered for consumption within the United States. [margin: "Refined petroleum product."]

(5) The Secretary may extend any time period prescribed in section 10 of this Act to the extent necessary to take into account delays in action by the Secretary on a license application on account of the provisions of this subsection. [margin: Time extension.]

(f) CERTAIN PETROLEUM PRODUCTS.—Petroleum products refined in United States Foreign Trade Zones, or in the United States Territory of Guam, from foreign crude oil shall be excluded from any quantitative restrictions imposed under this section except that, if the Secretary finds that a product is in short supply, the Secretary may issue such regulations as may be necessary to limit exports. [margin: Regulations.]

(g) AGRICULTURAL COMMODITIES.—(1) The authority conferred by this section shall not be exercised with respect to any agricultural commodity, including fats and oils or animal hides or skins, without the approval of the Secretary of Agriculture. The Secretary of Agriculture shall not approve the exercise of such authority with respect to any such commodity during any period for which the supply of such commodity is determined by the Secretary of Agriculture to be in excess of the requirements of the domestic economy except to the extent the President determines that such exercise of authority is required to carry out the policies set forth in subparagraph (A) or (B) of paragraph (2) of section 3 of this Act. The Secretary of Agriculture shall, by exercising the authorities which the Secretary of Agriculture has under other applicable provisions of law, collect data with respect to export sales of animal hides and skins. [margin: Export sales of animal hides and skins, data.]

(2) Upon approval of the Secretary, in consultation with the Secretary of Agriculture, agricultural commodities purchased by or

PUBLIC LAW 96-72—SEPT. 29, 1979   93 STAT. 520

for use in a foreign country may remain in the United States for export at a later date free from any quantitative limitations on export which may be imposed to carry out the policy set forth in section 3(2)(C) of this Act subsequent to such approval. The Secretary may not grant such approval unless the Secretary receives adequate assurance and, in conjunction with the Secretary of Agriculture, finds (A) that such commodities will eventually be exported, (B) that neither the sale nor export thereof will result in an excessive drain of scarce materials and have a serious domestic inflationary impact, (C) that storage of such commodities in the United States will not unduly limit the space available for storage of domestically owned commodities, and (D) that the purpose of such storage is to establish a reserve of such commodities for later use, not including resale to or use by another country. The Secretary may issue such regulations as may be necessary to implement this paragraph. [margin: Regulations.]

(3) If the authority conferred by this section or section 6 is exercised to prohibit or curtail the export of any agricultural commodity in order to carry out the policies set forth in subparagraph (B) or (C) of paragraph (2) of section 3 of this Act, the President shall immediately report such prohibition or curtailment to the Congress, setting forth the reasons therefor in detail. If the Congress, within 30 days after the date of its receipt of such report, adopts a concurrent resolution disapproving such prohibition or curtailment, then such prohibition or curtailment shall cease to be effective with the adoption of such resolution. In the computation of such 30-day period, there shall be excluded the days on which either House is not in session because of an adjournment of more than 3 days to a day certain or because of an adjournment of the Congress sine die.

(h) BARTER AGREEMENTS.—(1) The exportation pursuant to a barter agreement of any goods which may lawfully be exported from the United States, for any goods which may lawfully be imported into the United States, may be exempted, in accordance with paragraph (2) of this subsection, from any quantitative limitation on exports (other than any reporting requirement) imposed to carry out the policy set forth in section 3(2)(C) of this Act.

(2) The Secretary shall grant an exemption under paragraph (1) if the Secretary finds, after consultation with the appropriate department or agency of the United States, that—

(A) for the period during which the barter agreement is to be performed—

(i) the average annual quantity of the goods to be exported pursuant to the barter agreement will not be required to satisfy the average amount of such goods estimated to be required annually by the domestic economy and will be surplus thereto; and

(ii) the average annual quantity of the goods to be imported will be less than the average amount of such goods estimated to be required annually to supplement domestic production; and

(B) the parties to such barter agreement have demonstrated adequately that they intend, and have the capacity, to perform such barter agreement.

(3) For purposes of this subsection, the term "barter agreement" means any agreement which is made for the exchange, without monetary consideration, of any goods produced in the United States for any goods produced outside of the United States. [margin: "Barter agreement."]

(4) This subsection shall apply only with respect to barter agreements entered into after the effective date of this Act.

93 STAT. 521

PUBLIC LAW 96-72—SEPT. 29, 1979

(i) UNPROCESSED RED CEDAR.—(1) The Secretary shall require a validated license, under the authority contained in subsection (a) of this section, for the export of unprocessed western red cedar (Thuja plicata) logs, harvested from State or Federal lands. The Secretary shall impose quantitative restrictions upon the export of unprocessed western red cedar logs during the 3-year period beginning on the effective date of this Act as follows:

(A) Not more than thirty million board feet scribner of such logs may be exported during the first year of such 3-year period.

(B) Not more than fifteen million board feet scribner of such logs may be exported during the second year of such period.

(C) Not more than five million board feet scribner of such logs may be exported during the third year of such period.

Export terminations.

After the end of such 3-year period, no unprocessed western red cedar logs may be exported from the United States.

Export licenses, allocation.

(2) The Secretary shall allocate export licenses to exporters pursuant to this subsection on the basis of a prior history of exportation by such exporters and such other factors as the Secretary considers necessary and appropriate to minimize any hardship to the producers of western red cedar and to further the foreign policy of the United States.

(3) Unprocessed western red cedar logs shall not be considered to be an agricultural commodity for purposes of subsection (g) of this section.

"Unprocessed western red cedar."

(4) As used in this subsection, the term "unprocessed western red cedar" means red cedar timber which has not been processed into—

(A) lumber without wane;

(B) chips, pulp, and pulp products;

(C) veneer and plywood;

(D) poles, posts, or pilings cut or treated with preservative for use as such and not intended to be further processed; or

(E) shakes and shingles.

(j) EXPORT OF HORSES.—(1) Notwithstanding any other provision of this Act, no horse may be exported by sea from the United States, or any of its territories and possessions, unless such horse is part of a consignment of horses with respect to which a waiver has been granted under paragraph (2) of this subsection.

Regulations.

(2) The Secretary, in consultation with the Secretary of Agriculture, may issue regulations providing for the granting of waivers permitting the export by sea of a specified consignment of horses, if the Secretary, in consultation with the Secretary of Agriculture, determines that no horse in that consignment is being exported for purposes of slaughter.

FOREIGN BOYCOTTS

Regulations. 50 USC app. 2047.

SEC. 8. (a) PROHIBITIONS AND EXCEPTIONS.—(1) For the purpose of implementing the policies set forth in subparagraph (A) or (B) of paragraph (5) of section 3 of this Act, the President shall issue regulations prohibiting any United States person, with respect to his activities in the interstate or foreign commerce of the United States, from taking or knowingly agreeing to take any of the following actions with intent to comply with, further, or support any boycott fostered or imposed by a foreign country against a country which is friendly to the United States and which is not itself the object of any form of boycott pursuant to United States law or regulation:

(A) Refusing, or requiring any other person to refuse, to do business with or in the boycotted country, with any business concern organized under the laws of the boycotted country, with

93 STAT. 522

PUBLIC LAW 96-72—SEPT. 29, 1979

any national or resident of the boycotted country, or with any other person, pursuant to an agreement with, a requirement of, or a request from or on behalf of the boycotting country. The mere absence of a business relationship with or in the boycotted country with any business concern organized under the laws of the boycotted country, with any national or resident of the boycotted country, or with any other person, does not indicate the existence of the intent required to establish a violation of regulations issued to carry out this subparagraph.

Employment discrimination, prohibition.

(B) Refusing or requiring any other person to refuse, to employ or otherwise discriminating against any United States person on the basis of race, religion, sex, or national origin of that person or of any owner, officer, director, or employee of such person.

(C) Furnishing information with respect to the race, religion, sex, or national origin of any United States person or of any owner, officer, director, or employee of such person.

Business information.

(D) Furnishing information about whether any person has, has had, or proposes to have any business relationship (including a relationship by way of sale, purchase, legal or commercial representation, shipping or other transport, insurance, investment, or supply) with or in the boycotted country, with any business concern organized under the laws of the boycotted country, with any national or resident of the boycotted country, or with any other person which is known or believed to be restricted from having any business relationship with or in the boycotting country. Nothing in this paragraph shall prohibit the furnishing of normal business information in a commercial context as defined by the Secretary.

(E) Furnishing information about whether any person is a member of, has made contributions to, or is otherwise associated with or involved in the activities of any charitable or fraternal organization which supports the boycotted country.

Letter of credit.

(F) Paying, honoring, confirming, or otherwise implementing a letter of credit which contains any condition or requirement compliance with which is prohibited by regulations issued pursuant to this paragraph, and no United States person shall, as a result of the application of this paragraph, be obligated to pay or otherwise honor or implement such letter of credit.

Regulatory exceptions.

(2) Regulations issued pursuant to paragraph (1) shall provide exceptions for—

(A) complying or agreeing to comply with requirements (i) prohibiting the import of goods or services from the boycotted country or goods produced or services provided by any business concern organized under the laws of the boycotted country or by nationals or residents of the boycotted country, or (ii) prohibiting the shipment of goods to the boycotting country on a carrier of the boycotted country, or by a route other than that prescribed by the boycotting country or the recipient of the shipment;

(B) complying or agreeing to comply with import and shipping document requirements with respect to the country of origin, the name of the carrier and route of shipment, the name of the supplier of the shipment or the name of the provider of other services, except that no information knowingly furnished or conveyed in response to such requirements may be stated in negative, blacklisting, or similar exclusionary terms, other than with respect to carriers or route of shipment as may be permitted by such regulations in order to comply with precautionary requirements protecting against war risks and confiscation;

(C) complying or agreeing to comply in the normal course of business with the unilateral and specific selection by a boycotting country, or national or resident thereof, of carriers, insurers, suppliers of services to be performed within the boycotting country or specific goods which, in the normal course of business, are identifiable by source when imported into the boycotting country;

(D) complying or agreeing to comply with export requirements of the boycotting country relating to shipments or transshipments of exports to the boycotted country, to any business concern of or organized under the laws of the boycotted country, or to any national or resident of the boycotted country;

(E) compliance by an individual or agreement by an individual to comply with the immigration or passport requirements of any country with respect to such individual or any member of such individual's family or with requests for information regarding requirements of employment of such individual within the boycotting country; and

(F) compliance by a United States person resident in a foreign country or agreement by such person to comply with the laws of that country with respect to his activities exclusively therein, and such regulations may contain exceptions for such resident complying with the laws or regulations of that foreign country governing imports into such country of trademarked, trade named, or similarly specifically identifiable products, or components of products for his own use, including the performance of contractual services within that country, as may be defined by such regulations.

(3) Regulations issued pursuant to paragraphs (2)(C) and (2)(F) shall not provide exceptions from paragraphs (1)(B) and (1)(C).

(4) Nothing in this subsection may be construed to supersede or limit the operation of the antitrust or civil rights laws of the United States.

(5) This section shall apply to any transaction or activity undertaken, by or through a United States person or any other person, with intent to evade the provisions of this section as implemented by the regulations issued pursuant to this subsection, and such regulations shall expressly provide that the exceptions set forth in paragraph (2) shall not permit activities or agreements (expressed or implied by a course of conduct, including a pattern of responses) otherwise prohibited, which are not within the intent of such exceptions.

(b) FOREIGN POLICY CONTROLS.—(1) In addition to the regulations issued pursuant to subsection (a) of this section, regulations issued under section 6 of this Act shall implement the policies set forth in section 3(5).

(2) Such regulations shall require that any United States person receiving a request for the furnishing of information, the entering into or implementing of agreements, or the taking of any other action referred to in section 3(5) shall report that fact to the Secretary, together with such other information concerning such request as the Secretary may require for such action as the Secretary considers appropriate for carrying out the policies of that section. Such person shall also report to the Secretary whether such person intends to comply and whether such person has complied with such request. Any report filed pursuant to this paragraph shall be made available promptly for public inspection and copying, except that information regarding the quantity, description, and value of any goods or technology to which such report relates may be kept confidential if the Secretary determines that disclosure thereof would place the

United States person involved at a competitive disadvantage. The Secretary shall periodically transmit summaries of the information contained in such reports to the Secretary of State for such action as the Secretary of State, in consultation with the Secretary, considers appropriate for carrying out the policies set forth in section 3(5) of this Act.

(c) PREEMPTION.—The provisions of this section and the regulations issued pursuant thereto shall preempt any law, rule, or regulation of any of the several States or the District of Columbia, or any of the territories or possessions of the United States, or of any governmental subdivision thereof, which law, rule, or regulation pertains to participation in, compliance with, implementation of, or the furnishing of information regarding restrictive trade practices or boycotts fostered or imposed by foreign countries against other countries.

PROCEDURES FOR HARDSHIP RELIEF FROM EXPORT CONTROLS

SEC. 9. (a) FILING OF PETITIONS.—Any person who, in such person's domestic manufacturing process or other domestic business operation, utilizes a product produced abroad in whole or in part from a good historically obtained from the United States but which has been made subject to export controls, or any person who historically has exported such a good, may transmit a petition of hardship to the Secretary requesting an exemption from such controls in order to alleviate any unique hardship resulting from the imposition of such controls. A petition under this section shall be in such form as the Secretary shall prescribe and shall contain information demonstrating the need for the relief requested.

(b) DECISION OF THE SECRETARY.—Not later than 30 days after receipt of any petition under subsection (a), the Secretary shall transmit a written decision to the petitioner granting or denying the requested relief. Such decision shall contain a statement setting forth the Secretary's basis for the grant or denial. Any exemption granted may be subject to such conditions as the Secretary considers appropriate.

(c) FACTORS TO BE CONSIDERED.—For purposes of this section, the Secretary's decision with respect to the grant or denial of relief from unique hardship resulting directly or indirectly from the imposition of export controls shall reflect the Secretary's consideration of factors such as the following;

(1) Whether denial would cause a unique hardship to the petitioner which can be alleviated only by granting an exception to the applicable regulations. In determining whether relief shall be granted, the Secretary shall take into account—

(A) ownership of material for which there is no practicable domestic market by virtue of the location or nature of the material;

(B) potential serious financial loss to the applicant if not granted an exception;

(C) inability to obtain, except through import, an item essential for domestic use which is produced abroad from the good under control;

(D) the extent to which denial would conflict, to the particular detriment of the applicant, with other national policies including those reflected in any international agreement to which the United States is a party;

(E) possible adverse effects on the economy (including unemployment) in any locality or region of the United States; and

Transmittal to Secretary of State.

50 USC app. 2408.

Reports.

Public inspection and copying.

(F) other relevant factors, including the applicant's lack of an exporting history during any base period that may be established with respect to export quotas for the particular good.

(2) The effect of a finding in favor of the applicant would have on attainment of the basic objectives of the short supply control program.

In all cases, the desire to sell at higher prices and thereby obtain greater profits shall not be considered as evidence of a unique hardship, nor will circumstances where the hardship is due to imprudent acts or failure to act on the part of the petitioner.

PROCEDURES FOR PROCESSING EXPORT LICENSE APPLICATIONS

SEC. 10. (a) PRIMARY RESPONSIBILITY OF THE SECRETARY.—(1) All export license applications required under this Act shall be submitted by the applicant to the Secretary. All determinations with respect to any such application shall be made by the Secretary, subject to the procedures provided in this section.

(2) It is the intent of the Congress that a determination with respect to any export license application be made to the maximum extent possible by the Secretary without referral of such application to any other department or agency of the Government.

(3) To the extent necessary, the Secretary shall seek information and recommendations from the Government departments and agencies concerned with aspects of United States domestic and foreign policies and operations having an important bearing on exports. Such departments and agencies shall cooperate fully in rendering such information and recommendations.

(b) INITIAL SCREENING.—Within 10 days after the date on which any export license application is submitted pursuant to subsection (a)(1), the Secretary shall—

(1) send the applicant an acknowledgment of the receipt of the application and the date of the receipt;

(2) submit to the applicant a written description of the procedures required by this section, the responsibilities of the Secretary and of other departments and agencies with respect to the application, and the rights of the applicant;

(3) return the application without action if the application is improperly completed or if additional information is required, with sufficient information to permit the application to be properly resubmitted, in which case if such application is resubmitted, it shall be treated as a new application for the purpose of calculating the time periods prescribed in this section;

(4) determine whether it is necessary to refer the application to any other department or agency and, if such referral is determined to be necessary, inform the applicant of any such department or agency to which the application will be referred; and

(5) determine whether it is necessary to submit the application to a multilateral review process, pursuant to a multilateral agreement, formal or informal, to which the United States is a party and, if so, inform the applicant of this requirement.

(c) ACTION ON CERTAIN APPLICATIONS.—In each case in which the Secretary determines that it is not necessary to refer an application to any other department or agency for its information and recommendations, a license shall be formally issued or denied within 90 days after a properly completed application has been submitted pursuant to this section.

50 USC app. 2409.

(d) REFERRAL TO OTHER DEPARTMENTS AND AGENCIES.—In each case in which the Secretary determines that it is necessary to refer an application to any other department or agency for its information and recommendations, the Secretary shall, within 30 days after the submission of a properly completed application—

(1) refer the application, together with all necessary analysis and recommendations of the Department of Commerce, concurrently to all such departments or agencies; and

(2) if the applicant so requests, provide the applicant with an opportunity to review for accuracy any documentation to be referred to any such department or agency with respect to such application for the purpose of describing the export in question in order to determine whether such documentation accurately describes the proposed export.

(e) ACTION BY OTHER DEPARTMENTS AND AGENCIES.—(1) Any department or agency to which an application is referred pursuant to subsection (d) shall submit to the Secretary, within 30 days after its receipt of the application, the information or recommendations requested with respect to such application. Except as provided in paragraph (2), any such department or agency which does not submit its recommendations within the time period prescribed in the preceding sentence shall be deemed by the Secretary to have no objection to the approval of such application.

Recommendations, time extension.

(2) If the head of any such department or agency notifies the Secretary before the expiration of the time period provided in paragraph (1) for submission of its recommendations that more time is required for review by such department or agency, such department or agency shall have an additional 30-day period to submit its recommendations to the Secretary. If such department or agency does not submit its recommendations within the time period prescribed by the preceding sentence, it shall be deemed by the Secretary to have no objection to the approval of such application.

(f) ACTION BY THE SECRETARY.—(1) Within 90 days after receipt of the recommendations of other departments and agencies with respect to a license application, as provided in subsection (e), the Secretary shall formally issue or deny the license. In deciding whether to issue or deny a license, the Secretary shall take into account any recommendation of a department or agency with respect to the application in question. In cases where the Secretary receives conflicting recommendations, the Secretary shall, within the 90-day period provided for in this subsection, take such action as may be necessary to resolve such conflicting recommendations.

Conflicting recommendations.

(2) In cases where the Secretary receives questions or negative considerations or recommendations from any other department or agency with respect to an application, the Secretary shall, to the maximum extent consistent with the national security and foreign policy of the United States, inform the applicant of the specific questions raised and any such negative considerations or recommendations, and shall accord the applicant an opportunity, before the final determination with respect to the application is made, to respond in writing to such questions, considerations, or recommendations.

Applicant notification and opportunity for written response.

(3) In cases where the Secretary has determined that an application should be denied, the applicant shall be informed in writing, within 5 days after such determination is made, of the determination, of the statutory basis for denial, the policies set forth in section 3 of the Act which would be furthered by denial, and, to the extent consistent with the national security and foreign policy of the United States, the specific considerations which led to the denial, and of the availability

Applicant denial procedures.

PUBLIC LAW 96-72—SEPT. 29, 1979   93 STAT. 527

of appeal procedures. In the event decisions on license applications are deferred inconsistent with the provisions of this section, the applicant shall be so informed in writing within 5 days after such deferral.

(4) If the Secretary determines that a particular application or set of applications of exceptional importance and complexity, and that additional time is required for negotiations to modify the application or applications, the Secretary may extend any time period prescribed in this section. The Secretary shall notify the Congress and the applicant of such extension and the reasons therefor. *[margin: Time extension, notification to Congress and applicant.]*

(g) SPECIAL PROCEDURES FOR SECRETARY OF DEFENSE.—(1) Notwithstanding any other provision of this section, the Secretary of Defense is authorized to review any proposed export of any goods or technology to any country to which exports are controlled for national security purposes and, whenever the Secretary of Defense determines that the export of such goods or technology will make a significant contribution, which would prove detrimental to the national security of the United States, to the military potential of any such country, to recommend to the President that such export be disapproved. *[margin: Review.]*

(2) Notwithstanding any other provision of law, the Secretary of Defense shall determine, in consultation with the Secretary, and confirm in writing the types and categories of transactions which should be reviewed by the Secretary of Defense in order to make a determination referred to in paragraph (1). Whenever a license or other authority is requested for the export to any country to which exports are controlled for national security purposes of goods or technology within any such type or category, the Secretary shall notify the Secretary of Defense of such request, and the Secretary may not issue any license or other authority pursuant to such request before the expiration of the period within which the President may disapprove such export. The Secretary of Defense shall carefully consider any notification submitted by the Secretary pursuant to this paragraph and, not later than 30 days after notification of the request, shall— *[margin: Export transactions, review.]*

(A) recommend to the President that he disapprove any request for the export of the goods or technology involved to the particular country if the Secretary of Defense determines that the export of such goods or technology will make a significant contribution, which would prove detrimental to the national security of the United States, to the military potential of such country or any other country;

(B) notify the Secretary that he would recommend approval subject to specified conditions; or

(C) recommend to the Secretary that the export of goods or technology be approved.

If the President notifies the Secretary, within 30 days after receiving a recommendation from the Secretary of Defense, that he disapproves such export, no license or other authority may be issued for the export of such goods or technology to such country.

(3) The Secretary shall approve or disapprove a license application, and issue or deny a license, in accordance with the provisions of this subsection, and, to the extent applicable, in accordance with the time periods and procedures otherwise set forth in this section.

(4) Whenever the President exercises his authority under this subsection to modify or overrule a recommendation made by the Secretary of Defense or exercises his authority to modify or overrule any recommendation made by the Secretary of Defense under subsection (c) or (d) of section 5 of this Act with respect to the list of goods and technologies controlled for national security purposes, the Presi- *[margin: Presidential statement, transmittal to Congress.]*

93 STAT. 528   PUBLIC LAW 96-72—SEPT. 29, 1979

dent shall promptly transmit to the Congress a statement indicating his decision, together with the recommendation of the Secretary of Defense.

(h) MULTILATERAL CONTROLS.—In any case in which an application, which has been finally approved under subsection (c), (f), or (g) of this section, is required to be submitted to a multilateral review process, pursuant to a multilateral agreement, formal or informal, to which the United States is a party, the license shall not be issued as prescribed in such subsections, but the Secretary shall notify the applicant of the approval of the application (and the date of such approval) by the Secretary subject to such multilateral review. The license shall be issued upon approval of the application under such multilateral review. If such multilateral review has not resulted in a determination with respect to the application within 60 days after such date, the Secretary's approval of the license shall be final and the license shall be issued, unless the Secretary determines that issuance of the license would prove detrimental to the national security of the United States. At the time at which the Secretary makes such a determination, the Secretary shall notify the applicant of the determination and shall notify the Congress of the determination, the reasons for the determination, the reasons for which the multilateral review could not be concluded within such 60-day period, and the actions planned or being taken by the United States Government to secure conclusion of the multilateral review. At the end of every 60-day period after such notification to Congress, the Secretary shall advise the applicant and the Congress of the status of the application, and shall report to the Congress in detail on the reasons for the further delay and any further actions being taken by the United States Government to secure conclusion of the multilateral review. In addition, at the time at which the Secretary issues or denies the license upon conclusion of the multilateral review, the Secretary shall notify the Congress of such issuance or denial and of the total time required for the multilateral review. *[margin: Multilateral review process. Notification to Congress and applicant. Applicant status, report to Congress.]*

(i) RECORDS.—The Secretary and any department or agency to which any application is referred under this section shall keep accurate records with respect to all applications considered by the Secretary or by any such department or agency, including, in the case of the Secretary, any dissenting recommendations received from any such department or agency.

(j) APPEAL AND COURT ACTION.—(1) The Secretary shall establish appropriate procedures for any applicant to appeal to the Secretary the denial of an export license application of the applicant.

(2) In any case in which any action prescribed in this section is not taken on a license application within the time periods established by this section (except, in the case of a time period extended under subsection (f)(4) of which the applicant is notified), the applicant may file a petition with the Secretary requesting compliance with the requirements of this section. When such petition is filed, the Secretary shall take immediate steps to correct the situation giving rise to the petition and shall immediately notify the applicant of such steps. *[margin: Filing of petition by applicant.]*

(3) If, within 30 days after a petition is filed under paragraph (2), the processing of the application has not been brought into conformity with the requirements of this section, or the application has been brought into conformity with such requirements but the Secretary has not so notified the applicant, the applicant may bring an action in an appropriate United States district court for a restraining order, a temporary or permanent injunction, or other appropriate relief, to require compliance with the requirements of this section. The United

States district courts shall have jurisdiction to provide such relief, as appropriate.

VIOLATIONS

50 USC app. 2410.

SEC. 11. (a) IN GENERAL.—Except as provided in subsection (b) of this section, whoever knowingly violates any provision of this Act or any regulation, order, or license issued thereunder shall be fined not more than five times the value of the exports involved or $50,000, whichever is greater, or imprisoned not more than 5 years, or both.

(b) WILLFUL VIOLATIONS.—(1) Whoever willfully exports anything contrary to any provision of this Act or any regulation, order, or license issued thereunder, with knowledge that such exports will be used for the benefit of any country to which exports are restricted for national security or foreign policy purposes, shall be fined not more than five times the value of the exports involved or $100,000, whichever is greater, or imprisoned not more than 10 years, or both.

(2) Any person who is issued a validated license under this Act for the export of any good or technology to a controlled country and who, with knowledge that such a good or technology is being used by such controlled country for military or intelligence gathering purposes contrary to the conditions under which the license was issued, willfully fails to report such use to the Secretary of Defense, shall be fined not more than five times the value of the exports involved or $100,000, whichever is greater, or imprisoned for not more than 5 years, or both. For purposes of this paragraph, "controlled country" means any country described in section 620(f) of the Foreign Assistance Act of 1961.

"Controlled country."
22 USC 2370.

(c) CIVIL PENALTIES; ADMINISTRATIVE SANCTIONS.—(1) The head of any department or agency exercising any functions under this Act, or any officer or employee of such department or agency specifically designated by the head thereof, may impose a civil penalty not to exceed $10,000 for each violation of this Act or any regulation, order, or license issued under this Act, either in addition to or in lieu of any other liability or penalty which may be imposed.

(2)(A) The authority under this Act to suspend or revoke the authority of any United States person to export goods or technology may be used with respect to any violation of the regulations issued pursuant to section 8(a) of this Act.

(B) Any administrative sanction (including any civil penalty or any suspension or revocation of authority to export) imposed under this Act for a violation of the regulations issued pursuant to section 8(a) of this Act may be imposed only after notice and opportunity for an agency hearing on the record in accordance with sections 554 through 557 of title 5, United States Code.

(C) Any charging letter or other document initiating administrative proceedings for the imposition of sanctions for violations of the regulations issued pursuant to section 8(a) of this Act shall be made available for public inspection and copying.

Deferral or suspension.

(d) PAYMENT OF PENALTIES.—The payment of any penalty imposed pursuant to subsection (c) may be made a condition, for a period not exceeding one year after the imposition of such penalty, to the granting, restoration, or continuing validity of any export license, permission, or privilege granted or to be granted to the person upon whom such penalty is imposed. In addition, the payment of any penalty imposed under subsection (c) may be deferred or suspended in whole or in part for a period of time no longer than any probation period (which may exceed one year) that may be imposed upon such person. Such a deferral or suspension shall not operate as a bar to the

collection of the penalty in the event that the conditions of the suspension, deferral, or probation are not fulfilled.

(e) REFUNDS.—Any amount paid in satisfaction of any penalty imposed pursuant to subsection (c) shall be covered into the Treasury as a miscellaneous receipt. The head of the department or agency concerned may, in his discretion, refund any such penalty, within 2 years after payment, on the ground of a material error of fact or law in the imposition of the penalty. Notwithstanding section 1346(a) of title 28, United States Code, no action for the refund of any such penalty may be maintained in any court.

(f) ACTIONS FOR RECOVERY OF PENALTIES.—In the event of the failure of any person to pay a penalty imposed pursuant to subsection (c), a civil action for the recovery thereof may, in the discretion of the head of the department or agency concerned, be brought in the name of the United States. In any such action, the court shall determine de novo all issues necessary to the establishment of liability. Except as provided in this subsection and in subsection (d), no such liability shall be asserted, claimed, or recovered upon by the United States in any way unless it has previously been reduced to judgment.

(g) OTHER AUTHORITIES.—Nothing in subsection (c), (d), or (f) limits—

(1) the availability of other administrative or judicial remedies with respect to violations of this Act, or any regulation, order, or license issued under this Act;

(2) the authority to compromise and settle administrative proceedings brought with respect to violations of this Act, or any regulation, order, or license issued under this Act; or

(3) the authority to compromise, remit or mitigate seizures and forfeitures pursuant to section 1(b) of title VI of the Act of June 15, 1917 (22 U.S.C. 401(b)).

ENFORCEMENT

50 USC app. 2411.

50 USC app. 2021 note, 2401 note.

SEC. 12. (a) GENERAL AUTHORITY.—To the extent necessary or appropriate to the enforcement of this Act or to the imposition of any penalty, forfeiture, or liability arising under the Export Control Act of 1949 or the Export Administration Act of 1969, the head of any department or agency exercising any function thereunder (and officers or employees of such department or agency specifically designated by the head thereof) may make such investigations and obtain such information from, require such reports or the keeping of such records by, make such inspection of the books, records, and other writings, premises, or property of, and take the sworn testimony of, any person. In addition, such officers or employees may administer oaths or affirmations, and may by subpena require any person to appear and testify or to appear and produce books, records, and other writings, or both, and in the case of contumacy by, or refusal to obey a subpena issued to, any such person, the district court of the United States for any district in which such person is found or resides or transacts business, upon application, and after notice to any such person and hearing, shall have jurisdiction to issue an order requiring such person to appear and give testimony or to appear and produce books, records, and other writings, or both, and any failure to obey such order of the court may be punished by such court as a contempt thereof.

(b) IMMUNITY.—No person shall be excused from complying with any requirements under this section because of his privilege against self-incrimination, but the immunity provisions of section 6002 of

PUBLIC LAW 96-72—SEPT. 29, 1979   93 STAT. 531

title 18, United States Code, shall apply with respect to any individual who specifically claims such privilege.

(c) CONFIDENTIALITY.—(1) Except as otherwise provided by the third sentence of section 8(b)(2) and by section 11(c)(2)(C) of this Act, information obtained under this Act on or before June 30, 1980, which is deemed confidential, including Shippers' Export Declarations, or with reference to which a request for confidential treatment is made by the person furnishing such information, shall be exempt from disclosure under section 552 of title 5, United States Code, and such information shall not be published or disclosed unless the Secretary determines that the withholding thereof is contrary to the national interest. Information obtained under this Act after June 30, 1980, may be withheld only to the extent permitted by statute, except that information obtained for the purpose of consideration of, or concerning, license applications under this Act shall be withheld from public disclosure unless the release of such information is determined by the Secretary to be in the national interest. Enactment of this subsection shall not affect any judicial proceeding commenced under section 552 of title 5, United States Code, to obtain access to boycott reports submitted prior to October 31, 1976, which was pending on May 15, 1979; but such proceeding shall be continued as if this Act had not been enacted.

Information disclosure.

Access to boycott reports.

(2) Nothing in this Act shall be construed as authorizing the withholding of information from the Congress, and all information obtained at any time under this Act or previous Acts regarding the control of exports, including any report or license application required under this Act, shall be made available upon request to any committee or subcommittee of Congress of appropriate jurisdiction. No such committee or subcommittee shall disclose any information obtained under this Act or previous Acts regarding the control of exports which is submitted on a confidential basis unless the full committee determines that the withholding thereof is contrary to the national interest.

Information, availability to Congress.

(d) REPORTING REQUIREMENTS.—In the administration of this Act, reporting requirements shall be so designed as to reduce the cost of reporting, recordkeeping, and export documentation required under this Act to the extent feasible consistent with effective enforcement and compilation of useful trade statistics. Reporting, recordkeeping, and export documentation requirements shall be periodically reviewed and revised in the light of developments in the field of information technology.

(e) SIMPLIFICATION OF REGULATIONS.—The Secretary, in consultation with appropriate United States Government departments and agencies and with appropriate technical advisory committees established under section 5(h), shall review the regulations issued under this Act and the commodity control list in order to determine how compliance with the provisions of this Act can be facilitated by simplifying such regulations, by simplifying or clarifying such list, or by any other means.

Review of regulations.

EXEMPTION FROM CERTAIN PROVISIONS RELATING TO ADMINISTRATIVE
PROCEDURE AND JUDICIAL REVIEW

SEC. 13. (a) EXEMPTION.—Except as provided in section 11(c)(2), the functions exercised under this Act are excluded from the operation of sections 551, 553 through 559, and 701 through 706 of title 5, United States Code.

50 USC app. 2412.

(b) PUBLIC PARTICIPATION.—It is the intent of the Congress that, to the extent practicable, all regulations imposing controls on exports

---

PUBLIC LAW 96-72—SEPT. 29, 1979   93 STAT. 532

under this Act be issued in proposed form with meaningful opportunity for public comment before taking effect. In cases where a regulation imposing controls under this Act is issued with immediate effect, it is the intent of the Congress that meaningful opportunity for public comment also be provided and that the regulation be reissued in final form after public comments have been fully considered.

ANNUAL REPORT

SEC. 14. (a) CONTENTS.—Not later than December 31 of each year, the Secretary shall submit to the Congress a report on the administration of this Act during the preceding fiscal year. All agencies shall cooperate fully with the Secretary in providing information for such report. Such report shall include detailed information with respect to—

Report to Congress. 50 USC app. 2413.

(1) the implementation of the policies set forth in section 3;

(2) general licensing activities under sections 5, 6, and 7, and any changes in the exercise of the authorities contained in sections 5(a), 6(a), and 7(a);

(3) the results of the review of United States policy toward individual countries pursuant to section 5(b);

(4) the results, in as much detail as may be included consistent with the national security and the need to maintain the confidentiality of proprietary information, of the actions, including reviews and revisions of export controls maintained for national security purposes, required by section 5(c)(3);

(5) actions taken to carry out section 5(d);

(6) changes in categories of items under export control referred to in section 5(e);

(7) determinations of foreign availability made under section 5(f), the criteria used to make such determinations, the removal of any export controls under such section, and any evidence demonstrating a need to impose export controls for national security purposes notwithstanding foreign availability;

(8) actions taken in compliance with section 5(f)(5);

(9) the operation of the indexing system under section 5(g);

(10) consultations with the technical advisory committees established pursuant to section 5(h), the use made of the advice rendered by such committees, and the contributions of such committees toward implementing the policies set forth in this Act;

(11) the effectiveness of export controls imposed under section 6 in furthering the foreign policy of the United States;

(12) export controls and monitoring under section 7;

(13) the information contained in the reports required by section 7(b)(2), together with an analysis of—

(A) the impact on the economy and world trade of shortages or increased prices for commodities subject to monitoring under this Act or section 812 of the Agricultural Act of 1970;

7 USC 612c-3.

(B) the worldwide supply of such commodities; and

(C) actions being taken by other countries in response to such shortages or increased prices;

(14) actions taken by the President and the Secretary to carry out the antiboycott policies set forth in section 3(5) of this Act;

(15) organizational and procedural changes undertaken in furtherance of the policies set forth in this Act, including changes to increase the efficiency of the export licensing process and to fulfill the requirements of section 10, including an

PUBLIC LAW 96-72—SEPT. 29, 1979

93 STAT. 533

analysis of the time required to process license applications, the number and disposition of export license applications taking more than 90 days to process, and an accounting of appeals received, court orders issued, and actions taken pursuant thereto under subsection (j) of such section;

(16) delegations of authority by the President as provided in section 4(e) of this Act;

(17) efforts to keep the business sector of the Nation informed with respect to policies and procedures adopted under this Act;

(18) any reviews undertaken in furtherance of the policies of this Act, including the results of the review required by section 12(d), and any action taken, on the basis of the review required by section 12(e), to simplify regulations issued under this Act;

(19) violations under section 11 and enforcement activities under section 12; and

(20) the issuance of regulations under the authority of this Act, including an explanation of each case in which regulations were not issued in accordance with the first sentence of section 13(b).

(b) REPORT ON CERTAIN EXPORT CONTROLS.—To the extent that the President determines that the policies set forth in section 3 of this Act require the control of the export of goods and technology other than those subject to multilateral controls, or require more stringent controls than the multilateral controls, the President shall include in each annual report the reasons for the need to impose, or to continue to impose, such controls and the estimated domestic economic impact on the various industries affected by such controls.

(c) REPORT ON NEGOTIATIONS.—The President shall include in each annual report a detailed report on the progress of the negotiations required by section 5(i), until such negotiations are concluded.

REGULATORY AUTHORITY

Regulations.
50 USC app.
2414.

SEC. 15. The President and the Secretary may issue such regulations as are necessary to carry out the provisions of this Act. Any such regulations issued to carry out the provisions of section 5(a), 6(a), 7(a), or 8(b) may apply to the financing, transporting, or other servicing of exports and the participation therein by any person.

DEFINITIONS

50 USC app.
2415.

SEC. 16. As used in this Act,—

(1) the term "person" includes the singular and the plural and any individual, partnership, corporation, or other form of association, including any government or agency thereof;

(2) the term "United States person" means any United States resident or national (other than an individual resident outside the United States and employed by other than a United States person), any domestic concern (including any permanent domestic establishment of any foreign concern) and any foreign subsidiary or affiliate (including any permanent foreign establishment) of any domestic concern which is controlled in fact by such domestic concern, as determined under regulations of the President;

(3) the term "good" means any article, material, supply or manufactured product, including inspection and test equipment, and excluding technical data;

(4) the term "technology" means the information and know-how that can be used to design, produce, manufacture, utilize, or

PUBLIC LAW 96-72—SEPT. 29, 1979

93 STAT. 534

reconstruct goods, including computer software and technical data, but not the goods themselves; and

(5) the term "Secretary" means the Secretary of Commerce.

EFFECT ON OTHER ACTS

50 USC app.
2416.

SEC. 17. (a) IN GENERAL.—Nothing contained in this Act shall be construed to modify, repeal, supersede, or otherwise affect the provisions of any other laws authorizing control over exports of any commodity.

(b) COORDINATION OF CONTROLS.—The authority granted to the President under this Act shall be exercised in such manner as to achieve effective coordination with the authority exercised under section 38 of the Arms Export Control Act (22 U.S.C. 2778).

(c) CIVIL AIRCRAFT EQUIPMENT.—Notwithstanding any other provision of law, any product (1) which is standard equipment, certified by the Federal Aviation Administration, in civil aircraft and is an integral part of such aircraft, and (2) which is to be exported to a country other than a controlled country, shall be subject to export controls exclusively under this Act. Any such product shall not be subject to controls under section 38(b)(2) of the Arms Export Control Act. For purposes of this subsection, the term "controlled country" means any country described in section 620(f) of the Foreign Assistance Act of 1961.

22 USC 2370.

(d) NONPROLIFERATION CONTROLS.—(1) Nothing in section 5 or 6 of this Act shall be construed to supersede the procedures published by the President pursuant to section 309(c) of the Nuclear Non-Proliferation Act of 1978.

99 Stat. 141.
42 USC 2139.

(2) With respect to any export license application which, under the procedures published by the President pursuant to section 309(c) of the Nuclear Non-Proliferation Act of 1978, is referred to the Subgroup on Nuclear Export Coordination or other interagency group, the provisions of section 10 of this Act shall apply with respect to such license application only to the extent that they are consistent with such published procedures, except that if the processing of any such application under such procedures is not completed within 180 days after the receipt of the application by the Secretary, the applicant shall have the rights of appeal and court action provided in section 10(i) of this Act.

(e) TERMINATION OF OTHER AUTHORITY.—On October 1, 1979, the Mutual Defense Assistance Control Act of 1951 (22 U.S.C. 1611-1613d), is superseded.

AUTHORIZATION OF APPROPRIATIONS

50 USC app.
2417.

SEC. 18. (a) REQUIREMENT OF AUTHORIZING LEGISLATION.—Notwithstanding any other provision of law, no appropriation shall be made under any law to the Department of Commerce for expenses to carry out the purposes of this Act unless previously and specifically authorized by law.

(b) AUTHORIZATION.—There are authorized to be appropriated to the Department of Commerce to carry out the purposes of this Act—

(1) $8,000,000 for each of the fiscal years 1980 and 1981, of which $1,250,000 shall be available for each such fiscal year only for purposes of carrying out foreign availability assessments pursuant to section 5(f)(5), and

(2) such additional amounts, for each such fiscal year, as may be necessary for increases in salary, pay, retirement, other

PUBLIC LAW 96-72—SEPT. 29, 1979   93 STAT. 535

employee benefits authorized by law, and other nondiscretionary costs.

EFFECTIVE DATE

SEC. 19. (a) EFFECTIVE DATE.—This Act shall take effect upon the expiration of the Export Administration Act of 1969.

(b) ISSUANCE OF REGULATIONS.—(1) Regulations implementing the provisions of section 10 of this Act shall be issued and take effect not later than July 1, 1980.

(2) Regulations implementing the provisions of section 7(c) of this Act shall be issued and take effect not later than January 1, 1980.

[margin: 50 USC app. 2418. 50 USC app. 2401 note. 50 USC app. 2409 note. 50 USC app. 2406 note.]

TERMINATION DATE

SEC. 20. The authority granted by this Act terminates on September 30, 1983, or upon any prior date which the President by proclamation may designate.

[margin: 50 USC app. 2419.]

SAVINGS PROVISIONS

SEC. 21. (a) IN GENERAL.—All delegations, rules, regulations, orders, determinations, licenses, or other forms of administrative action which have been made, issued, conducted, or allowed to become effective under the Export Control Act of 1949 or the Export Administration Act of 1969 and which are in effect at the time this Act takes effect shall continue in effect according to their terms until modified, superseded, set aside, or revoked under this Act.

(b) ADMINISTRATIVE PROCEEDINGS.—This Act shall not apply to any administrative proceedings commenced or any application for a license made, under the Export Administration Act of 1969, which is pending at the time this Act takes effect.

[margin: 50 USC app. 2420. 50 USC app. 2021 note.]

TECHNICAL AMENDMENTS

SEC. 22. (a) Section 38(e) of the Arms Export Control Act (22 U.S.C. 2778(e)) is amended by striking out "sections 6(c), (d), (e), and (f) and 7(a) and (c) of the Export Administration Act of 1969" and inserting in lieu thereof "subsections (c), (d), (e), and (f) of section 11 of the Export Administration Act of 1979, and by subsections (a) and (c) of section 12 of such Act".

(b)(1) Section 103(c) of the Energy Policy and Conservation Act (42 U.S.C. 6212(c)) is amended—

(A) by striking out "1969" and inserting in lieu thereof "1979"; and

(B) by striking out "(A)" and inserting in lieu thereof "(C)".

(2) Section 254(e)(3) of such Act (42 U.S.C. 6274(e)(3)) is amended by striking out "section 7 of the Export Administration Act of 1969" and inserting in lieu thereof "section 12 of the Export Administration Act of 1979".

(c) Section 993(c)(2)(D) of the Internal Revenue Code of 1954 (26 U.S.C. 993(c)(2)(D)) is amended—

(1) by striking out "4(b) of the Export Administration Act of 1969 (50 U.S.C. App. 2403(b))" and inserting in lieu thereof "7(a) of the Export Administration Act of 1979"; and

(2) by striking out "(A)" and inserting in lieu thereof "(C)".

PUBLIC LAW 96-72—SEPT. 29, 1979   93 STAT. 536

INTERNATIONAL INVESTMENT SURVEY ACT AUTHORIZATIONS

SEC. 23. (a) Section 9 of the International Investment Survey Act of 1976 (22 U.S.C. 3108) is amended to read as follows:

"AUTHORIZATIONS

"SEC. 9. To carry out this Act, there are authorized to be appropriated $4,400,000 for the fiscal year ending September 30, 1980, and $4,500,000 for the fiscal year ending September 30, 1981.".

(b) The amendment made by subsection (a) shall take effect on October 1, 1979.

[margin: Effective date. 22 USC 3108 note.]

MISCELLANEOUS

SEC. 24. Section 402 of the Agricultural Trade Development and Assistance Act of 1954 is amended by inserting "or beer" in the second sentence immediately after "wine".

[margin: 7 USC 1732.]

Approved September 29, 1979.

LEGISLATIVE HISTORY:

HOUSE REPORTS: No. 96-200 accompanying H.R. 4034 (Comm. on Foreign Affairs) and No. 96-482 (Comm. of Conference).

SENATE REPORT No. 96-169 (Comm. on Banking, Housing, and Urban Affairs).

CONGRESSIONAL RECORD, Vol. 125 (1979):
  July 18, 20, 21, considered and passed Senate.
  May 31, July 23, Sept. 11, 18, 21, 25, H.R. 4034 considered and passed House; passage vacated and S. 737, amended, passed in lieu.
  Sept. 27, Senate agreed to conference report.
  Sept. 28, House agreed to conference report.

○